THE RED FOX

Symposium on Behaviour and Ecology

BIOGEOGRAPHICA

VOLUME 18

SPRINGER-SCIENCE+BUSINESS MEDIA, B.V. 1980

THE RED FOX

Symposium on Behaviour and Ecology

edited by

Dr. ERIK ZIMEN

SPRINGER-SCIENCE+BUSINESS MEDIA, B.V. 1980

Library of Congress Cataloging in Publication Data CIP

Main entry under title:

The red fox
(Biogeographica; v.18)

1. Red fox – behavior. 2. Red fox – ecology. 3. Mammals – behavior. 4. Mammals – ecology.
I. Zimen, Erik, 1941– II. Series: Biogeographica (The Hague) = v.18.
QL737.C22R42 599.74'442 80–18261
ISBN 978-90-6103-219-9 ISBN 978-94-017-5592-4 (eBook)
DOI 10.1007/978-94-017-5592-4

Originally published by Dr. W. Junk bv Publishers, The Hague in 1980
Softcover reprint of the hardcover 1st edition 1980

CONTENTS

INTRODUCTION

A SHORT HISTORY OF HUMAN ATTITUDES TOWARDS THE FOX

Erik Zimen*

Save for the wolf, probably no animal has stirred the fantasy of our forefathers more than the fox. In the old Greek fables of Aesop it was the cunning jackal who outwitted his colleagues, but in Europe, by the Middle Ages, time and repetition had gradually transformed him into a fox. Although the fox looks weak and slight contrasted with the wolf, bear and lion, his darting intelligence and impudent daring usually get him out of all the scrapes his insatiable appetite gets him into. Other fables tell how more "naive" animals managed to trick him – like the famous Aesop fable, retold by Hans Sachs, of the stork who took his revenge, after the fox offered him soup in a shallow dish, by inviting the fox to partake of food in a narrow-necked jug. Or the "Brer Rabbit" stories supposed to be told by an old Negro slave, and dating back through the centuries. Thc simpleton always prevails, and the whole fascination of these fables lies in the fox's age-old reputation for shrewdness far in excess of his fellows.

Whatever moral the fables offered, their detailed and apt descriptions of animal behaviour mirror our forefathers' sound basic knowledge of their natural environment, the red fox included, a knowledge that is our common cultural heritage. Our children think of the fox as smart, cunning and crafty – a well-deserved reputation if we consider his enormous geographical distribution covering almost the whole holarctic region (Fig. 1), the largest natural distribution of all mammals, again with the exception of the wolf. But the fox, not the wolf, has managed to survive over almost all his former range, in spite of extermination efforts and habitat destruction. His numbers may even have increased considerably in certain areas, and regions temporarily lost seem to be resettled. Today he even lives in urban areas of several large cities, including London.

Through the ages the adaptive success of the fox has had a mixed reception by man, best demonstrated perhaps in the British Isles. Some thousand years ago it was the wolf who was valued by the nobles for his sports utility, but detested by farmers and villagers alike for his destructive potential to sheep and cattle. In those days wood was the main fuel in Britain, not only for home use but for the early smelting industries, and demand for these purposes and for shipbuilding destroyed the forests and deprived the wolf of his last retreats during the Middle Ages. The remaining

* Lehrstühl für Biogeographie, Universität des Saarlandes, D-6600 Saarbrücken, G.F.R.

Biogeographica, Vol. 18: The Red Fox, ed. by E. Zimen

Fig. 1. The natural distribution of the Red Fox.

wolves were hunted down and killed at the end of this period, save for very remote areas of Scotland and Ireland, whereas on the continent the wolf managed to survive over a large part of his range up to the end of the last century.

The economic motives and their linked emotions in the British farmers and villagers triumphed with the eradication of the wolf, and the nobles turned rather reluctantly to foxhunting, considered to be only a poor substitute for the tougher wolf-hunts. But slowly new cultural ritualizations emerged, leading to the highly sophisticated modern foxhunt. The wolf had left a hiatus for the farmers too, who now adopted the next smaller carnivore – the fox – as the object of their detestation, and developed equally sophisticated methods of persecution. The nobles bred pack-hunting foxhounds, and the farmers bred small but tough little earthdogs (Fox Terriers) which were trained to burrow underground and flush the fox from its earth.

The climate of Britain was pretty well perfect for sheep-rearing, and we can read that "On the pleasant downs within a six-mile radius of Dorchester more than half a million sheep were feeding" in 1700, for example, and "the average weight of sheep... doubled between 1710 and 1795". Britain produced sheep with "the most valuable fleeces in the world", and peasants grew rich by selling wool. Most sheep were let alone grassing out on the range all year around. On the European continent, however, where the wolf survived so much longer, sheep were usually herded and guarded by shepherds and their dogs. Probably the fox had no opportunity to develop techniques of predation on lambs, for which he is condemned in Britain. On the continent the fox competed with the hunters of small game, and here again sophisticated methods of fox control were developed, including the breeding of the Teckel.

As so often happens, this hatred-induced cultural ritualization becomes divorced from the ultimate cause for its appearance; efforts of farmers and hunters to control the fox are out of all proportion to the animal's actual destructive potential. Are our conceptions of his size at fault? A fox is a small animal, seldom exceeding 10 kg in weight. Nor is he able to kill any animal considerably larger than himself. A popular German children's song is "Fox, you've stol'n the goose away..." We kept a goose in our fox enclosure for weeks, but the picture we wanted never materialized; in fact the goose tyrannized over the foxes. Nevertheless, a fox in a chicken pen can certainly be destructive. And even if the effects of fox predation on his wild-living prey species is mainly compensatory in the sense of Errington, the rabbit, hare or pheasant he kills is definitely not available to the hunter, so that naturally, to any hunter who wants to maximize his bag, the fox is a competitor for their mutual prey species.

The resulting widespread fox antipathy in small-game hunters is expressed in the "closed seasons" hunting law/omission. Of the seven Central European countries from which I could get information, only two have closed seasons for the fox. According to our data from the southwestern part of Germany, a high proportion of female foxes are shot in spring and early summer, when they are still lactating or at least have dependent young. In most cases the pups were not shot together with the mother. This is certainly not in accordance with most hunters' moral code. But such "mistakes" in dealing with a competing predator appear in their true unsavoury character when we consider how utterly the same hunter would be damned by his ilk if he made the mistake of shooting a roe deer mother away from her unweaned fawn.

Obviously our relationship to the fox is intense and multiform. Considering our cultural heritage regarding the fox, his enormous dispersion and adaptability, his high trophic level in almost all holarctic ecosystems and his success in spite of persecution in most man-utilized areas, it comes as a surprise to learn how little scientific interest he has received. A thorough taxonomic examination of the genus Vulpes is still lacking, and equally a revision of the over 40 subspecies described for *Vulpes vulpes* alone. We study some of the most exotic animal species assiduously, but our knowledge of many basic features of the fox's spatial and social organization still seems to be mainly circumstantial. Only recently has fox ecology and behaviour received more attention – one exception being Tenbrock's classical study on red fox ethology.

The motivation of this increased research, however, was often not interest in the species itself, but its danger to human health through the spread of vulpine rabies over central Europe and North America. This basic motive in recent fox research seems even to have inverted many of the purist fox ecologies. At the Saarbrücken Fox Colloquium in January 1979, at which the following papers were presented, it was the Veterinarian amongst us who finally cautioned that we should not look upon the fox solely in respect to rabies control.

Whatever the individual researcher's motives, the increase in behavioural and ecological research on the fox over the last 10 years is certainly mainly the merit of the WHO project on wildlife rabies in Europe. But other founders, especially in the rabies-free areas of Scandinavia, England and the western United States, have also sponsored research, from an interest in the effect of the fox on game, or similar reasons, or pure interest in the fox itself. At the Institute of Biogeography at Saarbrücken, interest in the fox was sparked by his high trophic level and the resulting accumulation in his body of chemicals introduced into the environment by man. For our findings to have meaning, however, we must gain a much deeper insight into the nutrient transfer in the food chain, and this includes the fox's feeding habits. We must also comprehend his spatial behaviour and population biology, and these again can only be understood if we learn more about the basics of fox social behaviour, as well as rabies epidemiology.

Starting a new project, therefore, we were most fortunate in having, as well as the rabies experts, some of the world's leading experts on fox ecology and behaviour to come to Saarbrücken for a workshop on the object of our common interest – the red fox. Four days were of course insufficient to cover all aspects of fox biology; some prominent researchers were unfortunately unable to come, others had already done so much that their reports could only highlight their findings. This book must therefore not be considered a comprehensive summary of our present-day knowledge of the fox; such a publication would be premature anyway, considering the many newly-opened fundamental questions on fox biology, as well as on rabies control methods. But, like those attending the meeting, the reader may find this report stimulating for future research.

I do not want to close without expressing my sincere gratitude to all who made this workshop possible, especially Prof. Paul Müller who sponsored the meeting in his unparalleled, uncomplicated way. Financial support was given by the University of Saarbrücken, the Umweltbundesamt in Berlin, and environment-conscious M.Ps of a certain political Party which shall be nameless. Finally, Mr. Henderson of Dr. Junk bv publishers proved himself a most patient editor.

2 HABITAT REQUIREMENTS OF THE RED FOX

Huw Glen Lloyd*

HABITAT SUITABILITY AND DISTRIBUTION

Choosing a place to live is a feature that the more mobile animals can enjoy compared with the more passive or accidental ways of securing suitable habitats by sessile species.

Some mobile species may have very restricted distributions, for example the coypu, *Myocastor coypus*, in the marsh-land area of East Anglia, England; others, such as the red fox are widely distributed not only over a large geographical range, but widely within that range. The habitat and habitat requirements of the coypu can be more easily described, even if incompletely, than for the fox because it tends to be more specialized in its way of life. The way of life of the fox is such that it can live in a wide range of habitats; consequently its habitat requirements would be much more difficult to define even if habitats could be readily described, which they are not. It cannot be shown that the fox exercises choice in selecting a place to live, but it almost certainly does within the restrictions imposed by its social behaviour and the suitability of areas on the margin of its present day geographical and local range.

Habitat suitability or preference is most easily revealed by the densities of animals found in different places. With the fox however there is one major obstacle; it is hunted by man, probably throughout much of its range. Thus the densities revealed for any area where such a restriction or influence on numbers exists may only reflect the ability of the fox to withstand the effects of man (which vary considerably in terms of effort) in that area, and not the true suitability of the area. It may be that it reflects the suitability of the area in terms of harbourage and not of food – though food may be an important feature in determining reproductive performance as a response to a non-natural depression of numbers below that which the area could support. Since the fox is able to withstand even the best efforts of man to eliminate it (for disease or pest control) and can well tolerate intermittent, haphazard or regulated interference (for disease and pest control, for sport or for its fur) and also occurs locally where interference by man is slight or none – consideration needs to be given to habitat requirements and suitability in a range of circumstances but more especially in the middle range of man's activity since this is the more common throughout the distribution of the fox.

---

* Ministry of Agriculture, Fisheries and Food, Agricultural Science Service, Worplesdon Laboratory, Spa Road, Llandrindod Wells, Powys, U.K.

In the absence of interference by man the bald statement that the distribution and abundance of foxes are determined directly or indirectly by availability of food and cover would not be challenged and it is the ecological and behavioural adaptations of the fox to meet these requirements in different habitats or environmental situations that need examination. Mammals are not usually limited directly by food shortage but by biological mechanisms which prevent increase to the point where the food supply might be exploited to the extreme. Though the red fox has a highly varied diet, not confined to predated species, and can survive in diverse types of habitats, its potential for increase would probably not be limited directly by availability of food.

It does not have a specialist niche and competes, or at least takes foods that other more specialized feeders enjoy, such as voles (taken by hawks, owls and small mustelids), earthworms (badgers), rabbits (buzzards, domestic and feral cats, polecats, stoats and weasels) and carrion (corvids especially). It competes with other species for individual food items but it does not compete overall, nor can it logically occupy identical habitats with any other species. However, in the evolution of the species and in its adaptation to changing environments the fox must have competed with other species. Zeuner (1963) and Kurten (1968) describe the occurrence of the antecedent species of the red fox (*Vulpes alopecoides*) and of the corsac fox (*Vulpes praecorsac*) to occur contemporeaneously in the early pleistocene in Europe – and in the late pleistocene, the coexistence of the arctic and corsac and red foxes. Following the last glaciations the arctic fox withdrew with the receding ice to northern latitudes and the corsac fox to the open steppes of Russia. The red fox expanded to occupy a wide range of habitats in the more temperate regions. Its adaptability is such that it occurs throughout the Holarctic region, except for Greenland, most arctic islands and the drier zones of western North America and North Africa. Introduced successfully to Australia in 1871 (for hunting with horse and hound) it colonized most of its present day range from about 1900 to 1940 and can occur in areas enjoying no more than 8 millimetres annual rainfall. Thus its geographical distribution includes the barren tundra up to about 75° north (in Canada and USSR) at the one extreme, and the hot dry deserts of Australia at the other. In the drier zones of north America and north Africa it appears to be replaced by other foxes, but for sheer adaptability it is without peer. The enormous geographical range of the red fox is evidence of the adaptability of an unspecialized mammal and, because of the diversity of habitats in which it can survive, also a token of its ability to learn whatever food finding and gathering skills it needs wherever it occurs. Though variable in size the fox is a comparatively small mammal which undoubtedly confers advantages of concealment and also makes it less conspicuous as a predator of domestic animals than if it were much larger and capable of taking larger prey. In areas of deep winter snows it can move through deep and soft snow by virtue of its long legs and physical endurance – yet it is light enough to be able to walk on thinly frozen snow. It is not adapted to digging yet it can dig dens effectively, but because of its real and proverbial inclination to exploit it often prefers to usurp dens abandoned by other species.

ALTERATIONS OF RANGE

Few areas of Europe within the geographical range of the red fox are without foxes today, and many biologists consider that despite the current rabies epizootic, foxes are probably more numerous now than at any time during the last 400–500 years. Historical evidence in Britain suggests strongly that the red fox expanded its range and became more numerous from about 1750–1850, and circumstantial evidence suggests another expansion of range and numbers from 1950 to about 1965. This implies, correctly, that foxes were absent from parts of Britain during the past 200 years. Indeed it is only recently (1950–1965) that they have colonized parts of Scotland (Hewson and Kolb, 1973) and parts of some eastern counties of England – such as Norfolk. These areas are eminently suitable habitats for foxes and the point of interest is why they were absent from these areas at these times.

Hewson and Kolb (1973) suggest that the expansion of range in Scotland followed directly upon a temporary super-abundance of food which enhanced the survival of juveniles (and possibly adults). Dead and dying rabbits of the myxomatosis epizootic of 1954–56 provided the fox with this super abundance – short lived, since the supply failed within 18 months when rabbits became scarce compared with pre-myxomatosis numbers. There was evidence that foxes had begun to spread before myxomatosis was introduced (a feature which might have been associated with increased plantings of coniferous woodlands and the consequent provision of cover and of voles during the 10–15 year post-planting period) but following the sharp rise in numbers in 1954–55 the population declined when rabbits became scarce. Later, in 1960, fox numbers began to increase but by this time they had spread to many hitherto fox-free areas. The decline in rabbits resulted in vegatative changes which, after a delay, promoted an increase in vole numbers. Hewson and Kolbs' observations are interesting because they might indicate that the sudden movement to new areas was precipated by a higher number of foxes (following myxomatosis) being confronted with a sudden and dramatic food shortage which forced them to disperse. An ancillary factor might have been the very hard and prolonged frosts of 1963 which, as was reported in Wales, might have forced foxes off the higher land to scavenge in the more populous lower lands. Dispersion to new areas may occur only, or mainly, in times of adversity and not when foxes are living in the midst of plenty. This does not explain why foxes were absent formerly from those parts of Scotland however, nor the comparatively recent movement of foxes into parts of Norfolk, though, there, some historical documentation is available.

The fox was common in most parts of Norfolk in the 15th, 16th and 17th centuries as testified by the number of foxes recorded in Churchwarden's account of foxes killed for bounty from 1533. The numbers of foxes killed were small by comparison with kills obtained today; moreover the bounty was generous which suggests that foxes, though common, were not numerous and it is not improbable that they were eliminated by man. Lubbock reported in 1845, when writing his "Fauna of Norfolk" that "the fox is so rare as to be unworthy of mention and fox hounds have recently been given up in West Norfolk for lack of sport". However the hounds were re-established in West Norfolk by 1880. In the 1920's the fox was relatively common in the south western parts of the county and any fox seen east of a north–south line through East Dereham was considered to be sufficiently remarkable to be reported in the local press (R. P. Bagnall - Oakley, *pers comm.*). The fox continued to spread eastwards, slowly in the 1930's, but received some added impetus at the time of myxomatosis. Although still spreading, their numbers declined sharply from about 1958 to 1962 which coincided with many cases of secondary insecticidal poisoning (chlorinated hydrocarbons) in foxes, resulting from feeding on birds which had eaten dressed cereal seed, (Taylor and Blackmore, 1961; Rothschild, 1963; Blackmore, 1964). This was but a temporary set-back and foxes are now widespread and numerous throughout the county.

To those concerned with fox control as an anti-rabies measure in Europe it may seem improbable that the fox could be eliminated by traditional control efforts, and whilst there is no direct evidence that this happened in Norfolk, two examples are known in Wales, in North Pembrokeshire and the Lleyn Peninsula in North Wales. These two areas were the scene of intensive rabbit trapping from about 1910 when nearly all farms were visited by commercial rabbit trappers every year. The fox was harmful to their interests since, by mutilating trapped rabbits, the prices paid by dealers was less. In consequence a determined effort to eliminate the fox all but succeeded and foxes were very few from about 1920. A rapid recolonization occured from 1955 to 1960, following myxomatosis and the absence of rabbit trapping, and foxes became exceedingly numerous in both areas by 1963. This illustrates that foxes can survive in the virtual absence of the wild rabbit, and can rapidly establish large numbers if foxes, although few, are already in the area. A different situation occurred in the Isle of Anglesey where as in parts of Norfolk, foxes died out some time in the early 19th century. Here, three - probably no more - foxes were introduced in 1965 or 1966. Despite efforts to contain them foxes increased but their spread to all parts of the island (of about 1100 sq. km) took 10 to 11 years. It seems that rather than spread quickly to adjacent fox-free areas, dispersing juvenile foxes in increasing fox populations move only short distances away from the main body of the population. It required about 7 years for the population to occupy areas 16 km distant from the points of introduction.

Extreme wintry conditions causing food shortages might promote greater dispersal into uncolonized areas, but while food is adequate it seems that foxes find it advantageous to stay close to their own kind even though the then uncolonized habitat later proves to be highly suitable for them. This of course merely illustrates the obvious that the most important requirement within any habitat to a fox, is another fox.

FOOD AND COVER

A quick glance at the results of investigations on fox diet in different regions of its geographical range will show that its range of foods is enormously varied, which not only reflects the availability of different prey species, for example, but also the fox's ability to secure a livelihood in different ways in different places. Its ubiquity denies the possibility of defining the habitats it occupies precisely since it is found in widely divergent habitat types such as boreal forest, open agricultural land, mixed woodland or, in Britain, urban areas. Clearly there are some components of the habitats which are common to all and provide the fox with the ability to live there – but no attempts have been made to establish this qualitatively or quantitatively, comparing one area with another. Subjective appraisal of habitat suitability and of potential carrying capacity does however highlight features which are related to abundance of foxes in different areas, but true preferences reflected by relative numbers are difficult to derive because natural fox communities untrammeled by by man are not common and may not cover the broad scope of habitats where foxes are to be found.

Except where fox control is very intensive, areas where foxes are few in number probably reflect the least suitable or preferred habitats. But in the middle range of fox numbers, say one to three foxes per sq km the picture is clouded because man may or may not have depressed the population significantly, and there is little opportunity to determine this. Even less is known about habitat preferences of individual foxes. Fox diet investigations reveal the opportunistic feeding behaviour of the fox as a species but individual behaviour is not thus revealed. Circumstantial evidence suggests that some foxes may show preferential or specialist feeding activity – on litter bins, at picnic sites, sheep carrion, or cattle afterbirths for example. If, as adults, foxes, like polecats (Apfelbach, 1973), tended to select those prey foods – or other foods – upon which they were reared, such behaviour could be more plausible. But there are seasonal differences in food supply and uptake and it is commonly believed that it is the fox's ability to switch foods that enables it to subsist over such widely divergent habitats. Nevertheless, individual foxes may show preferences for certain food items, or may have developed individual methods of taking prey.

The fox's ability to take a wide range of prey and its ability to exploit alternative foods raises questions concerned with decisions that a fox makes when settling in an area new to it, when dispersing from its natal area and when faced with a variety of potential prey or other foods in its range. It has to learn or decide where to look for food, even which prey or food is the best for it to exploit at a given time in terms of reward for effort. Determination of the size of its home – or territory – which will provide it with all its requirements throughout the four seasons of the year would be critical to its survival. Perhaps the time of dispersal is the key in this context since, if the final choice of area is made at the period of year when food is least abundant, and if the fox survives that period, then things can only get better; or, perhaps, the area selected is much larger than is really necessary to maintain a fox. Observations on movements by radio tracking in mid-Wales reveal, for some foxes occupying large ranges (2.5 to 15 sq km) in upland and hill land, of fragmented and discontinuous habitats, show that nocturnal – presumably foraging – activity is confined to parts of the range only, as if there were islands of potential foraging places within its entire range. In smaller ranges, a less obvious tendency still occurs. Sometimes foraging areas, like a string of beads, lead out from a fox's centre of activity.

The fox has a strong tendency to investigate the unfamiliar – a trait which may be fundamental to its ability to exploit urban areas successfully – but it must also be of value to it when foraging, since it is unlikely to encounter all situations during its juvenile life in its family group. Connected with this would be its ability to switch from one purposeful activity to another – say from hunting voles to chasing a winged pheasant – and by learning, to ignore signals that are of no value to it as food indicators, e.g. the squeaks of shrews. Whilst the concept of optimal foraging is not entirely relevant to fox habitat requirements it has relevance to dispersal behaviour and movement into fox-free areas.

The fox is known to feed upon earthworms, particularly *Lumbricus terrestris*, at times heavily in some situations. This does not necessarily imply that *Lumbricus* – rich areas are *per se* ideal fox habitat. Indeed the intake of large amounts of earthworms may indicate occupation of a poor habitat to which the fox has been relegated by social pressure, for example. This and other preferences of foxes are poorly understood, as is their uptake of available food in any habitat. Thus its habitat requirements can only be expressed crudely and could be mistaken when particular or circumscribed areas are being considered. Furthermore, in examining suitability of habitats it is implied that the habitat reflects availability of food and cover. But this cannot always be so even in the gross definition of apparently similar biotopes, since biological systems are in a perpetual state of imbalance. The rabbit in Britain today can be found at low and high population densities, in apparently similar habitats and also at the same densities in grossly different habitats. Thus, like the rabbit, though a high density of foxes may obtain in one kind of habitat, it does not follow that they will be as numerous concurrently in another similar area. There must be many variable features – in addition to the influence of man – which will render estimates of numbers of foxes dubious if based upon habitat classification. Only an estimate of the potential carrying capacity of any habitat can be possible.

Habitats which support small mammals, especially *Microtus*, are potentially highly suitable for foxes, as are those with an abundance of hares and rabbits. But there are many examples of habitats which support high numbers of foxes where these species are of minor importance in the diet of the fox. It is its ability to exploit a wide variety of foods – flesh, carrion, human waste and fruits – that enables the fox to do well even in areas which on cursory examination do not appear to provide it with the means to thrive.

In addition to food the fox needs cover of some sort for daytime refuge, whether it be a pile of timber in an urban factory, a dry drainpipe running beneath a road, a couch in a thick conifer wood or the more usual den or earth, in rock, boulders or soil. Ample food with poor cover, or ample cover with little food are not highly suitable habitats, but if there is an abundance of food the fox can in many circumstances learn to adapt itself to scarcity of cover, as is the case in urban areas in Britain. The fox is a surface dweller which requires secure cover, usually below ground, for rearing cubs. At other times, dens or earths may only be used when the weather is harsh or very wet, or where surface cover is either not secure or is scarce.

Since Macdonald, (1977) revealed some of the spatial and social patterns of foxes and showed that they do not always exist as mated pairs with their own exclusive range or territory, but sometimes as groups with the male or males outnumbered by vixens, ideas about home ranges or territories shrinking or expanding with changes in population density need to be re-examined, in addition to the examination of the social relationships of foxes in areas where they are few. There may be a range of social groupings or spatial patterns ranging from the exclusive ranges held by a pair of foxes (and their juvenile offspring) as observed by Sargeant, (1972) in north America to that described by Macdonald, (1977), possibly with other arrangements outside that range. Radio tracking of foxes in mid-Wales in an area where foxes are under moderate but intermittent, density-dependent pressure from hunters from October to May suggest, because home ranges are far from exclusive to either sex, that the causes of fox spatial characteristics are complex – and are complicated in many areas by interference by man, before, during and after the juvenile dispersal period. Similarly, Niewold's, (1973) observations on the foraging movements of vixens to distant places well outside their range, highlights our lack of knowledge on many aspects of fox behaviour in relation to habitat. Notwithstanding that the quality of the habitat in terms of food is but poorly understood, some relationships as shown in Figs. 1 to 5 can be broadly outlined.

Surface cover which the fox seeks for much of the year, except when it is insecure in such places or where cover is sparse, is of considerable importance to the fox especially where man pursues it, (for whatever reason). Where man tolerates the fox, it finds surface cover in unlikely places – such as small ornamental bushes where it is hardly concealed and is apparently unperturbed by the activities of garden attendants in the nurseries of the Botanic Gardens, Edinburgh (Hugh Kolb, *pers comm.*).

It is not possible to assess the importance of denning places to the survival of the fox without knowing the disturbance imposed by man's hunting activities and the methods employed to kill foxes in different habitats. Where the effort is considerable and sustained – as for rabies control – inaccessible or impenetrable cover assume greater importance if traditional methods of control only are used – such as shooting or hunting with dogs.

The fox is not prey to many species today, but it tends to shun otherwise suitable areas where coyotes, for example, exist (Dennis Voigt, *pers. comm.*) as in parts of Ontario Province, Canada. In Europe erstwhile predators are largely absent but some predators of fox cubs may still obtain. Secure cover, above and below ground is only of significance where man or other predators exert some pressure upon the fox. Simple disturbance is of course

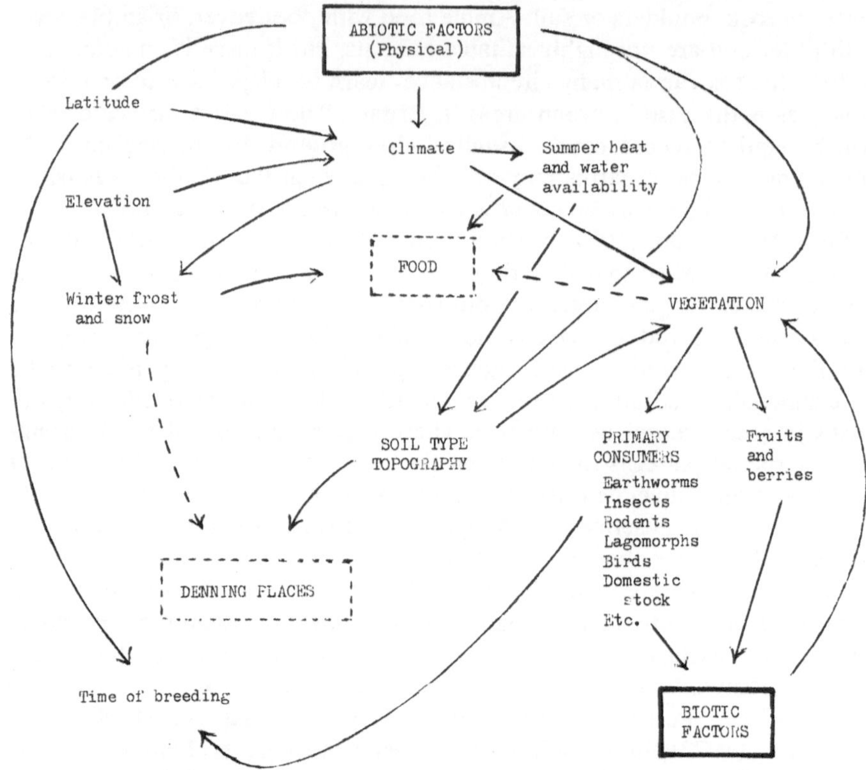

Fig. 1. Physical, abiotic factors are ultimate in determining the potential for food and cover. Climate and soil determines vegetation type, which in its turn largely determines the kind and availability of food in any area. Similarly, soil type and topography affect the potential for denning places. The time of breeding, though governed largely by latitude, is a response to biotic factors.

inimical, and were it not for the presence of domestic dogs urban foxes in Britain might be even more bold in their choice of places to lie up in daylight.

Disturbance by man is probably one of the more important influences upon choice of denning places, but much will depend upon the kind of disturbance and whether or not it is regular or intermittent. If foxes are disturbed by casual hunting, secure cover or dens are not at a premium, nor if hunting is regulated by close seasons, voluntary or otherwise, as might occur where pelt hunting is an important incentive. Where foxes of all ages are killed for control purposes, the necessity for cover will depend upon the kind of control methods used, and the nature of the terrain.

The active pursuit of foxes, that is seeking them out, as with gassing or shooting, would be less effective where secure or inaccessible cover predominate. Methods not involving pursuit – as by interception, using snares, traps or poisoned bait – would succeed irrespective of security of den and cover. In addition, if control incentive varies according to density of foxes (density dependent) topography and den and cover security are important to the survival of the fox, but if control effort is constant, and is conducted

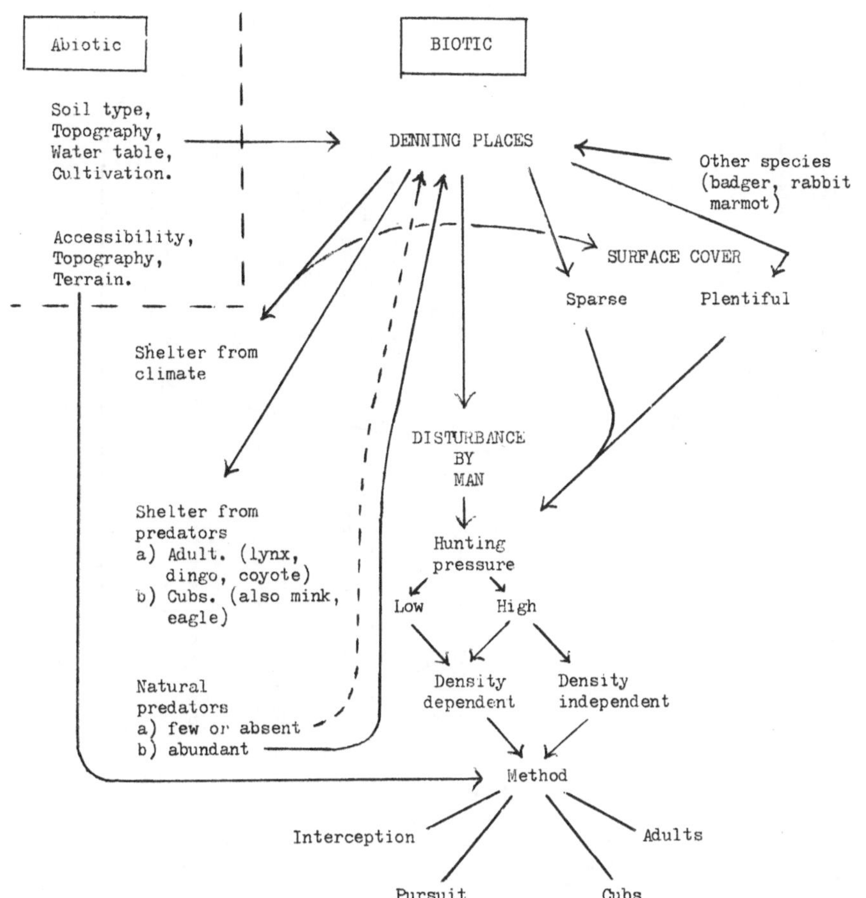

Fig. 2. Denning places – whether surface cover or below-ground harbourage in soil or rock are determined both by abiotic and biotic features of the environment. Some features, such as a high water table, may be limiting but not necessarily exclusively for breeding places since the fox can tailor its needs according to availability, for example by digging burrows in slight elevations, or in man-made dykes in drained marshland – as in the Norfolk fens. Below-ground denning places are often provided by physical features – in rocks or boulders, but though the fox can dig its own burrows it often exploits old or extant burrows of other species – or even man-made artifacts such as the spaces beneath barns, or garden sheds, dry drainpipes or piles of factory rubbish.

persistantly, irrespective of the densities of foxes (density independent), den security only then becomes important to those few foxes which survive. Full time, paid, fox control staff (density independent control), using methods involving interception would be a combination that foxes in many habitats could not withstand, but this is rarely done.

Competition with other species is of two kinds – for food, with non-canids which, with regard to prey, is probably only exceptionally of importance, but perhaps for carrion could be depriving; and for space. Coyotes and dingoes both kill foxes and since in many circumstances foxes are deprived of space to live by these species, or occur at lower densities, by implication food is also denied. Still on shakey ground, intra-specific

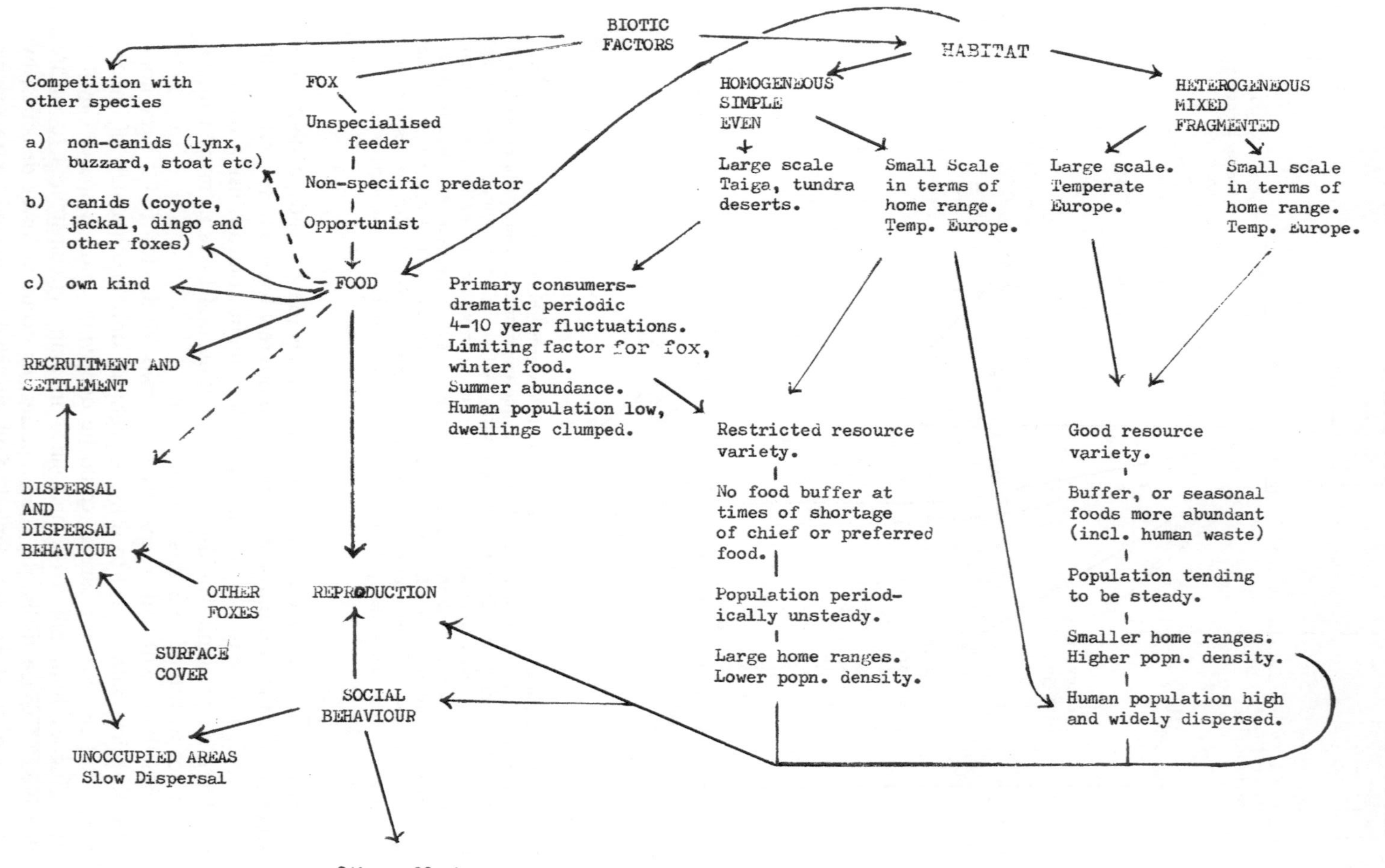

Fig. 3. Deals entirely with biotic factors which influence the number of foxes found in any area where physical factors are not limiting.

competition logically exists but few observations have been made (Vincent, 1958; Macdonald, 1977) and there is much to be known about social organisation, food utilisation and spatial organisation at different densities in different habitats.

In general terms, mixed, heterogeneous, fragmented or discontinuous habitats provide the fox with greater resource variety and in the long run more steady fox populations than homogeneous habitats. Habitat variety within a fox's range provides it with alternative foods in time of shortage and a variety of potential denning places. Homogeneous habitats can however support very high densities of foxes, especially in good rodent years in the higher latitudes – or even in Britain in newly planted forestry plantations, but buffer foods against hard times are not so plentiful. Food availability, in addition to social factors, stimulates dispersal and settlement in any area.

Observations by radio tracking of foxes in different parts of Wales show that foxes frequent the vicinity of human habitation in rural areas more frequently in high fox density populations than elsewhere. In Switzerland human waste has been found in over 50% of autopsied rural foxes, (Alexander Wandeler, *pers. comm.*). Thus, widely scattered or small groups of residences seem to offer advantages to the fox providing it is tolerated and that it does not, by its activity, attract the attention and opprobrium of occupants.

Throughout its range the fox subsists heavily upon small rodents, though it can do well without them if other foods are available. Table 1 (based upon Golley, Ryszkowski and Sokur 1975) shows a range of availability of food and proportions consumed by small mammals in different habitats. From this can be estimated the relative consumption of food by rodents which to some extent also reflects the relative abundance of small mammals in the habitats indicated. For example, availability of food in beech forests is high but consumption by small mammals is low, probably because other species

Table 1. Energy consumption of small mammals as the percentage of available primary production consumed yearly in temperate ecosystems. (After Golley, Ryszkowski and Sokur, 1975)

	Available primary production 10^6 Kcal/ha/yr	% Consumption by small mammals	Relative consumption and relative numbers
Agricultural fields			
Rye grass	41	0.5	3.3
Alfalfa	40	1.4–21.4	9–138
Grasslands			
Grass field	41	1.3	8.6
Old field, seeds only	0.5	12	1
Desert shrubs	2.5	5.5	2.2
Forest plantation	6.7	3	3.5
Oak/pine	13	0.7	1.5
Mixed forest	2.1	4.6	1.6
Beech forest	16.2	0.6	1
Spruce grass	17.9	2.9	8.2

(possibly birds) are better adapted to feeding there. The relative consumption by small mammals is low in such areas (at the relative base, 1). The extremely high alfalfa relative consumption represents, at the upper limit, an example of the effect of a locally high rodent population, perhaps of plague proportions. Elsewhere the higher relative consumption and rodent abundance is associated with permanent grass in fields, and with spruce woodlands where grass and tree seed is available, but not all species of small rodents are equally preferred by foxes and Table 1 does not take account of this. However, the most preferred of the small rodent species *Microtus*, found throughout much of Europe, is primarily a rough grassland species. Less preferred species, such as *Apodemus*, tends to occupy wooded areas and *Clethrionomys*, woodland and overgrown areas. On this basis rough permanent grassland offers the fox a supply of its preferred or widely dispersed (geographically) prey food; if cover is also available in the form of woodland, thicket or rock, one of the best habitats for the fox would be rough grassland with small parcels of woodland or scrub cover, amounting to a small proportion of the total area distributed throughout. Additional prey species (hares, rabbits, winged game and earthworms) and carrion (livestock, livestock afterbirths), fruit bearing shrubs or trees and human waste (at picnic sites, caravan parks, farmhouses, factory canteens etc) would enhance the area by ensuring a wide range of foods against a shortage of some.

Where rabbits exist at high densities (up to 8–10 adult rabbits per hectare in the winter) the fox is assured of a food supply since large fluctuations of winter numbers of rabbits are not common. The reproductive productivity of the rabbit at high densities requires a mortality of about 80% of juveniles (that is, about 30 individuals per hectare – the biomass of which is not known) to ensure a steady over-winter population – and the fox would have its share of this.

In areas of endemic myxomatosis, (in Britain the mortality rate of myxomatous rabbits lies in the range 70–95%, Vaughan and Vaughan, 1968), not only mortally affected rabbits but also those which would survive the disease are vulnerable and available to predators and scavengers. Englund (1970) has shown that fox populations in the southern parts of Sweden where rabbits are found, are more steady in number and breeding success than foxes in the northern, predominantly rodent areas. While the small rodent is the staple of the fox other prey can enhance conditions for it, but the distribution and abundance of alternative prey are localised and, as with the rabbits, cannot be deduced by a potential suitability of the habitat since other environmental features over-ride this.

Areas least suitable for foxes, whether harassed or not, would have some or all of the following characteristics; flat, open country; few woodlands or very open deciduous woodland; neat or simple field boundaries (fences or ditches, e.g.); no scrub or uncultivable land; large fields, mainly arable and a high water table.

Whilst habitat requirements in terms of food and cover can be defined in broad terms and certain gross features of the environment which have some influence on this can be identified from field experience, there will be many

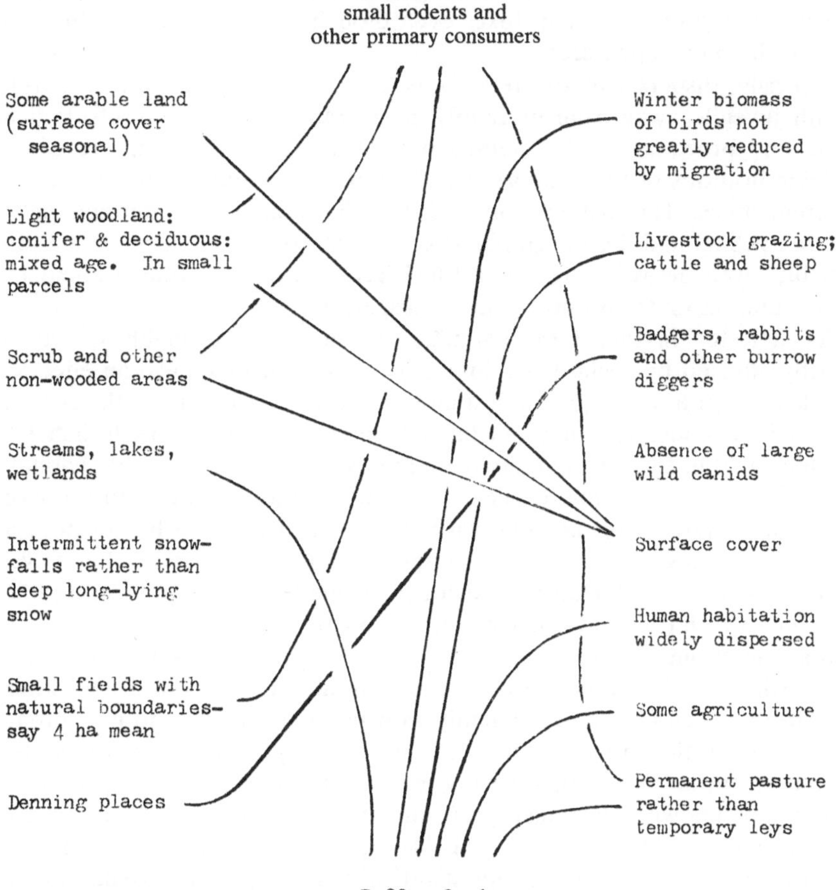

Fig. 4. Suggests habitat components best suited to foxes where they are not harassed by man. Where foxes are harassed and killed food availability will be more or less important according to the reduction of number of foxes achieved. The followng features will provide some insurance against successful control by hunting or by gassing:- much surface cover; large areas of dense coniferous woodland; rough terrain; inaccessible earths or dens and little snow to aid tracking.

inexplicable exceptions where the numbers of foxes are incompatible with these observations. Difficulties arise because the fox is not a prey specific predator and since no rule of thumb can be applied, caution is needed when attempting to assess the probable range of fox populationn densities from habitat classification alone. The potential suitability of an area can be assessed with greater confidence but, as in parts of Britain in the very recent past, highly suitable areas may have no foxes at all.

Misleading indications of fox abundance can arise from the numbers of foxes killed by hunting, because foxes are not equally vulnerable to certain kinds of hunting in all areas; for example, if cover is not abundant foxes may tend to lie below ground more often than in areas where there is abundant surface cover, yet hunting by shooting in daylight in the latter area could produce a greater number of foxes even though they may be less abundant than in the more open area.

Typically, dispersal of the red fox is an emigrant movement of a 6–10 month juvenile away from its family group range to an area outside it to settle and breed at 10–12 months old. Dispersal movements are highly variable in terms of the proportions of animals that apparently emigrate in different areas. The distances moved, the duration of the itinerant phase before settling, and the geographical orientation of the movements are also variable, even between the sexes. Much remains to be revealed about the stimulation, motivation and tactics of dispersal.

As indicated in Fig. 3 the stimulus to emigrate is probably closely or directly affected by social behaviour but the habitat may have an effect on whether or not a fox stays and, if it moves where it decides to settle, It is not known if juveniles are more inclined to emigrate from habitats liable to much change in food availability compared with more dependable habitats, but some dispersal seems to occur in all areas studied though the proportions that emigrate appear to be variable. Nor is it known whether dispersing foxes seek a vacant area or, by competition, forge a place for themselves. It can however be inferred from evidence gained by ear tagging and radio-tracking – slender as it is – that vacant areas are sought.

If a habitat deteriorates adults also might be forced to emigrate temporarily, as when harsh weather forces foxes from hill areas to lower land where food can be secured, or permanently as when, explicably – and possibly for reasons other than food – some foxes move away from areas where they have been settled for a few or many months. The causation of a permanent shift is not known. The tagging of adult foxes has revealed that most of those recaptured, even several years later, are in much the same area as when they were first handled, but slight shifts in range not revealed by this method certainly occur (Sargeant, 1972). There is however insufficient information to permit generalisations on motivation or causes of emigration of adults. It is clear however that there is a strategic advantage for adults to stay put, once settled, and for emigration to be confined mainly to juveniles, but until studies on the effects of manipulation of well-studied natural populations are made, changes within local populations that can promote or affect dispersal will not be uncovered.

In the few cases of adult dispersal observed by radio-tracking in mid-Wales it appears that emigration is not a response to widespread habitat deterioration because other foxes in the same areas do not move away, and others – juveniles – move in. In addition, a few foxes appear to be intermittently itinerant. For example, an adult male living in an area where home ranges spanned 1.3 to 5 sq km moved frequently to other areas up to 16 km distant, but sooner or later it would return to areas known to be familiar to it. Often it could not be found, perhaps on some occasions because it had moved well beyond the area which could be physically covered by the radio-tracking team.

Usually one expects foxes to occupy ranges as pairs, whether or not their ranges are entirely exclusive to them but, since the description of group ranges by Macdonald, (1977) this supposition clearly cannot apply throughout. It is possible that the groups observed by Macdonald are characteristic of fox populations unaffected adversely by man and would occur widely but for this. Field evidence gathered in Wales and north America, (Lloyd, 1980; Sargeant, 1972) suggest strongly that most foxes exist as pairs, and since most information on dispersal of juveniles relate to such populations these circumstances only can be considered.

Because the fox is comparatively short lived in many situations and suffers a high mortality and turnover of population, the occupancy of ranges or territories must be in a state of flux throughout much of the year. Account can be taken of this in seeking explanations for the different dispersal movements observed.

Adult mortality and age structures of fox populations studied in different areas are variable (see Table 2). In mid-Wales, in one period, adult mortality was a mean of 57% annually, (Lloyd, 1976); here much of the mortality – but an unknown proportion – is attributable to man and tends to be seasonal – from November to March. If these data on mortality are used, probabilities of change of paired territory holders from one year to another can be represented as follows:

Annual adult mortality 57%
Probability of

both members dying	$= 0.57 \times 0.57 = 0.325$
one member of a pair dying	$= 0.57 \times 0.43 = 0.245$
both members of a pair surviving	$= 0.43 \times 0.43 = 0.185$

Between seasons (100 pairs of foxes)

32.5 pairs lose both members	65 dead
24.5 pairs lose the male	24.5 dead
24.5 pairs lose the female	24.5 dead
18.5 pairs survive	37 alive
plus 49 single survivors	49 alive

Of the 18.5 pairs that survive, 3.5 will survive as a pair to the next season.

An adult mortality of 57% represents, in a steady population, a replacement by an equivalent number of juveniles. Thus, among foxes occupying fifty home ranges, 28.5 juvenile males and 12.25 adult males will be competing for the corresponding number of unattached adult and juvenile vixens, and the proportion seeking mates, at 57% mortality, is 81.5% which represents a considerable change of tenancy from one year to another. For purposes of these calculations it is assumed that foxes remain paired so long as both survive – a supposition supported by the observations of fox ranchers and others that the wild fox is monogamous (Tembrock, 1957).

The calculation of the annual state of change is shown schematically in Fig. 5. Since tagging studies have not revealed large shifts in range of adults throughout their lives, the surviving member of a pair is shown to occupy its original area, and the gaps left by the loss of a single animal or of a pair

Table 2. Age structures; percentage in year groups; selected examples

Year group	1	2	3	4	5	6	7	8	9	10
West Wales	43.0	23.0	13.0	11.0	2.5	4.5	0.6	1.3	0.6	
Mid Wales	59.0	24.5	10.5	4.5	2.0	0.8				
Mid Wales (by tagging returns)	60.0	22.0	11.0	5.0	1.5					
Isle of Skye	67.0	19.0	9.5	3.4	0.8					
Ireland (Fairley, 1969, assuming 50% mortality annually)	50.0	25.0	13.0	6.0	3.0	1.5				
Denmark (Jensen and Nielsen, 1968)	75.0	11.0	7.0	3.0	1.0	1.5	0.8			
London (Harris, 1977)	52.0	26.0	10.0	5.0	3.5	2.5	0.5	0.5		
Switzerland (Wandeler et al., 1974)	70.0	13.0	7.0	4.0		6.0		?		
USSR (Chirkova, 1955; by tagging)	90.0		10.0	?						
France (Artois, pers. comm.)	49.0	24.0	10.0	11.0	7.0					
USA (Phillips, 1970)	72.0		28.0		?	(In area of little control)				
	85.0		15.0		?	(In area of intensive control)				

Animals aged by dental or skeletal examination except where indicated, and figures rounded to nearest decimal place.
Year group 1 represents juveniles and sub-adults up to twelve months old, and the other groups correspondingly.

remain until the dispersal period. Minor shifts or encroachment upon parts of vacated adjacent ranges need not concern us. On this basis dispersing juveniles, though seeking both mates and suitable habitats in areas already colonised by foxes, are provided with ready-made previously selected habitats which can support foxes; thus habitat requirements are already more or less assured where a member of a pair survives. Olfactory and other clues probably indicate to itinerant foxes moving into a completely vacant range that the area was occupied by foxes formerly, and that it is consequently propitious.

Several features are relevant to dispersal and uptake of home ranges by juveniles. Broadly they are related to high, and moderate to low density fox populations where the juveniles live. The high population densities of 3.5 to 4.5 foxes/km^2 of parts of Pembrokeshire where there was little interference by man, are nearer to a state of balance with the habitat than those of mid-Wales where the population is subject to considerable interference. In the latter area fox numbers are below (at 1 to 2.5 foxes/km^2) that which the habitat can support. Elsewhere, to make the distinction, foxes may be at low densities (as on the more isolated Scottish hills) and be near the carrying capacity of the habitat, but little is known of the biology of foxes in such areas. In high density fox populations individual or home range sizes of pairs of foxes are small. The annual adult mortality, recruitment rate of juveniles and distances moved during dispersal are also low compared with these features in populations of low density: but the converse obtains with the

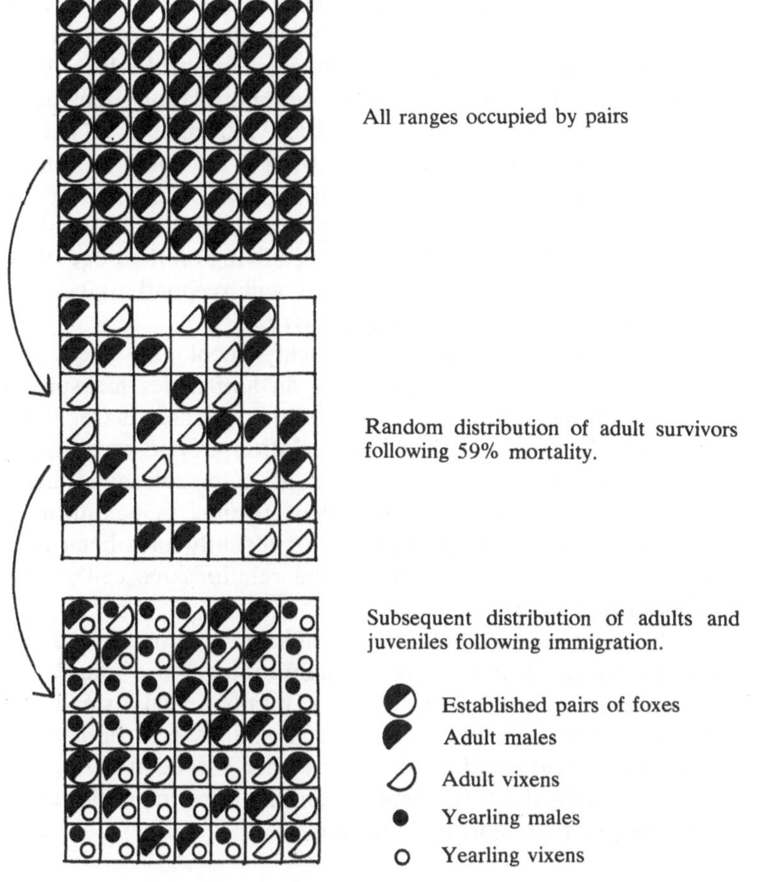

Fig. 5. Schematic representation of vacancies occurring in home ranges occupied by pairs of foxes when mortality of adults is 59% annually. For ease of display the figure of 59% is used instead of 57% as quoted in the text.

incidence of non-productive vixens and the mean number of cubs per vixen in the population. These features are all reflected indirectly in Fig. 5. For example, if the recruitment rate and adult mortality is low, say at 33%, 45% of pairs will remain intact from one year to another, and only 55% of foxes will be seeking mates compared with 81.5% in the mid-Wales example. Since home range sizes are smaller in the high density groups emigrant foxes will not need to move so far as those foxes in low density populations. It is simply a matter of scale. In both areas the number of encounters with prospective vacancies will be the same, per emigrant fox, since the number of emigrant animals and of vacancies are roughly similar locally, but the distances involved can be very different. This is oversimplification not least because of the possibility of kinship group ranges occurring in high density populations, but at least the idea fits most field observations where sufficient is known of the populations being studied. For example, emigrant movements in the high density populations of Pembrokeshire are a mean of 4.7

and 1.8 km for males and females respectively, and in the lower density populations of mid-Wales 13.7 and 2.25 km, with 11% of male foxes moving more than 20 km compared with none in Pembrokeshire, (Lloyd, 1980). Similarly in north America, where fox population densities are generally much lower than in Britain, emigrant movements in excess of 20 km predominate, and movements beyond 100 km are common (Storm et al., 1976; Dennis Voigt, *pers. comm.*). Such dispersal movements – except in north America – all refer to emigration from and to similar habitat or population density areas, but high density groups living on the edge of areas where foxes are limited in numbers by man will respond to conditions obtaining in the areas into which the foxes move.

It is commonly expressed that foxes quickly recolonise areas where clearance of foxes has been made, usually as an anti-rabies measure. The observation is not in dispute but the implication that the area is free of foxes is not qualified, nor the status of the fox population around or adjacent to such areas. Observations on the slow rate of colonization of hitherto fox-free areas suggest that foxes are unadventuresome in colonising new areas and will only do so readily if foxes are, or recently have been present in the area, which is a strong hint that social relationships can over-ride habitat requirements in some situations.

The generalisations presented here, based on inferences made in the field, which are part of the stock in trade of the field biologist are founded on a limited number of particular examples which can only reveal a small part of the complicated relationships between the fox and its environment. It is hoped that however rashly and inadequately this has been done it may stimulate further investigations, in particular on the experimental manipulation of wild fox populations and their resources.

REFERENCES

Apfelbach, R. 1973. Olfactory sign stimulus for prey selection in polecats. *Z. Tierpsychol.*, 33: 270–3.

Blackmore, D. K. 1964. A survey of disease in British wild foxes (*Vulpes vulpes*). *Vet. Rec.*, 33: 527–33.

Chirkova, A. F. 1955. Fox tagging studies. *Voprosy Biologii Pushnykh Zverery*, Moscow, 14: 191.

Englund, J. 1970. Some aspects of reproduction and mortality rates in Swedish foxes. *Viltrevy*, 8: 1–82.

Fairley, J. S. 1969. Tagging studies of the red fox, (*Vulpes vulpes*) in north-east Ireland. *J. Zool., Lond.*, 159: 527–32.

Golley, F. B., Ryszkowski, L. and Sokur, J. T. 1975. The role of small mammals in temperate forests, grasslands and cultivated fields in *Small mammals, their productivity and population dynamics*; Golley, F. B. et al. eds. London: Cambridge University Press.

Harris, S. 1977. Distribution, habitat utilization and age structure of a suburban fox (*Vulpes vulpes*) population. *Mammal Rev.*, 7: 25–39.

Hewson, R. and Kolb, H. 1973. Changes in numbers and distribution of foxes (*Vulpes vulpes*) killed in Scotland from 1948–70. *J. Zool., Lond.*, 171: 285–92.

Jensen, B. and Nielsen, L. B. 1968. Age determination of the red fox (*Vulpes vulpes* L.) from canine tooth sections. *Dan. Rev. Game Biol.*, 5: (6), 3–15.

Kurten, B. 1968. Pleistocene mammals of Europe. London: Weidenfeld and Nicolson.

Lloyd, H. G. 1976. Wildlife rabies in Europe and the British situation. *Trans. Roy. Soc. Trop. Med. Hyg.*, 70: 179–87.

Lloyd, H. G. 1980. The red fox. London: Batsford.
Lubbock, Richard. (1845) *Observations on the fauna of Norfolk.* Norwich: Charles Muskett.
Macdonald, D. W. 1977. The behavioural ecology of the red fox in Kaplan, ed., Rabies the facts. London: Oxford University Press.
Niewold, F. J. J. 1973. Various movements of the red fox (*Vulpes vulpes*) determined by radio-tracking. *Proc.* 11*th Congress Int. Union of Game Biologists,* Stockholm.
Phillips, R. L. 1970. Age ratios of Iowa foxes. *J. Wildl. Mgmt.*, 34: 52–6.
Sargeant, A. B. 1972. Red fox spatial characteristics in relation to waterfowl predation. *J. Wildl. Mgmt.*, 36: 225–36.
Storm, G. L., Andrews, R. D., Phillips, R. L., Bishop, R. A., Siniff, D. B. and Tester, J. R. 1976. Morphology, reproduction, dispersal and mortality of mid-western red fox populations. *Wildl. Monographs,* (*The Wildlife Society*). 49: 82pp.
Taylor, J. C. and Blackmore, D. K. 1961. A short note on the heavy mortality in foxes during the winter 1959–60. *Vet. Rec.*, 73: 232–3.
Tembrock, G. 1957. Zur Ethologie des Rotfuches (*Vulpes vulpes* L.) unter besondere Beruck-sichtigung der Fortpflanzung. *Zool. Gart. Lpag.*, 23: 289–532.
Vaughan, H. E. N. and Vaughan, J. A. 1968. Some aspects of the epizootiology of myx-omatosis. *Symp. zool. Soc. Lond.*, 24: 298–309.
Rothschild, M. 1963. A rise in the flea index of the hare, with relevant notes on the fox and woodpigeon, at Ashton, Peterboroguh. *Proc. zool. Soc. Lond.*, 140: 341–6.
Vincent, R. E. 1958. Observations of red fox behaviour. *Ecology*, 39: 755–7.
Wandeler, A., Muller, J., Wachendorfer, G., Schale, W., Forster, U. and Steck, F. 1974. Rabies in wild animals in Central Europe. *Zbl. Veter. Med.*, B. 21: 765–73.
Zeuner, F. E. 1963. A history of the domesticated animals. London: Hutchinson.

3 THE RED FOX – *Vulpes vulpes* (L., 1758) – IN EUROPE

Michael Stubbe*

INTRODUCTION

Relative to the total size of the individual countries, not including the European part of the Soviet Union, the red fox populates a territory of about 4,738,000 km^2. It is not present in Iceland, Crete and on the Balearic Islands. Its ecological prevalence is impressive. In the European highlands the red fox can be found as high as 2,500 m, in the summer months up to 3,500 m NN.

Despite very severe persecution, the existence of the red fox is not endangered in any European country. This has to be stressed particularly since the fox, like other beasts of prey, is deeply embedded in myth, legend and fable. Time and again this gives rise to emotional arguments for various, often irrational, reasons when necessary reductions have to be implemented. The fact that hundreds of thousands of red foxes are shot in Europe every year is often ignored.

Nevertheless, in an era of profound ecological transformation, it seems more imperative than ever to have an exact survey of the areas of each individual animal species, to assess their population density and to work out stock prognoses. This should not be the privilege or task of one country but calls for systematic coordinated international cooperation.

By calculating the relative fox density, the hunting indicator of population density (HIPD, according to Bögel et al., 1974), i.e. the number of animals shot annually per 100 ha (=1 km^2) hunting-ground, important comparative data may be obtained for the assessment of rabies incidence as well as of the economic and ecological role of the red fox. Moreover, international comparison helps determine the sites to which reduction efforts should be directed and also the actual stock in a given country.

Since in the epidemiology of rabies the fox plays an important role (though this is not its only one!) it is suggested that standardized statistics are established of the annual shootings, captures and other losses in all countries. By bringing the shooting numbers into relation with the known reproduction rates of red foxes the annual average spring stock may be determined.

Thanks to contributions from a number of foreign colleagues it was possible to record the annual fox bag for 60% of the European distribution area, though not including the European Soviet Union. By means of

* Wissenschaftsbereich Zoologie der Sektion Biowissenschaften, Martin-Luther-Universität, Halle-Wittenberg, DDR (WB-Leiter: Prof. Dr. sc. J. Schuh), DDR-402 Halle, Domplatz 4.

proportional conversion further data on the mean annual spring stock in Europe from 1971 to 1975 could be obtained for the remaining 40 per cent. We are especially indebted to Dr. Z. Pielowski (Čempin), Dr. P. Suminski (Warsaw), Dr. N. Erdei (Hódmezövásárhely), Dr. K. Tschiderer (Schwechat), M. F. Broggi (Vaduz), Dr. H. Sägesser (Bern), to the centre of the Austrian Hunting Societies (Vienna), to the Eidgenössisches Oberforstinspektorat (Bern), Dr. S. Myrberget (Trondheim), Dr. B. Jensen (Kalö), Prof. Dr. L. Bencze (Sopron), O. Aukrust (Oslo), Dr. P. Dragoev (Sofia), Prof. Dr. E. Schäfer (Göhr) Dr. J. van Haaften (Arnhem), H. G. Lloyd (Llandrindod Wells), Prof. Dr. K. Borg (Stockholm), T. Mörner (Stockholm), Dr. P. Hell (Zvolen), Dr. J. Kučera (Prague), Dr. Combe (Paris), Dr. G. Heidemann (Munich), Prof. Dr. T. Lampio (Helsinki), Dr. M. Delibes (Sevilla), to the Deutscher Jagdverband (Bonn), to the Direktion des Eaux et Forests (Luxemburg).

ANNUAL FOX BAGS AND ASSESSMENT OF THE RED FOX STOCK IN EUROPE

An attempt to assess the annual fox bags of all European countries revealed that statistic records are well organized and carried on continuously over a longer period of time only in slightly more than 50% of the relevant countries (cf. Tables 1 and 2). In several countries (among them Spain, Italy, Ireland and Great Britain) the recording of the annual fox shootings has been completely neglected so far, which is partially attributed to the complicated hunting-licence systems. It is, however, imperative also under these conditions to compile hunting statistics for predators. From other countries no data could be obtained in spite of intensive effort. Thus the existence of statistics may be assumed also for Romania and Albania. Within the group of countries with good hunting statistics different key-days are used. Taking into account the annual population dynamics, the most efficient method from the biological point of view would be to follow the annual population cycle or else the hunting year (from April 1 to March 31 of the following year). Since, however, in most countries records are kept according to the calendar year (January 1 to December 31) the tables show the calendar years, while for countries with hunting year statistics the fox bags are summarized under the year first mentioned (e.g. 1975/76 under 1975). When mean values of many years are calculated this mistake is not of very great relevance.

The fox statistics (Tables 1 and 2) show minimum numbers, i.e. they do not include found carcases, traffic accident losses and the non-recordable victims of gas campaigns. Another, quite considerable proportion disappears in the "private sphere", thus defying calculation in some countries. It seems most likely that the non-recorded losses amount to 50 per cent of the total bag. Thus e.g. Pielowski (in lit.) estimates the fox bag in Poland to be twice to three times as high as the official data. Since our analysis comprises the official data the final result represents a minimum stock and the minimum annual losses, respectively. We succeeded in recording the hunting

Table 1. Fox bags in some European countries between 1961 and 1970.

Country	1961	1962	1963	1964	1965	1966	1967	1968	1969	1970
Bulgaria	59,028	61,690	42,279	39,128	30,343	28,205	42,915	46,006	50,468	60,888
FRG (apart from Bavaria)	(129,079)	(116,017)	(125,896)	(130,590)	(104,322)	(99,393)	(96,355)	(91,584)	114,398	113,015
Denmark	52,000	54,700	58,000	59,300	56,500	56,000	55,000	62,000	59,000	54,000
GDR	42,070	48,375	45,041	49,412	53,193	65,210	84,112	70,758	57,149	57,890
Finland	?	?	?	25,000	20,000	22,000	20,000	18,000	22,000	22,000
France	?	?	?	?	?	?	95,300	102,608	110,895	110,064
Liechtenstein	88	100	111	124	115	132	182	147	82	97
Luxemburg	2,529	2,341	2,450	2,671	1,965	2,976	2,572	1,148	734	957
Norway	?	?	?	?	?	?	?	?	?	?
Austria	42,068	40,378	39,670	43,635	39,940	42,064	42,398	40,030	37,026	37,197
Poland	19,639	24,916	22,499	27,234	18,179	24,160	24,614	23,789	21,235	25,438
Sweden	73,500	75,000	76,790	71,727	71,130	79,411	75,926	71,634	67,749	76,744
Switzerland	19,628	17,207	19,616	22,087	22,433	25,615	23,310	28,697	23,244	19,650
Czechoslovakia	?	?	?	?	?	41,419	40,385	28,732	30,259	30,684
Hungary	28,238	24,827	30,625	29,143	?	?	?	?	31,393	28,356

Table 2. Fox bags in some European countries between 1971 and 1975 and relations to hunting areas.

Country	1971	1972	1973	1974	1975	∅ 1971–1975	Total size in km²	Hunting area in km²	Percentage of agricultural area from total area	Percentage of forest area from total area	Killed foxes per km² hunting area
Bulgaria	72,744	76,796	82,053	90,518	82,473	80,917	110,910	97,610	53.9	34.1	0.83
FRG	121,803	117,597	129,278	177,248	219,550	153,275	247,140	205,890	54.3	30.0	0.74
Denmark	56,000	53,000	50,000	48,000	53,000	52,000	43,070	34,580	69.3	11.0	1.50
GDR	57,630	57,223	50,580	54,255	61,607	56,259	108,330	92,370	58.0	27.2	0.61
Finland	17,000	18,000	17,000	15,000	20,000	17,400	337,010	214,630	8.2	55.5	0.08
France	109,322	114,263	109,964	102,528	106,739	108,563	547,030	470,490	59.3	26.7	0.23
Liechtenstein	93	37	28	38	102	60	158	(125)	(52.0)	(27.0)	(0.48)
Luxemburg	1,013	1,284	1,633	1,732	1,897	1,512	2,590	2,180	51.0	33.2	0.69
Norway	44,000	48,000	52,000	67,000	38,000	49,800	324,220	92,340	2.8	25.7	0.54
Austria	37,338	39,671	36,279	37,149	41,870	38,461	83,850	70,430	45.2	38.8	0.55
Poland	31,948	30,551	29,871	31,725	33,672	31,553	312,680	278,320	61.8	27.2	0.11
Sweden	71,976	68,657	74,777	75,569	68,988	71,993	449,960	264,310	8.3	50.5	0.27
Switzerland	17,168	20,041	18,004	19,443	24,973	19,926	41,290	30,690	48.8	25.5	0.65
Czechoslovakia	32,060	30,533	30,243	32,743	38,473	32,810	127,880	115,260	55.2	35.0	0.28
Hungary	31,459	29,015	26,253	24,017	25,434	27,236	93,030	83,250	73.5	16.0	0.34
total	702,454	704,668	707,963	776,965	816,778	741,765	2,829,148	2,052,425	—	—	0.36

bag from 1971 to 1975 for 15 European countries, comprising 60 per cent of the distribution area of the fox in Europe (without the Soviet Union) (Table 2). According to these findings, a five-year annual average of 741,765 foxes were killed. Starting with 702,454 specimens in 1971, there was a continuous increase up to 816,778 foxes in 1975.

A proportional conversion to the remaining 40 per cent of the fox area leads to the conclusion that theoretically there had been an annual fox bag of 500,772 foxes within the same period of time. Thus, an annual minimum loss of about 1,242,500 foxes is obtained for the evaluated European area for the years 1971 to 1975.

According to the summarizing data and findings obtained by Wandeler (1968), Lloyd (1968), Fairley (1970), Englund (1970), Lloyd et al. (1976) as well as by Stubbe (1973, 1977) an average reproduction rate of 4.7 to 4.8 surviving (bred) young animals may be assumed per litter. Minor series of investigations from Czechoslovakia (Soviš, 1967) and from Poland (Pielowski, 1976) yield values between 5.5 and 6.3. Large-scale tests covering several climatically differing years are likely to reveal relatively high litters of an average of nearly 4.7, too.

Dividing the average annual shooting rate of 1,242,500 foxes by 4.7 we obtain a mean annual number of fox litters for Europe amounting to 264,362. Large-scale tests showed a predominance of males in the fox populations (cf. Lit). Moreover, a certain percentage of female animals do not reproduce. Lloyd et al. (1976) mention a figure of 10 to 12 per cent, which is sometimes assumed to be even higher. Considering these two phenomena, one may say that the share of old foxes per litter is 2.4 to 2.5. The optimum conclusion from the above mentioned considerations for the minimum stock would thus be 2.5 old foxes per litter. When this figure is multiplied by the mean number of litters an annual basic or spring stock, on April 1, of about 660,900 foxes is obtained for Europe (average of the years 1971 to 1975).

Taking into account the theoretical character of this calculation these figures represent minimum values, which, however, do provide a real background for the stock assessment.

Red fox shooting numbers known from traditional fur countries, such as the Soviet Union, Mongolia and from North America show that the fox has retained its considerable economic importance. In the Soviet Union an annual average of 487,300 red foxes were flayed (skinned) from 1924 to 1958 (Heptner et al., 1967). As in the Asian part of the Soviet Union there also still exist in central Asia many natural enemies which have a controlling effect on the population density. In the Mongolian People's Republic an average of 30,000 to 40,000 foxes are killed (Stubbe, 1975).

Data concerning the relative population density of the fox in the evaluated European countries can be seen in Table 2. It is very difficult to assess the data since the statistics in the individual countries are unlikely to share the same objective value, and it is unknown what percentage of the population losses is actually recorded. Some informations point to considerable differences.

With regard to fox density (∅ of the years 1971 to 75) one can roughly distinguish between three categories of countries:

(1) Countries with a low relative fox density (HIPD = 0.08 to 0.40 killed foxes/100 ha hunting district): Finland, France, Poland, Sweden, Czechoslovakia, Hungary.
(2) Countries with an average fox density (HIPD 0.41 to 0.80 killed foxes/100 ha hunting district): GDR, Norway, Switzerland, Liechtenstein, Luxemburg, FRG.
(3) Countries with a relatively high fox density (HIPD >0.81 killed foxes/100 ha hunting district): Denmark, Bulgaria.

All density data refer to hunting areas consisting mainly of agricultural and forest land (cf. Tables 2 and 3). International statistics include the category "other areas". In spite of intensive efforts it was impossible to find out in what way these other areas are further subdivided. Among others they include traffic roads, inland waters, residential areas including industrial premises, glacial and débris areas as well as waste land etc. However, as far as the fox is concerned, certain waste land areas have to be included in the hunting district, which would mean a slight decrease in the density values given for some countries, especially Sweden (Norway).

In the future further material should be provided in order to conscientiously investigate the reasons for the remarkable differences in population density, since this could possibly lead to important conclusions for epidemiologists and hunters. The whole set of problems contains many unsolved and important questions such as: What kind of hunting methods predominate in the individual countries? – What is the role of natural enemies? – Is the number of hunters of any influence? – What degrees of preference are there in the occupation of different biotopes, including space structure and levels? – What is the influence of prey supply on fox density? – In what way do fox populations in non-hunting areas adjust?

At first sight the share of agricultural and forest areas in the hunting district do not reveal any clear tendency for the differences in fox population density (Table 2). In order to elucidate these relations a great number of factors have to be analysed.

We suggest an international programme for the investigation of the red fox, the objective of which would be to provide clearly defined basic data over a period of five to ten years using the same methods. This project should evaluate and use the experience gathered from the analysis of wild-animal rabies in Europe carried out by WHO and FAO in some West European countries with a view to achieving a coordinated fox research scheme throughout Europe.

SUMMARY

Based upon the shooting record and the established rates of reproduction of the red fox in some European countries a mean spring stock of at least 660,900 foxes in the total area (excluding the USSR) can be calculated.

Table 3. Areas and main forms of land usage or cultures in European countries, including the USSR (according to Yearbook of the GDR).

Country	Year	Area (in 1,000 hectars) Total	Agricultural area Total	Arable land[1]	Meadows + pastures	Forest area	Other areas
Albania	1974	2,875	1,240	640	600	1,200	435
Belgium	1973	3,051	1,568	835	733	601	882
Bulgaria	1973	11,091	5,982	4,502	1,480	3,779	1,330
FRG	1973	24,714	13,425	8,078	5,347	7,164	4,125
Denmark (apart from Farö Islands)	1973	4,307	2,986	2,668	318	472	849
GDR	1973	10,833	6,287	4,858	1,429	2,950	1,596
Finland	1973	33,701	2,766	2,714[3]	52	18,697	12,238
France	1974	54,703	32,441	18,844	13,597	14,608	7,654
Greece	1974	13,194	9,155	3,905	5,250	2,615	1,424
Great Britain and Northern Ireland	1973	24,404	18,683	7,164	11,519	1,984	3,737
Ireland	1973	7,028	4,841	1,179	3,662	216	1,971
Iceland	1973	10,300	2,280	1	2,279	120	7,900
Italy	1973	30,123	17,484	12,235	5,249	6,226	6,413
Yugoslavia	1973	25,580	14,431	8,087	6,344[2]	8,858	2,291
Liechtenstein	1973	16	?	?	?	?	?
Luxemburg	1974	259	132	62	70	86	41
Netherlands	1973	3,685	2,101	834	1,267	303	1,281
Norway	1973	32,422	904	790	114	8,330	23,188
Austria	1973	8,385	3,793	1,612	2,181	3,250	1,342
Poland	1973	31,268	19,326	15,107	4,219	8,506	3,395
Portugal	1969	8,769	4,900	4,370	530	3,109	760
Romania	1973	23,750	14,904	10,426	4,478	6,309	2,537
Sweden	1973	44,996	3,718	3,018	700	22,713	18,565
Switzerland	1973	4,129	2,017	384	1,633	1,052	1,060
Spain	1973	50,475	?	?	?	?	?
Czechoslovakia	1973	12,788	7,060	5,311	1,749	4,466	1,262
USSR	1973	2,240,220	550,085	230,684	319,401	?	?
Hungary	1973	9,303	6,835	5,555	1,280	1,490	978

[1] According to the international nomenclature, this column comprises, together with arable land: garden land, fruit-cultures, land planted with vines, osier land, rubber plantations, temporary fallow land and temporary meadows (rotation utilization).
[2] 1971.
[3] Only agricultural enterprises.

In spite of severe persecution and reduction measures the existence of the red fox in any European country is not endangered. A minimum of 1,242,500 foxes were killed every year in the above mentioned area.

By assessing the relative population density – the hunting indicator of population density – important comparative data may be obtained for the examined countries, the values being between 0.08 and 1.50 killed foxes annually per 100 ha hunting district.

Since the red fox plays an important role in the epidemiology of rabies it is suggested that standardized records are kept of the annual fox bag and other losses in all countries. A proposal is made for setting up an international programme for the investigation of the red fox.

REFERENCES

Bögel, K., et al. 1974. Recovery of Reduced Fox Populations in Rabies Control. *Zbl. Vet. Med.* B 21: 401–412.

Englund, J. 1970. Some aspects of reproduction and mortality rates in Swedish Foxes (*Vulpes vulpes*) 1961–1963 and 1966–1969. *Viltrevy* 8: 1–82.

Fairley, J. S. 1970. The food, reproduction, form, growth and development of the fox *Vulpes vulpes* (L.) in Northeast Ireland. *Proc. Royal Irish Acad.* 69 (B, No. 5): 103–137.

Heptner, V. G. and Naumov, N. P. 1967. Die Säugetiere der Sowjetunion. Moskau, Bd. II.

Lloyd, H. G. 1968. The control of foxes (*Vulpes vulpes* L.). *Ann. appl. Biol.* 61: 334–345.

Lloyd, H. G., Jensen, B., Van Haaften, J. L., Niewold, F. J., Wandeler, A., Bögel, K. and Arata, A. A. 1976. Annual Turnover of Fox Populations in Europe. *Zbl. Vet. Med.* B. 23: 580–589.

Pielowski, Z. 1976. The Role of Foxes in the Reduction of the European Hare Population. Ecology and Management of European Hare Populations (Proc. int. symp. in Poznań on Dec. 23–24, 1974), Warszawa, 135–148.

Soviš, B. A. 1967. A contribution to the food ecology and population dynamics of the common fox with regard to its economic importance. *Acta Zootechn. Univ. Agric. Nitra* 15: 159–171.

Stubbe, M. 1973. Der Fuchs (*Vulpes vulpes* L.). In Stubbe, H.: Buch der Hege. Berlin, Bd. I: 182–212.

Stubbe, M. 1975. Die Jagd in der Mongolischen Volksrepublik. *Jagdinformationen* 4 (2): 20–28.

Stubbe, M. and Stubbe, W. 1977. Zur Populationsbiologie des Rotfuchses *Vulpes vulpes* (L.) III. *Hercynia N. F.* 14: 160–177.

Wandeler, A. 1968. Einige Daten über den bernischen Fuchsbestand. *Revue Suisse Zool.* 75: 1071–1075.

4 COMPARISON OF THE DIET OF THE RED FOX (Vulpes vulpes L., 1758) IN GELDERLAND (HOLLAND), DENMARK AND FINNISH LAPLAND

Darrell M. Sequeira*

INTRODUCTION

During this century the ecology of the red fox (*Vulpes vulpes*, L.) has been studied throughout its biogeographic (Holarctic, Oriental, Australasian, Northern Neotropical and African) range. In recent decades interest in the European red fox has increased because of its occurrence in practically all biotopes and due to its role as a predator and scavenger, its importance in the transmission of rabies, and to the economy of farmers and hunters.

In order to determine the extent of our knowledge about the food ecology of the European red fox and the methods used to investigate its diet, I first present a review of the literature.

The food ecology of the red fox in Europe has been studied extensively (Table 1) by various methods. Most of the workers have attempted to answer the question: *What do foxes eat and in what relative quantities?* They have usually analysed stomach contents, but also: faecal samples, stomach and intestine contents, stomach contents and faecal samples, and remains outside fox earths.

But only food remains which resist digestion will occur in the faecal samples and this bias must be acknowledged when considering the results, as only the principle foods will be indicated in faecal analysis.

The possibility of using faecal analysis as a method for estimating the quantities of prey consumed has been evaluated by Scott (1941) and Lockie (1959). On the basis of results obtained from experiments with captive foxes Scott (op. cit.) suggested that the frequency of occurrence multiplied by a conversion number is a reliable method for calculating the numbers and weights of prey animals eaten. But Lockie (1959) concluded that measurement of frequency of occurrence was frought with serious bias as it depended on the amount of food eaten and the order in which it is eaten. He recommended the use of dry undigested matter in the faeces for estimating bulk, the relative proportion, and daily consumption in weight of different prey eaten.

Based on the analysis of faeces, Lockie (1964), Ryszkowski et al. (1971) and Goszczynski (1974) have obtained information about the numbers and weights of prey eaten in addition to general information about the diet of

* Wildlife Research Unit, Department of Agriculture & Forest Zoology, University of Helsinki, Viikki, 00710 Helsinki 7, Finland.

Table 1. Studies on the food ecology of the red fox in Europe (with the exception of Estonia, excluding U.S.S.R.).*

Country	References	No. of stomachs with contents analysed	No. of intestines with contents analysed	No. of faecal samples analysed	Other information
Norway	Lund, 1962	484	—	984	Snow tracking (274 km) remains at 18 earths and captive foxes
Sweden	Englund, 1965a	1,166	—	—	—
	Englund, 1965b	120	—	—	—
	Englund, 1969	243	—	—	—
	Palm, 1970	—	—	—	Snow tracking (162 km)
Finland	Lampio, 1953	325	—	—	—
	Österholm, 1964	—	—	—	Captive foxes
Estonia USSR	Naaber, 1967	—	—	229	42 remains
Poland	Rzebik-Kowalska, 1972	496	—	—	—
	Ryszkowski et al., 1971	—	—	523	—
	Goszszynski, 1974	—	—	ca. 1,000	—
	Mateljka, Röben and Schröder, 1977	—	—	329	—
Denmark	Bistrup, 1890	40	—	—	—
	Jensen and Sequeira, 1978	285	—	—	Remains at 275 earths
West Germany	Behrendt, 1955	128	—	—	Remains at 4 earths, captive foxes
	Lessmann, 1971	235	—	—	—
	Witt, 1976	131	131	—	—
	Frisch and Frisch, 1971	—	—	—	Captive foxes
	Klenk, 1971	—	—	—	Captive foxes
	von Lutz, 1978	—	—	806	—
France	Brosset, 1975	—	—	450	—
Switzerland	Wandeler and Hörning, 1972	623	—	—	—
	Fuchs, 1972	—	—	—	3,021 bones
Czechoslowakia	Sovis, 1967	86	—	—	Remains at 11 earths
Rumania	Hellwing, 1960	8	—	—	—
Bulgaria	Atanassov, 1958	192	—	142	—
	Peshev, 1965	262	—	—	—
Italy	Leinati et al., 1960	—	—	5,280	—
Spain	Amores, 1975	121	121	—	—
Portugal	Magalhaes, 1974	—	—	157	—
Holland	Sequeira, 1978	153	141	28	Captive foxes and age determination
United Kingdom	Southern and Watson, 1941	40	—	18	—
	Lever, 1959	420	—	123	—
England	MacDonald, 1977a	—	—	2,872	Captive foxes
	MacDonald, 1977b	—	—	—	Captive foxes
Scotland	Lockie, 1956	14	—	98	—
	Lockie, 1964	—	—	178	—
	Douglas, 1965	14	—	—	—
	Hewson et al., 1975	—	—	523	—
	Watson, 1976	—	—	219	—
	Richards, 1977	—	—	186	64 feeding remains
Ireland	Fairley, 1970	503	—	—	Remains at 23 earths, captive foxes
	Forbes and Lance, 1976	—	—	53	—
	Total	6,089	393	14,198	

* After Jensen and Sequiera 1978.

foxes. Ryszkowski et al. (op. cit.) attempted to relate the prey consumed by foxes to the availability of prey by quantitive analysis of faeces and estimation of the population density of prey in the study area.

Quantification of the diet was also attempted from the results of stomach analysis by Lampio (1953) and Behrendt (1955), who both considered it reasonable to quantify the diet from its occurrence in the contents of stomachs and more recently from gut contents (Amores, 1975). Englund (1965a) had previously concluded that this was not possible.

The investigations (Table 1) cover the analysis of the contents of about 6,000 stomachs and more than 14,000 faecal samples. Although different biotopes are represented and comparison between studies is difficult because different methods of presenting the results were used, general features of the foxes' diet emerge.

What do foxes eat?

The principal food, according to frequency, are rodents, rabbits, hares, birds, carcasses, offal, carrion and domestic livestock. During certain seasons invertebrates and fruit are a principal food.

The most important rodent genus was *Microtus*. In areas where rabbits and hares were present, rabbits were usually more numerous and were eaten in larger numbers (Englund 1965a). *Galliformes* were the most important birds eaten. Most of these were domestic poultry waste. Pigs offal is another agricultural byproduct which was eaten. A significant portion of the flesh eaten was derived from carcasses and carrion. During winter there was a tendency to eat more wild ungulates such as roe deer (*Capreolus capreolus*), red deer (*Cervus elaphus*), moose (*Alces alces*), and fallow deer (*Dama dama*). Other domestic livestock eaten were sheep, lambs and cattle.

Items of secondary importance in the diet were: *Insectivora,* such as shrews and hedgehogs; invertebrates, mostly insects, particularly beetles, but also earthworms, slugs and snails; berries, fruits and seeds, particularly bilberries, apples and pears.

Insectivorous mammals were not common although they occur in most biotopes in Western Europe. Insects were eaten in summer and berries, fruits and seeds towards the end of summer and in autumn. The incidence of insects has been mentioned in most studies but its importance has not been evaluated.

In addition frogs, lizards, snakes, fish, squirrels, muskrats, weasels, cats, badgers and garbage have been reported in the diet occasionally.

How do foxes respond to a changing availability of prey, to the availability of prey in different habitats and to a virtual disappearance of a principal prey?

Faecal analysis has been used to investigate the diet of fox populations in relation to a changing availability of prey (Lockie, 1964, Hewson et al., 1975, Richards, 1977, Forbes and Lance, 1976, Brosset, 1975, Behrendt,

1955, Ryszkowski et al., 1971, Goszczynski, 1974); or for comparing the foxes diet in different habitats: (Hewson et al. (1975) between Scottish forests and open hills and between forests in West Scotland (Argyll) and North East Scotland (Kincardineshire), Southern and Watson (1941) between hill districts in Wales and Mid-England plains, Lever (1959) between lowlands and hills in England, MacDonald (1977a) between home ranges in Oxford, Fairley (1970) between Northern Ireland and Britain, Leinati et al. (1960) at different altitudes, Peshev (1965) between foothills, mountainous regions and flat country, Ryszkowski et al. (1971) between forests and cultivated fields, Rzebik-Kowalska (1972) between south and north Poland.

A consensus of opinion is that the food remains recovered in the faeces reflect the availability of prey according to the season of the year and the habitat where the foxes feed.

A catastrophic reduction in the population of rabbits in some areas of Western Europe due to myxomatosis provided an opportunity to test the response of foxes to the virtual disappearance of a principal prey. The expected change in diet was investigated by Lockie (1956) and Ingleby (1956) in Scotland; Cobnut (1955) and Lever (1959) in England; and Englund (1965b) in Gotland, Sweden.

The decline of the rabbit population was advantageous to farmers (Ingleby 1956) and resulted in an increase in the population of voles and hares (Lockie 1956), which provided an alternative food supply for foxes in Scotland. The shortage of rabbits was compensated with an increased consumption of the short-tailed vole (*Microtus agrestis*) and the brown rat (*Rattus norvegicus*) in England (Lever 1959), and by hare, mice and pheasants in Gotland (Englund 1965b). Englund (op. cit.) reported that there was no evidence of food shortages but Lockie (1956) claimed that there was a reduction in the number of predators.

In spite of the reduction in the availability of rabbits foxes still continued to hunt them; Englund (1965b) found that the rabbit population had decreased to $\frac{1}{20}$th of its original size but the occurrence of rabbits in the stomachs decreased by 50%. This suggests active prey selection. No increase in the depredation by foxes on livestock and game was found by Lockie (1956) and Ingleby (1956) but hares and pheasants were hunted in Gotland (Englund, 1965b).

How does the diet differ between age classes and sexes?

In order to answer this question it is clearly necessary to determine the ages and sexes of foxes so that they may be classified into different age groups for purposes of comparison. Morris (1972) has reviewed mammalian age determination methods.

Previous work on age determination of foxes according to eye lens weights has been done by Lord (1961), Friend and Linhart (1964), van Haaften (1970), Brömel and Zettl (1974). Jensen and Nielson (1968) and Grue and Jensen (1973) have determined age from annular cementum of canine tooth sections, Harris (1978) has used the first premolar, and van Haaften (op. cit.), Lüps et al. (1972) and Brömel and Zettl (1974) have determined age from dental development. Wandelar (1976) has used both methods. Geiger et al. (1977) has compared the three methods and Harris (1978) has evaluated the efficiency of age determination methods of the red fox.

Special studies of the diet of cubs were made by Douglas (1965), Lever (1959), Fairley (1970), Behrendt (1955), Englund (1969). Behrendt (op. cit.) concluded that the growth of captive cubs depends on the nutritive value, quantity and choice of nourishment. There were slight differences in the diets of cubs, juveniles and adults, according to Englund (1965a). None of the workers considered the nutritional and energetic value of the food eaten.

Lampio (1953) suggested that there was a difference in the weight and identity of the food eaten by males and females. Subsequent studies by Lund (1962), Englund (1965a), Fairley (1970) and Rzebik-Kowalska (1972) concluded that this was not so. The method of obtaining the foxes did not have any effect on the weight and prey contents of stomachs (Englund, 1965a).

What kind of endoparasites occur in the foxes' gut?

Some workers complimented stomach analysis with records of the incidence of gut parasites (Douglas, 1965, Magalhäus, 1974, Brosset, 1975, Behrendt, 1955, Wandeler and Hörning, 1972). In addition special studies of gut parasites have been made by Bereford-Jones (1961), Thompson (1976) in Britain; Ross and Fairley (1969) in Northern Ireland; Lamina and Main (1964) in West Germany, Lozanic (1967) in Yugoslavia, Prokopic (1960) in Czechoslovakia and Persson and Christenson (1971) in Sweden. Parasitic nematodes (e.g. Toxascara canis, leonina) and cestodes (e.g. Mesocestoides spp. and Taenia spp.) were most common.

What do foxes prefer to eat?

Lund (1962) investigated food preferences of captive foxes. He concluded that *Microtidae* was most preferred. MacDonald (1977b) did several experiments on the responses of foxes to *Microtidae*, *Muridae*, *Soricidae*, and *Carnivora*.

In addition estimates of the amounts eaten per day were also obtained (Lockie, 1959, Lund, 1962). Fairley (1970) reported that 90% of the food fed to foxes in Edinburgh Zoo had been passed within two days. According to Nessini et al. (1955) 10% of a meal may remain for more than 26 hours. In field trials Ryszkowski et al. (1971) estimated that a red fox defaecates 5 times in 24 hours.

When do foxes forage for food?

Österholm (1964) concluded, from experiments on captive foxes, that foxes forage at twilight in autumn and at night in winter. In the northern latitudes it forages during the light hours in summer. Haacke et al. (1973) found foxes to be most active at dawn and dusk. Accordingly foxes might be considered to be crepuscular animals which forage in the evening and early morning, in synchrony with the diurnal rhythm of their prey, for example, of mice, voles and leporids. But in fact foxes often forage throughout the night as revealed by radio tracking and infra red equipment (MacDonald per. comm.). When foraging the distance receptors in order of importance were, hearing, sight and smell. Sight had the highest potential for releasing the catch response but was insignificant in practice because most of the prey were not easily seen. Another indication was that foxes are monochromatic and are only capable of seeing movement and shapes (Haltenorth and Roth (1968). The remarkable powers of hearing and smell possessed by foxes is explained by Dudley (1977).

How do foxes behave while feeding?

Insight into the feeding behaviour of foxes has been obtained from snow-tracking observations (Lund, 1962) and by radio tracking and observation with infra-red equipment (MacDonald, 1977a). Lund (op. cit.) observed that foxes mostly dug into snow for voles, but also leave caches of food which they exploit later on, when passing a routine food trail. MacDonald (1976) has made experiments which have revealed much about caching behaviour. Montgomery's (1974) studies have revealed that foxes often forage for food individually. Other workers have reported foxes visiting rubbish heaps (Niewold, 1974) or exploiting special concentrations of food resources such as nesting bird colonies on islands (Bergman, 1966, Naaber, 1967).

What is the economic impact of fox food ecology on domestic and game animals?

Its depredations on domestic animals are of interest to farmers (Pursor and Young (1959), Fairley, 1969b) and on game animals to hunters (Bergman, 1966, Jensen, 1970, Spittler, 1972, Hewson et al., 1975, Mateljka et al., 1977, Jensen and Sequeira, 1978). Lund (1962) investigated the latter with 4,914 questionnaires and concluded that foxes exert little influence on stocks of small game. Domestic animals are killed when available, poultry and lambs especially, when rodent populations have declined.

Ingleby (1956) pointed out that the absence of rabbits is a substantial advantage to farmers and there was no indication of any increased depredation by foxes in spite of their increased population and wider distribution. Lockie (1956) also reported that there was little damage to livestock and game by wild predators following the almost complete disappearance of rabbits. According to Rzebik-Kowalska (1972) foxes have practically no influence on populations of game fowl and mammals. Poultry might be taken when under careless management, but foxes, together with polecats (*Putorius putorius*) and pine martens (*Martes martes*) play an important role in the destruction of harmful field and forest rodents.

But it seems that foxes are capable of influencing the bag of hunters (Jensen, 1970, Jensen and Sequeira, 1978). Foxes might be a serious menace to concentrations of breeding birds (Bergman, 1966, Naaber, 1967) and might significantly reduce the population of some species. Sovis (1967) has related the food eaten by foxes to agriculture and hunting.

Some workers take a rather dismal view of the value of foxes. Peshev (1965) recommends its further extermination and Naaber (1966) considers it to be a "noxious" animal and wishes that it should be killed by any means possible.

Two special studies of the relations between foxes and their prey were made by Ryszkowski et al. (1971) and Goszczynski (1974). All the above mentioned studies give an insight into the role of foxes as predators and scavengers and consequently their impact on prey populations in terrestrial ecosystems.

How do we accurately identify prey remains?

The results which have been reviewed in this introduction depend on accurate identification of prey remains. This needs considerable practical experience.

Reference works on the subject are as follows: (Hausman, 1920a and b, 1924, 1930, Williamson, 1938, Stoves, 1944a, Oyer, 1946, Lebland, 1951, Noback, 1951, Mandelli, 1961, Day, 1966, Ryder, 1973) on the structure and classification of mammalian hairs; (Rudall, 1941, Appleyard, and Greyville, 1950, Williamson, 1951), on the structure and classification of the hair cuticle; (Williams, 1934, Dearborne, 1938, Mathiak, 1938, Stoves, 1944b) on sectioning and identifying mammalian hair from cross-sectional outline. Mathiak (1938), Mayer (1952), Stains (1958), Dziurdzik (1973) have published keys on the identification of hairs.

Two useful guides on the identification of mammalian skulls are by Husson (1962) and Corbet (1969), and one of birds by Peterson et al., (1969). The identification of feathers is reviewed by Lillie (1942) and Chandler (1960). Day (1966) has also considered feather identification. In addition Svenson (1970) has published a detailed guide on the identification of passerine feathers.

Research on foxes has increased our knowledge about the ecology of foxes and perhaps of ways of controlling their populations. From this point of view the influence of food supply on the population density is of special interest (Lockie, 1964, Englund, 1970).

In general, a good knowledge of the diet of foxes in most European biotopes has been obtained but much needs to be learned about the relative quantities eaten in relation to the availability of prey species in defined study areas; within the foraging area of the fox population under study. Information is also short on preferred foods; food sharing among members of the same family group, and the nutritional value of the food. Some studies have given too little attention to sample size and many of the conclusions were not valid on the basis of available evidence (Cochran, 1953, pp. 50–59, Hanson and Graybill, 1956).

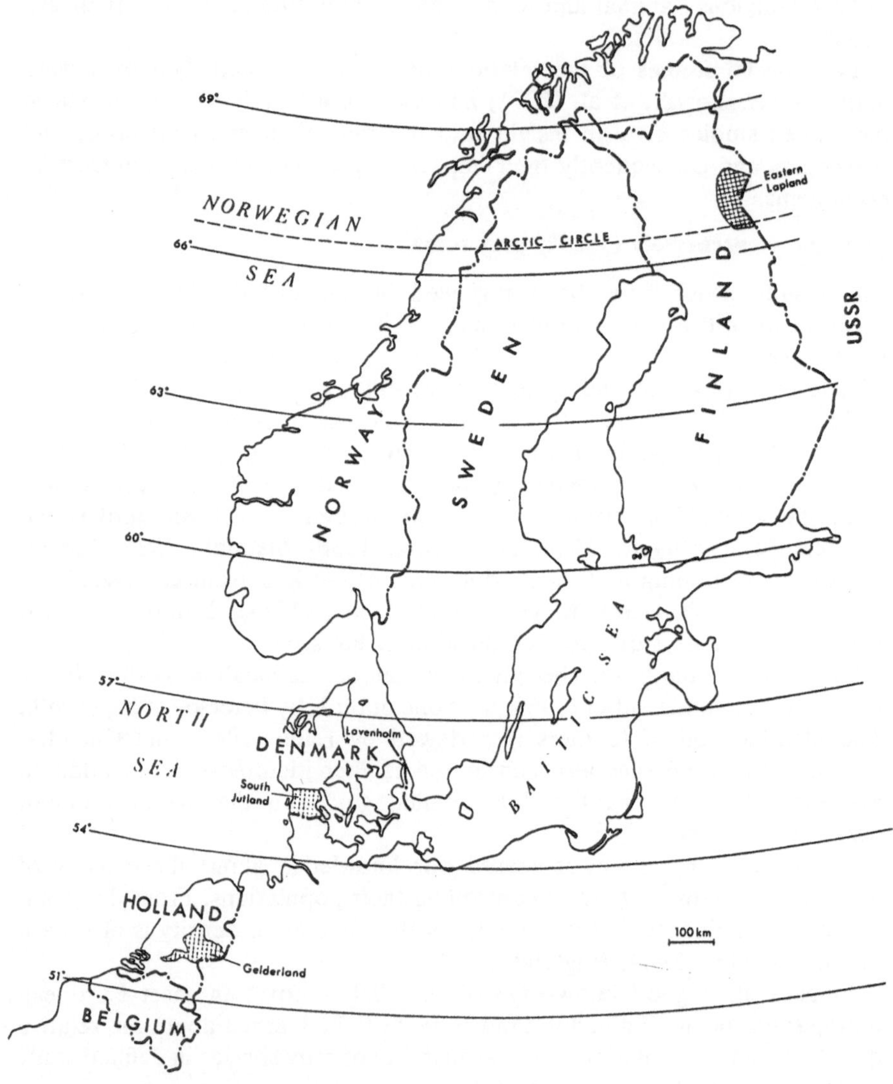

Fig. 1. Areas in Holland, Denmark and Finland where foxes were collected.

RED FOX FOOD ECOLOGY IN HOLLAND, DENMARK AND FINLAND

My studies of the red foxes' food ecology in Holland, Denmark and Finland (Fig. 1) has been the result of opportunities which arose in these countries, quite by chance, and does not embody any underlying scientific reasons for this combination of areas.

I have studied the ecological implications of the diet of the red fox in the province of Gelderland, Holland, (Sequeira, 1978) an area of 513,000 hectares of diverse habitats, including farmland, coniferous and deciduous forests, and heathlands; in 3 sample areas in Denmark (Jensen and Sequeira, 1978) the Lövenholm forest of about 30 Km^2 in Jutland, in South Jutland, and other areas of Denmark; and thirdly in Finland (unpublished) in an area of boreal forests in Eastern Finnish Lapland, in the municipalities of Salla and Savukoski.

153 stomachs and 141 intestines with contents, collected during 1969–70 in Gelderland, 285 stomachs with contents of >6 months old foxes obtained during 1965–70 in Denmark and 155 stomachs, with contents, collected during September–May 1967–73 in Finnish Lapland were analysed. In addition feeding experiments were done on 4 captive foxes in Holland.

The research in Gelderland and Denmark (1969–70), was part of a programme to control the transmission of rabies. It was important to know how diet might influence feeding behaviour and territory size of foxes and whether any prey animals might be responsible for the transmission of the rabies virus.

What and how much do foxes of different age classes and sexes eat during different seasons and annually?

The primary vertebrate foods eaten annually by foxes in Gelderland, Denmark and Finnish Lapland is shown in Table 2. The summer diet of fox cubs in Gelderland is shown in Table 3 and the summer diet of adult foxes in Gelderland and South Jutland is shown in Table 4.

No statistically significant difference was found in the diet of males and females and the sexes were therefore not separated. The prey have been listed in the same order because a chi-squared test for independence showed no significant difference between the occurrence of primary foods. But in reality differences might exist depending on the season and local availability of prey. However, such differences were not evident in the present study.

An important question relating to fox cubs is: *At what age are fox cubs weaned?*

In the case of fox cubs in Gelderland 86% of the cubs with only milk in their stomachs were under 1 month old. No milk was found in any cubs' stomachs over 2 months old. This suggests that fox cubs in Gelderland are weaned by the age of 8 weeks.

Table 2. Primary vertebrate foods eaten annually in Gelderland, Denmark and Finnish Lapland

Gelderland	Denmark	Finnish Lapland
152 s. 141 i.	285 s.	155 s.
Rabbits and hares	Hares and rabbits	Blue hare
Voles	Voles	Voles
Mice	Mice	—
Poultry	Pigs and poultry offal	—
Song birds	Song birds	—
—	Pheasants and partridges, and other bigger birds	*Tetraonidae*
		Reindeer offal

s. = stomachs, i. = intestines.

In the first month of life they are almost totally dependent on milk but by the age of 2 months they have been introduced to the important items of food and have become acquainted with a variety of food species. It is not clear for how long the parents supplement the cubs' diet after weaning.

In addition to the primary foods, cubs <2 month old ate some wood mice (*Apodemus sylvaticus*). No invertebrates and plants were found in the stomachs but remains of beetles were found in intestines.

The 2–6 month old cubs ate a variety of secondary foods which were as follows: Doves and pigeons, red deer, wild boar, domestic pigs, moles, shrews, pheasants and eggs. In addition one case of a garden snail (*Helix pomatia*) and some remains of beetles and flies (*Diptera*) were found. Also, evidence for pears, bilberries, grain, and prunes was found.

Not surpisingly there was a close similarity in the diet of adults and cubs <2 month old as they were fed by the adults. Cubs 2–6 months old ate a more varied diet.

The primary vertebrate foods eaten by foxes in Gelderland and South Jutland in summer were very similar. This implies that foxes occupy a similar ecological niche in these two ecosystems which are about 600 km apart. But unlike foxes in Gelderland, Danish foxes ate domestic pigs' offal, and pheasants and partridges.

Secondary foods eaten by adult foxes in Gelderland in summer were wild boar offal and pheasants. Small frequencies of shrews, mostly common

Table 3. Primary vertebrate foods eaten by fox cubs in Gelderland in summer

<2 m	2–6 m
48 s. 53 i.	15 s. 19 i.
Rabbits and hares	Rabbits and hares
Voles	Voles
—	Mice
Poultry	Poultry
Song birds	Song birds

Table 4. Primary vertebrate foods eaten by adult foxes in Gelderland and South Jutland in summer

Gelderland	South Jutland
44 s. 44 i.	59 s.
Rabbits and hares	Hares and a few rabbits
Voles	Voles
Mice	Mice
Poultry	Poultry and pigs' offal
Song birds	Song birds
(including undetermined mammals and birds)	Pheasants and partridges

shrews (*Sorex araneus* – 1.4%), red deer and roe deer (2.5%) occurred in the stomachs of Danish foxes. Also in Denmark 2 cases of fish (1%), a lavaret (*Coregonus lavaretus*) and 2 pheasants and 4 domestic hens eggs were found.

Important non-vertebrate foods in late summer and autumn in both the areas were fruit and insects. In Gelderland foxes ate apples, bilberries (*Vaccinium myrtillus*) cowberries (*Vaccinium vitis-idaea*) and cranberries (*Vaccinium oxycoccus*). In South Jutland foxes ate cherries, cherry plums, and plums (9%). In most cases up to 10 stones were found but as many as 48, 31 and 27 stones were recorded. Also apples and pears were found regularly in autumn and winter (4%).

In both areas beetles and maggots common on carcasses and dung were found. For example in Gelderland the frequency of insects was 24% of the total sample. Most were dung beetles (*Geotrupes*). In Gelderland 1 stomach contained 18 g of cuticle parts, but also ground beetles (*Carabidae*), sexton beetles (*Silphidae*), rove beetles (*Staphylinidae*) and maggots of flies (*Diptera*) were found.

Probably the most nutritious insects eaten were larvae of *Agrotinae* e.g. (*Apamea monoglypha*) – and cutworms agrotid moth. In Gelderland a stomach contained 11 g of the body walls of 110 apamea larvae. In South Jutland 25 cases of cutworms from Aug.–Nov. were recorded. There were a few in most cases but in 3 cases, about 100, and in 1 case 200 individuals.

This suggests that foxes actively seek fruit and insects particularly in late summer and autumn. Fruit and insects could be an important food for cubs and sub-adults. The implications of the occurrence of sarcosaprohagous insect fauna which are common on carcasses will be considered separately.

The primary vertebrate foods eaten in winter by foxes in Gelderland, Denmark and Finnish Lapland are shown in Table 5.

There was a great similarity in the diet of foxes in Gelderland and Denmark as already observed with the summer diet. The 4 common primary foods in both areas were hares, voles, mice and poultry. Unlike foxes in Gelderland which probably did not eat domestic pigs' offal in winter, foxes in Denmark ate lots of remains of piglets and offal from slaughtered pigs. Even those foxes living in a relatively large wood for Danish conditions,

Table 5. Primary vertebrate foods eaten by foxes in Gelderland, Denmark (Lövenholm forest, South Jutland, Various other localities) and Finnish Lapland in winter

Gelderland	Lövenholm	South Jutland	Other localities	Finnish Lapland
25 s. and 25 i.	60 s.	105 s.	56 s.	155 s.
Rabbits and hares	Hares	Hares and a few rabbits	Hares	Blue hare
Voles	Voles	Voles	Voles	Voles
Mice	Mice	Mice	Mice	—
Poultry	Poultry	Poultry	Poultry	—
—	Domestic pigs	Domestic pigs	Domestic pigs	—
—	—	Pheasants, partridges, song birds, and other bigger birds	Pheasants, partridges, song birds, and other bigger birds	*Tetraonidae*
—	—	—	—	Reindeer offal

s. = stomachs, i. = intestines.

Lövenholm forest (30 km^2), travelled to agricultural land in search of domestic pigs and in fact scored the highest frequency. The latter foxes did not eat pheasants, partridges and song birds in the same frequencies as foxes in South Jutland and other localities in Denmark.

Foxes in Gelderland ate a variety of secondary foods: Roe deer and possibly other artiodactyls, shrews, hedgehogs, song birds, *Anseriformes*, pheasants and undetermined birds. But foxes in Denmark apparently ate less variety of secondary foods as only 9 shrews (1.4%) and 7 cases of roe deer (2.5%) were found.

Once again in both areas fruit and insects were important non-vertebrate foods eaten in early winter (autumn). Apples and pears were eaten regularly and bilberries were found in a Gelderland fox stomach. In Gelderland, in addition to the beetles already mentioned, chafers and dung beetles (*Scarabaeidae*) e.g. *Aphodius* species common on cow's dung and *Typhoeus typhoeus* which is associated with the dung of rabbits and sheep was found. Also blowfly (*Lucilia*) and dipteran maggots were found.

In contrast, while foxes in Finnish Lapland exploited prey populations of blue hare (*Lepus timidus*) (brown hare do not survive in Lapland), and voles as in the other more southern areas, they do not seem to eat mice, poultry and domestic pigs in large frequencies, but eat *Tetraonidae* and reindeer offal instead. Of course poultry and domestic pigs' offal is very likely to be quite scarce as the area is sparsely populated with *Homo sapiens.*

The tendency to concentrate on a few prey items is probably the optimal feeding strategy for exploiting the available prey in balance with the energy budget which promoted survival.

It is reasonable to ask: *What do foxes prefer to eat and how does this compare with the diet actually eaten and the species of prey available in the habitat?* The results of my experiments with 4 captive foxes and other studies on this question are in preparation.

REFERENCES

Amitrov, V. K. 1956. Foxes, transmitters of rabies. *Veterinariya* 2: 33–34.

Amores, F. 1975. Diet of the Red Fox (*Vulpes vulpes*) in the Western Sierra Morena (South Spain). *Donana Acta Vertebrata* 2: 221–239.

Appleyard, H. M. and Greville, C. M. 1950. The cuticle of mammalian hair. *Nat. Lond.* 166: 1031, 1 fig.

Attanassov, N. 1958. Der Fuchs (*Vulpes vulpes* crucigera, Bechstein) in Bulgarien (Foxes in Bulgaria *V. vulpes* c.B.). Bulgarische Akademie der Wissenschaften, Arbeiten des Zoologischen Instituts No. 5: 324 p. (Bulgarian with German summary).

Behrendt, G. 1955. Beiträge zur Ökologie des Rotfuches (*Vulpes vulpes*, L.) (Observations on the ecology of the Red Fox (*V. vulpes*, L.). *Z.F. Jagdwissenschaft* 1: 113–145 and 161–183.

Bereford-Jones, W. P. 1961. Observations on the helminths of British wild red foxes. *Vet. Rec.* 1961, 73 (36): 882–883.

Bergman, G. 1966. Saaristomme ketuista (Occurrence and living habits of the foxes, *Vulpes vulpes*, in Finnish island areas and its effects on birds. *Suomen Riista* 18: 30–41.

Bistrup, E. 1890. Raeven og Musene (Foxes and rodents). *Tidsskrift for Skovvaesen* 2A: 37–43. (Cited in Jensen and Sequeira 1978).

Brosset, A. 1975. Regime alimentaire d'une population suburbaine de renards aú cours d'un cycle annúel (The food habits of a surburban population of foxes in an annual cycle). *La Terre et. la Vie* 29: 20–30.

Brömel von, J. and Zettl, K. 1974. Beitrag zur Altersbestimmung beim Rotfuchs (*Vulpes vulpes*, L. 1758); (Estimation of age in Red Foxes). *Z.F. Jagdwissenschaft* 20 (2). 96–104.
Chandler, A. C. 1960. A study of the structure of feathers with reference to their taxonomic significance. *Zool. Vol.* 13, No. 11, Univ. of California Publ. 243–446.
Cochran, W. G. 1953. Sampling Techniques. John Wiley and Sons, New York: 50–59.
Collett, R. 1911–12. Norges Hvirveldyr 1, Norges Paltedyr. 1–744, Kristiania (as cited in Lund 1962).
Corbet, C. B. 1969. The identification of British Mammals. Brit. Mus. (Nat. Hist.).
Croft, J. D. and Hone, L. J. 1978. The stomach contents of Foxes, *Vulpes vulpes*, collected in New South Wales. *Aust. Wildl. Res.* 5: 85–92.
Day, M. G. 1966. Identification of hair and feather remains in the gut and faeces of stoats and weasels. *J. Zool.* 148. 201–217.
Dearborne, N. 1938. Sections aid in identifying hair. *J. Mammal. Balt.* 20: 346–348, 1 fig.
Douglas, M. J. W. 1965. Notes on the Red Fox (*Vulpes vulpes*) near Braemar, Scotland. *J. Zool.* 147: 228–233.
Dudley, E. 1977. Our unknown wildlife, The Fox. Frederick Muller Ltd.
Dziurdzik, B. 1973. Key to the identification of hairs of mammàls from Poland. *Acta Zool. Crac. Tom.* XVIII, No. 4: 73–91, 11 figs.
Englund, J. 1965a. Studies on the food ecology of the red fox (*Vulpes vulpes*) in Sweden. *Viltrevy* 3 No. 5: 375–485.
Englund, J. 1965b. The diet of foxes (*Vulpes vulpes*) on the island of Gotland since myxomatosis. *Viltrevy* 3, No. 6: 505–530.
Englund, J. 1969. The diet of fox cubs (*Vulpes vulpes*) in Sweden. *Viltrevy* 6 No. 1: 1–39.
Englund, J. 1970. Population dynamics of the Swedish red fox (*Vulpes vulpes* L.). Research Dept. Section for Vertebrate Zoology. Swedish Museum of Natural History, Stockholm.
Fairley, J. S. Oct. 1969b. The fox as a pest of agriculture. *Irish Nat. J.* 16, No. 8: 216–219.
Fairley, J. S. 1970. The food, reproduction, form, growth and development of the fox, *Vulpes vulpes*, L., in North-East Ireland. *Proc. Royal Irish Acad.* 69 (B) No. 5: 103–137.
Forbes, T. O. A. and Lance, A. N. 1976. The contents of fox scats from Western Irish Blanket bog. *J. Zool. Lond.* 179: 224–226.
Friend, M. and Linhart, S. B. 1964. Use of the eye lens as an indicator of age in the red fox. *N.Y. Fish and Game J.* 11. 1: 58–66.
Frisch, O. and Frisch, H. von, 1971. Beobachtungen bei der Handaufzucht und spaeteren Aussetzung einer Fuchsfaeke *Vulpes vulpes* (Observations of a hand-raised female fox V.v.) *Z. Tierpsychol.* 28 (5): 534–541.
Fuchs, F. 1972. Uber die Nahrung des Rotfuches (*Vulpes vulpes* L.) im bernischen Hugelland (Diet of Red Foxes (*V. vulpes* L.) in Bern's Hugelland). *Jahrbuch des Naturhistorischen Museums der Stadt Bern* 5: 119–131.
Geiger, G., Brömel, J. and Habermehl, K. H. 1977. Konkordanz verschiedener Methoden der altersbestimmung beim Rotfuchs (*Vulpes vulpes* L. 1758). (Concordance of various methods of determining the age of red fox). *Z. Jagdwissenschaft* 23 (2): 57–64.
Goszczynski, J. 1974. Studies on the food of foxes. *Acta Theriologica* 19: 1–18.
Grue, H. and Jensen, B. 1973. Annular structures in canine tooth cementum in Red Foxes (*Vulpes vulpes* L.) of known age. *Dan. Rev. Game Biol.* Vol. 8. No. 7.
Haacke, H. R., Hirneiss, R., Hommel, R., Kraft, H. and Landauer, G. 1973. Funkortung Freilebender Füchse. (Telemetric studies on foxes). *Z. Jagdwissenschaft* 19 (2): 90–98.
Haltenorth, T. (Munich) and Roth, H. H. (Rome) 1968. Short review of the Biology and Ecology of the red fox. *Saugetrierk Mitt.* 16, No. 4: 339–352.
Hanson, R. G. 1973. An index of food quality. *Nutrition Reviews.* Vol. 31, No. 1: 1–7.
Hanson, W. R. and Graybill, F. 1956. Sample size in food habits analysis. *J. Wildl Manage.* 20: 64–68.
Harris, S. 1977. Age determination in the Red Fox *Vulpes vulpes* – an evaluation of technique efficiency as applied to a sample of suburban foxes. *J. Zool.* 184 (1): 91–117.
Harris, S. 1978. Age determination in the Red Fox (*Vulpes vulpes*) – an evaluation of technique efficiency as applied to a sample of suburban foxes. *J. Zool. Lond.* 184: 97–117.
Hausman, L. A. 1920a. Structural characteristics of the hair of mammals. *Amer. Nat.* 54: 496–523.
Hausman, L. A. 1920b. Classification of mammalian hairs on configuration of cuticle scales and medulla. *Amer. J. Anat.*
Hausman, L. A. 1924. Further studies on structural characteristics of mammalian hairs. *Amer. Nat.* 58: 544.

Hausman, L. A. 1930. Recent studies of hair structure relationships. *Sc. Month.* 30: 258–277.
Hellwing, S. 1960. Einiges über die Nahrung des Fuchses, *Vulpes vulpes* L. (About the diet of foxes, *V. Vulpes* L.). *Travaux du Mus. d'Hist. Nat. "Gr. Antipa"* 2: 415–417.
Hewson, R., Kolb, H. H. and Knox, A. G. 1975. The food of foxes (*Vulpes vulpes*) in Scottish forests. *J. Zool. Lond.* 176: 287–292.
H.M.S.O. June 1971. Report of the committee of inquiry on rabies. Final Report.
Husson, A. M. 1962. Het determinera van Zoogdieren in Braakballen van Uilen (The determination of mammals in the pellets of owls). *Zool. Bij. Uit. Rij. Mus. Nat. Hist. Leid.* No. 5: 1–63.
Ingleby, J. A. 1956. Sheep, foxes and rabbits. *Scott. Agric.* 36: 69–70.
Jensen, B. 1969. The migration of foxes and the rabies situation in South Jutland. *Mamm. Soc. Bull. Lond.* 32: 6–8.
Jensen, B. 1970. Effect of a fox control programme on the bag of some other game species. *Trans. IX Int. Congr. Game Biol. Mosk.* 480.
Jensen, B. and Nielsen, L. B. 1968. Age determination in the red fox (*Vulpes vulpes* L.) from canine tooth sections. *Dan Rev. Game Biol.* Vol. 5 No. 6: 1–16.
Jensen, B. and Sequeira, D. M. 1978. The diet of the Red Fox (*Vulpes vulpes* L.) in Denmark. *Dan. Rev. Game Biol.* Vol. 10. No. 8: 1–16.
Kaplan, C., Brown, F., Crick, J., Warrell, D. A., Haig, D. H., MacDonald, D. W., Lloyd, H. G. and Turner, G. S. 1977. Rabies The Facts. Oxf. Univ. Press: 148 pp.
Klenk, K. 1971. Das Activitaetsmuster des Rotfuchses (*Vulpes vulpes* L.) in einen Freilandgehage mit kunstlichem Bau (The activity of the Red Fox (*Vulpes vulpes* L.) in an enclosure with an artificial den). *Z. Sauget.* 36 (5): 257–279.
Lampio, T. 1953. Tutkimuksia ketun ravinnosta (On the food of the fox). *Suomen Riista* 8: 156–164. English summary on page 229–230.
Lamina, J. and Main, F. 1964. Der Parasitenbefall bei Rotfüchsen in Südhessen (The parasite burden among Red Foxes in South Hesse). *Z. Jagdwiss.* 10 (4): 137–142.
Lebland, C. P. 1951. Histological structure of hair with a brief comparison of other epidermal appendages and the epidermis itself. *Ann. N.Y. Acad. Sc.* 53: 464–475, 23 figs.
Leinati, L., Mandelli, G., Videsott, R. and Grimaldi, E. 1960. Indagini sulle abitudini alimentari della volpe (*Vulpes vulpes* L.) del Parco Nazionale del Gran Paradiso. *La Clin. Vet.* 83: 305–328. (Italian, summarised in German by Leinati, L.: Research on the diet of foxes in Italy's Gran Paradiso National Park). *Monatsheft für Vet. Med.* 15: 641–643, 1960.)
Lessman, F. J. 1971. Wildtiertollwut, (Rabies, Inaugaral Dissertation) Tierärzlichen Fakultät. Univ. München: 115 pp.
Lever, R. A. 1959. The diet of the fox since myxomatosis. *J. Anim. Ecol.* 28: 359–375.
Lillie, F. R. 1942. On the development of feathers. *Biol. Rev.* 17: 247–266.
Lockie, J. D. 1956. After myxomatosis-notes on the food of some predatory animals in Scotland. *Scott. Agric.* 36: 65–69.
Lockie, J. D. 1959. The estimation of the food of foxes. *J. Wildl. Manage.* 23: 224–227.
Lockie, J. D. 1964. The breeding density of the golden eagle and fox in relation to food supply in Wester Ross, Scotland. *Scott. Nat.* 71: 67–77.
Lord, R. D. 1961. The lens as an indicator of age in the gray fox. *J. Mammal.* 42: 109–111.
Lozanic, B. M. 1967. Contribution a la connaissance de la faune des Helminthes chez le Renard de nos regions (*Vulpes vulpes* L.). Contribution to the knowledge of the helminthological fauna of the red fox living in our regions. *Int. Congr. Game Biol.* 7: 473–474.
Lund, M. K. 1962. The red fox in Norway 11. The feeding habits of the red fox in Norway. *Medd. Fra Sta. Vilt. Ser.* 2, No. 12: 1–79.
Lüps, P., Neuenschwander, A. and Wandeler, A. 1972. Gebissentwicklung und Gebissanomalien bei Füchsen (*Vulpes vulpes* L.) aus dem Schweizerischen Mittelland. (Normal and abnormal dental development of foxes in Swiss Mittelland.) *Rev. Swiss. Zool.* 79: 1090–1103.
MacDonald D. W. 1976. Food caching by red foxes and some other carnivores. *J. Tierpsychol.* 42, 170–185.
MacDonald, D. W. 1977a. The behavioral ecology of the Red Fox, *Vulpes vulpes*: a study of social organisation and resource exploitation. D. Phil. Thesis, Oxford.
MacDonald, D. W. 1977b. On food preference in the red fox. *Mammal Rev. vol.* 7, No. 1: 7–23.
Magalhaes, C. M. P. 1974. Habitos alimentares da raposa (*Vulpes vulpes*, silacea) et da Geneta (*Genetta genetta*) na Tapada de Mafra. (Food habits of the Red Fox (*V. vulpes*, s.) and the Genet (*G. genetta*) of Tapada in Mafra). *Estudos Divulgacao Tecnica, Grupo A. Seccao Zool, Flor. e. Cinegetica* 18: 5–17.

Mandelli, G. 1961. Research on the food habits of foxes (*Vulpes vulpes* L.) in Gran Paradiso National Park: Preliminary investigations on the macroscopic characteristics of the hairs of their principal prey. *Clin. Vet.* 83, 8: 235–243.

Mateljka, H., Röben, P. and Schröder, E. 1977. Zur Ernahrung des Rotfuchses, *Vulpes vulpes* (Linné 1758) im offinen Kulturland. (On the feeding of Red Fox, *Vulpes vulpes* (Linné 1758) in open cultivated areas.) *Z. Säugetierkunde* 42: 347–357.

Mathiak, H. A. 1938. A rapid method of x-sectioning mammalian hairs. *J. Wildl. Manage.* 2 (3): 162–164.

Matthews, L. H. 1960. British Mammals. Second edition. Collins, London.

Mayer, W. V. 1952. The hair of California mammals with keys to the dorsal guard hairs of California mammals. *Amer. Midl. Nat.* 48: 480–512.

Montgomery, G. G. 1974. Communication in red fox dyads: A computer simulation study. Smith. Contr. to *Zool.* No. 187: 1–30.

Morris, P. 1972. A review of mammalian age determination methods. *Mammal Review* No. 3: 69–104.

Naaber, J. 1967. Materials for the diet of the red fox in the western parts of the Matsalu National Park. *Ornitoloogiline Kogumik* 4: 172–176 (Estonian with English summary).

Nesseni, R., Lecht, M. and Scheven, B. 1955. Uber die Durchgangszeit des Futters beim Silverfuchs. (Time taken for the food to pass through the gut of silver foxes). *Archs. Tier.* 5: 26–32.

Niewold, F. J. J. 1974. Irregular movements of the Red Fox (*Vulpes vulpes*) determined by radio tracking. XI Int. Congr. Game Biol. Stock. Sept. 3–7: 331–337.

Noback, C. R. 1951. Morphology and phylogeny of hair. *Ann. N.Y. Acad. Sci.* 53: 476–492, 24 figs.

Oyer, E. R. 1946. Identification of mammals from studies of hair structure. *Trans. Kansas Acad. Sci.* 49. No. 2: 155–160.

Palm, P. 1970. The food ecology of Swedish red foxes. *Zool. Revy* 32: 43–46. (English summary).

Persson, L. and Christensen, D. 1971. Endoparasiten hos Röd Rave i Sverige. *Zool. Revy* 33: 17–28.

Peshev, Z. 1965. The food of fox (*Vulpes vulpes* L.) in some parts of Bulgaria. *Ann. de l'Univ. de Sofia, Fac. de Biol.* 58: 87–119 (Bulgarian with English summary).

Peterson, R. T., Mountfort, G. and Hollom, P. A. D. 1969. A field guide to the birds of Britain and Europe. Collins. London.

Petrides, G. A. 1975. Principle foods vs preferred foods and their relations to stocking rate and range condition. *Biol. Cons.* Vol. 7, No. 3: 161–169.

Prokopic, J. 1960. The helminth fauna of wild and ranch foxes in Czechoslovakia. *Zool. Listy* 23 (3): 239–244.

Pursor, F. and Young, C. B. 1959. Lamb survival in two hill flocks. *J. Anim. Prod.* 1: 85–91.

Richards, D. F. 1977. Observations on the diet of the Red Fox (*Vulpes vulpes*) in South Devon. *J. Zool. Lond.* 183: 495–504.

Ross, J. G. and Fairley, J. S. 1969. Studies of disease in the Red Fox in Northern Ireland. *J. Zool.* 157 (3): 375–381.

Rudall, K. M. 1941. The structure of the hair cuticle. *Proc. Leeds Philos. Lit. Soc. Sci. Sect.* Vol. 4: 13–18.

Ryder, M. L. 1973. Hair. Edward Arnold: 56 pp.

Ryszkowski, L., Wagner, C.K., Goszczynski and Truszkowski, J. 1971. Operations of predators in a forest and cultivated fields. *Ann. Zool. Fenn.* 8: 160–168.

Rzebik-Kowalska, B. 1972. Studies on the diet of the carnivores in Poland. *Acta Zool. Crac.* 17: 415–506. (In Polish with English summary).

Scott, T. G. 1941. Methods and computation in faecal analysis with reference to the red fox. *Iowa State Coll. J. Sci.* 15: 279–285.

Sequeira, D. M. 1978. Age determination and food ecology of the red fox (*Vulpes vulpes* L.) in the province of Gelderland, Holland. (Licenciate thesis, University of Helsinki).

Southern, H. N. and Watson, J. S. 1941. Summer food of the red fox (*Vulpes vulpes*) in Gt. Britain. *J. Anim. Ecol.* 10: 1–11.

Sovis, B. 1967. A contribution to the food ecology and population dynamics of the common fox with regard to its economic significance. *Acta Zootech. Univ. Agric. Nitra-Czech.* 15: 159–171. (Czechoslovakian with English summary).

Spittler, H. 1972. Ueber die Auswirkung der durch die Tollwut hervorgerufenen Redusierung der Fuchspopulation auf den Niederwildbesatz in Nordrhein-Westfalen. (On the effect of the

reduction of foxes due to rabies on the small game populations in North Rhein Wetsfalen). *Z. Jagdwiss.* 18 (2): 76–95.
Stains, H. J. 1958. Field guide to guard hairs of Middle Western furbearers. *J. Wildl. Manage.* 22: 95–97.
Stoves, J. L. 1944a. Morphology of mammalian hair. *Nature Lond.* 153: p. 285.
Stoves, J. L. 1944b. The appearance in cross-section of the hairs of some carnivores and rodents. *Proc. Roy. Soc. Edin. Vol.* 62, Sect. B: 99–103, 3 p. of phot.
Surdan, C., Dragonescu, N., Gheorghiu, V. L., and Enache, A. 1969. Bovine rabies enzootic transmitted by foxes. *Rev. Roum Inframicro.* 6 (1): 75–82.
Svensson, L. 1970. Identification guide to European passerines. K. L. Bechman's Trykeri: 152 pp.
Thompson, R. C. A. 1976. The occurrence of meso-cestoides spp. in British wild Red Foxes (*Vulpes vulpes*, crucigera). *J. Helminth.* 50 (2): 91–94.
van Haaften, J. L. 1970. Food ecology studies in the Netherlands. Trans. IX International Cong. Game. Biol. Moscow, 539–543.
von Lutz, W. 1978. Beitrag zur Nahrung des Rotfuchses (*Vulpes vulpes*, L.) im "National park Bayerischer Wald". (The diet of the Red Fox (*Vulpes vulpes*, L.) in the "Bayericher Wald National Park"). *Z. Jagdwiss.* 24: 1–9.
Wandelar, A. I. 1976. Alterbestimmung bei Füchsen. (Age determination of Red Foxes). *Rev. Suisse Zool.* 83 (4): 956–963. (In German with English summary).
Wandelar, A. I. and Hörning, B. 1972. Aspekte des Cestodenbefalles bei bernischen Füchsen. (Aspects of cestode parasitism in Bern's foxes). *Jahrbuch des Nat. Mus. der Stadt Bern* 4: 231–252.
Watson, K. 1976. Food remains in the droppings of foxes (*Vulpes vulpes*, L.) in the Cairngorms. *J. Zool.* 180: 495–496.
WHO-FAO 1970. Co-ordinated research programme on wildlife rabies in Central Europe. Report on informal discussions held at Munich (unpublished):
Williams, C. C. 1934. A simple method of sectioning mammalian hairs for identification purposes. *J. Mammal. Vol.* 15: 251–252.
Williamson, C. S. 1938. Aids to the identification of mole and shrew hairs with general comments on hair structure and hair determination. *J. Wildl. Manage.* 2 (4): 239–250.
Williamson, C. S. 1951. Determinations of hairs by impressions. *J. Mamm.* 32: 80–84.
Witt, H. 1976. Untersuchungen zur Nahrungswahl von Füchsen (*Vulpes vulpes*, Linné 1758) in Schleswig-Holstein. (Research on the food preference of the Red Fox in Schleswig-Holstein). *Zool. Anz. Jena* 197: 377–400.
Österholm, H. 1964. The significance of distance receptors in the feeding behaviour of the fox (*Vulpes vulpes*, L.). *Acta Zool. Fenn.* 106: 1–31.

5 PREY CONSUMPTION OF A RED FOX POPULATION IN SOUTHERN SWEDEN

Torbjörn von Schantz*

INTRODUCTION

This study was carried out as a part of an integrated research project concerning relationships between vertebrate predators and their prey animals in an open field area in southernmost Sweden. The analysis of the fox diet was based on the analysis of 1,028 scats, and 162 prey remains found at fox dens. The percentage of weight of the prey items in the diet was calculated; *Lagomorphs* was the most important prey throughout the year. The number of foxes in the population and their monthly prey demands were estimated and analysis of the annual predation showed that 10 to 15 percent of the annual production of each of the dominant prey populations was consumed by the foxes.

The diet of the red fox has been studied during the last five decades (for review of literature see; Jensen and Sequeira, 1978, Lutz, 1978, and Korschgen, 1959). Most studies have described the qualitative composition of the diet. Only in a few cases have a quantitative analysis been made and there have been few attempts to estimate the impact of fox predation on the prey populations (Pils and Martin, 1978, Goszczyński 1977, Goszczyński et al., 1976, Ryszkowski et al., 1971, and Korschgen, 1959).

This study evaluates the fox diet composition, and the impact of fox predation on the prey populations, calculated from data on the number of foxes in the population and their food demands.

THE STUDY AREA

The Wildlife Research Group at the University of Lund examines a community of vertebrate predators and their prey animals in the Revinge area, southernmost Sweden (55.42°N, 13.25°E). The project aims to examine factors regulating numbers of predators in the community and also to examine the impact of predators on the prey populations.

The study presented in this paper was carried out between August 1974 and March 1977 in the Revinge area. The study area, about 45 km^2, was earlier agricultural, but since 1965 it has been used for military training. The area contains open fields grazed by cattle interspersed by coniferous plantations and copses of deciduous trees. Dry parts of the area are densely inhabited by rabbits (*Oryctolagus cuniculus*). The area also contains extensive marshes populated by voles, primarily *Microtus agrestis* and *Arvicola terrestris.*

* Department of Animal Ecology, University of Lund, S-223 62 Lund, Sweden.

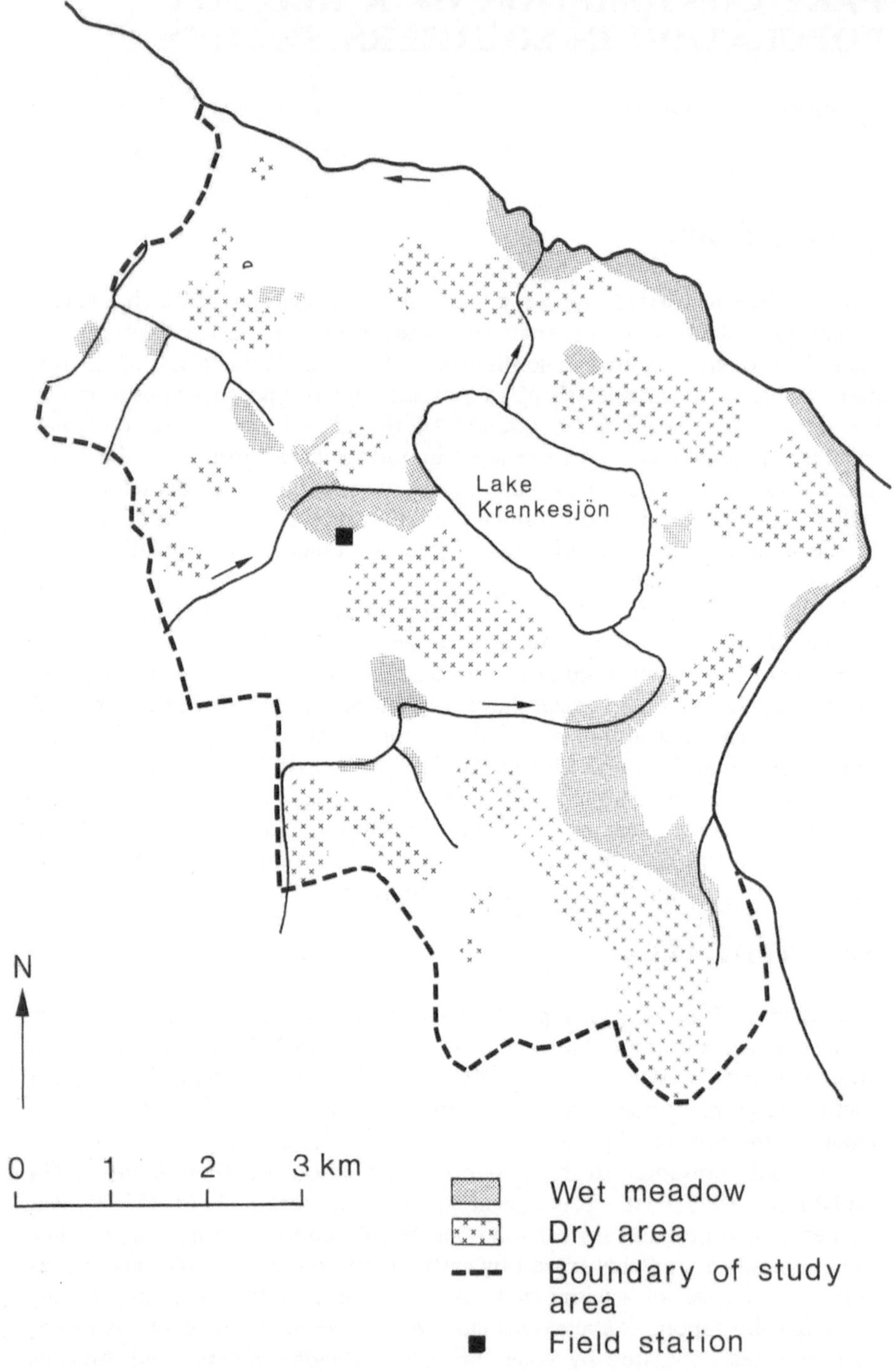

Fig. 1. Study area and distribution of main habitats.

Methods

The food habits of adult foxes were determined by analysis of scats, collected twice each month on a special scat-collecting-route, approximately 7 km long, covering all types of habitats. The scat analysis was performed as described by Lockie (1959) and Goszczyński (1974). Determination of mammalian hair was made according to Brunner and Coman (1974). When calculating the percentage of weight in the diet for the prey items I used the correction factors of Lockie (1959).

The diet of cubs during the denning period was determined by registration of prey remains at the dens. Mean weight of found prey species was estimated and multiplied by the number of occurrences to yield the percentage of weight for each prey type. The calculation of food demands of the foxes was based on the results of Sargeant (1978).

To determine the number of foxes in the population, I counted, using a searchlight, foxes at night in 140 plots, in all covering 300 hectares. This was done five times in late autumn and five times in late winter every year. Telemetry data (von Schantz unpubl.) showed how long the foxes spent in open habitats where they could be seen. The counting plots were distributed only in such habitats. That is, if the foxes for example spent 80 percent of their time in "visible" habitats during night, I consequently assumed that the population size calculated on the basis of the counts was 80 percent of the true population. Further details on this method and the assessment of its significance will be published elsewhere. The number of cubs was determined by counting the number of litters in the study area, and the number of cubs in a sample of litters during May and June.

RESULTS AND DISCUSSION

The diet

The percentage of total number of occurrences of fox prey items, based on scat analysis, is given in Table 1. There was no significant difference between

Table 1. Percentage of total number of occurrences of prey items in the fox diet, based on analysis of 1,028 scats. The figures are approximated to closest unit. *Lagomorphs* consist of rabbit (*Oryctolagus cuniculus*) and European hare (*Lepus europeus*), other small rodents consist of *Clethrionomys glareolus* and *Apodemus* sp. Thrushes and bigger birds have been considered as large birds. Only vegetables and fruit with nutritional value for the fox have been recorded. Carcasses, eggshells, garbage and occasional prey items, e.g. fish, cat, squirrel, etc. have been assign to miscellaneous.

Period	Lagomorphs	Arvicola terrestris	Microtus agrestis	Other small rodents	Large birds	Small birds	Insects	Veget fruit	Miscel-laneous	Total number of occurrences	$p (x^2)$	
I Dec-Mar	48	2	26	3	5	6	3	1	6	435	$p < 0.001$ (36.0)	I-II
II Apr-May	46	7	16	1	4	7	11	2	7	249	$p < 0.01$ (23.6)	II-III
III Jun-Sep	40	7	10	1	5	7	18	7	5	856	$p < 0.001$ (92.5)	III-IV
IV Oct-Nov	48	4	21	1	2	3	6	11	4	584	$p < 0.001$ (68.5)	IV-I

the years, so I pooled the data from the same month each year. Consecutive months, with similar (no significant differences) diet, were put together in food seasons. *Lagomorphs* and, to a certain extent, small rodents dominated in all periods. Birds occurred less frequently in October and November, insects occurred most frequently in April to September, while vegetables and fruit were more common in the diet from June to November than in winter and spring.

Analysis based on frequency of occurrence does not give any accurate information on the relative importance of the different prey items. Large prey, such as *Lagomorphs* are underestimated, while small prey, such as *Microtus* will be overestimated. This is well illustrated by comparing Table 1 with Table 2, where I have used Lockie's correction factors and calculated the percentage of weight for the prey items. Insects and vegetables are excluded as their weight was negligible. This type of data illustrates better the relative importance of the different prey categories. It is also the only data that can be used, when working with scat analysis, to calculate the amount of consumed prey biomass or number of prey animals taken, during a specified period of time.

Lagomorphs were far the most important prey throughout the year. The smallest percentage (77) of *Lagomorphs* was in April (Fig. 2). In late May and in the beginning of June the rabbit kittens began to leave the burrows. The amount of *Lagomorphs* in the diet increased during this period to the peak value in July (91%). From April to October *Arvicola* was the second most important prey. *Arvicola* is less available for foxes during winter than *Microtus*, as *Arvicola* during this season seldom leaves its tunnels in the frozen ground (Jeppsson pers. comm.). *Microtus* make its tunnels in the litter (Hansson pers. comm.) and consequently is within reach of foxes also in winter.

Cub diet in may and June, based on prey remains found at dens, was dominated by *Lagomorphs* and large birds (Table 3). No remains of voles were found, probably because they are eaten whole. The percentage of occurrences of other prey items corresponds well to Englund's (1969) data from southern Sweden, based on stomach analysis. Therefore, I used his figures regarding the percentage of occurrences of voles in Table 3. Among small rodents, *Arvicola* was the main prey in the diet of adult foxes in May and June (Fig. 2). I therefore conclude that voles in the cub diet consisted of *Arvicola* only.

Table 2. Percentage of weight of prey items (corrected according to Lockie, 1959) in the fox diet, based on analysis of 1,028 scats. The figures are approximated to closest unit.

Period	Lagomorphs	Arvicola terrestris	Microtus agrestis	Other small rodents	Large birds	Small birds	Number of scats
Dec–Mar	86	2	9	1	2	1	230
Apr–May	78	11	7	>1	3	1	124
Jun–Sep	89	6	2	>1	2	1	374
Oct–Nov	87	5	6	>1	1	>1	300

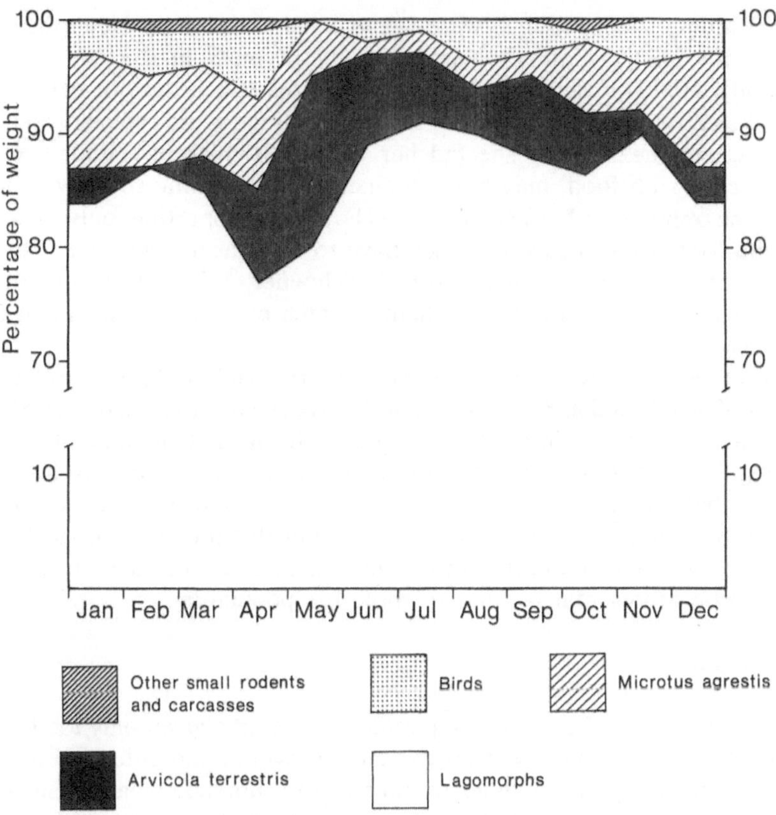

Fig. 2. Percentages of weight, on monthly basis, of prey items in diet of adult foxes, and of cubs from July and onwards (see p. 14). $N = 1{,}028$ scats.

The diet of adults and cubs in May and June, excluding small rodents, insects and vegetables, showed a significant difference ($p < 0.001$, $X^2 = 24.6$). The most important difference was the great amount of large birds in the cub diet. A possible overrepresentation of birds in remains outside dens, compared with mammals, is suggested by Pils and Martin (1978). They found, when digging out fox dens, that 50 percent of all "large bird" remains found were found outside the dens, while only 34 percent of all mammal remains

Table 3. Diet of the fox cubs during May and June, based on 162 prey remains collected at dens. The figures are approximated to closest unit.

Prey item	Percentage of total occurrences	Percentage of weight
Lagomorphs	40	52
Large birds	28	33
Small birds	4	<1
Voles*	15	2
Miscellaneous	13	13

* For voles, see the text.

found, excluding small rodents, were found outside the dens. Nevertheless this does not fully explain the large amount of birds in the cubs' diet. Macdonald's (1977) observations on a female fox, which did not eat chickens after she was 3 months old, suggest that adult foxes are reluctant to eat birds. However, next spring she fed her cubs with chickens. Adult foxes, supplying cubs with food, may have to change their hunting strategy, from being "time minimizers" (Schoener, 1971) when supporting only themselves with food, to a stage where they have to catch as much prey per unit of time as possible, "energy maximizers" (Schoener, 1971), when supporting cubs also. This probably drives them to catch even birds, which earlier were rejected.

In the study area, fox cubs were born in the end of March and the beginning of April, and as they are weaned 4 weeks later (Sargeant, 1978), I have assumed that the cubs started to eat prey from the beginning of May. According to Sargeant the postdenning period begins when the cubs are 13 weeks old, that is in the beginning of July. This corresponds to my own field observations, which also indicate that the cubs at this time manage to feed on their own. I have therefore considered the scats collected in July and onwards to be representative for the whole fox population.

The fox population

Data on the numbers of foxes in the population are here given only for 1976 (Table 4). In 1975, the figures were the same in winter and autumn, but no data was obtained on the number of cubs in the summer. I have assumed that all cubs were born on the first of April and that there was a constant mortality and emigration rate of juveniles from June, when the last cubs were counted at the dens, to September and of adults and subadults from

Table 4. Number of foxes, and estimated prey consumption by foxes in the Revinge area in 1976. The figures on the amount of prey biomass consumed are approximated to the closest ten. The prey biomass consumed during May and June is separated for adults and juveniles as their diet was different in this period.

Month	Number of foxes	Prey biomass consumed (kg)
Jan	52 ad	840
Feb	35 ad	510
Mar	35 ad (20 pregnant ♀♀)	590
Apr	35 ad, 76 juv (preweaned)	830
May	35 ad, 76 juv	560 (ad) 920 (juv)
Jun	35 ad, 76 juv	550 (ad) 1,020 (juv)
Jul	35 ad, 70 juv	1,720
Aug	35 ad, 65 Juv	1,910
Sep	35 ad, 58 Juv	1,450
Oct	87 ad	1,400
Nov	80 ad	1,250
Dec	70 ad	1,130
		Σ 14,680

October to February. It was further assumed that there was no adult mortality from February to September.

Food demands

Sargeant (1978) made a study on prey demands of adult and juvenile red foxes during the spring and summer cub-rearing period. Based on his paper, I calculated the prey demands for the foxes in the Revinge population. As the latter are approximately 60 percent heavier than those studied by Sargeant, their prey demands are accordingly higher. Therefore, estimated prey demands were 0.52 kg per fox and day for adult foxes and for subadults (25 weeks and older). I assumed that this figure was valid throughout the year. One week before whelping the vixens needed 0.69 kg per fox and day. During their first four weeks, the cubs had a weekly demand of 0.93 kg per cub, attributed to the vixen for providing milk to the cubs. At the age of 5–8 weeks, when the cubs are weaned, they needed 2.25 kg per cub and week, during the weeks 9–12 they needed 3.10 kg per cub and week, and during the weeks 13–24 they needed 4.14 kg per cub and week. The figures above include nonconsumed prey parts.

Impact of predation

When calculating the number of animals taken of the different prey species it is important to know whether the foxes preferred a certain size category of the animals in any of the prey populations, or if the various size categories were equally exploited. A good approximation for calculating the number of animals taken could be achieved by using mean weights of the dissappearing fraction of the prey populations. However, this has been possible only for the hare (*Lepus europeus*), and the calculated number of hares taken, presented in Table 5, is based on such data (Frylestam unpubl.). Consequently, the calculated number of individuals taken of the other prey species, presented in Tables 5–8, are based on the mean weights of individuals in the prey populations, as computed from trappings.

The separation of rabbits and hares in Table 5 is based on a sample of 250 identifications of *Lagomorph* hair according to Brunner and Coman (1974). The largest number of hares in the diet occurred from February throughout May ($p<0.05$, $X^2=4.3$, D.F. $=1$). The earliest hare litters were born in February and the latest in October (Frylestam unpubl.). Such early and late litters probably were especially exposed to predation (Frylestam unpubl.), and this may explain the peaks in these months. The large number of hares in the fox diet from February throughout May may be explained by the foxes' thorough scanning of the ground when hunting for small rodents, which they did to a high extent during this period (Fig. 2). This probably caused many hare litters to become predated. In June and July there is a great production of rabbit kittens. During this period the foxes change to rabbit hunting which is reflected in the diet as a decrease in the amount of small rodents (Fig. 2) and a decrease in the amount of hares (Table 5).

Table 5. Consumed biomass, and calculated number of rabbits and hares eaten by the fox population in 1976. The figures on numbers of rabbits and hares eaten are approximated to closest ten. Mean weight of rabbits and hares were supplied by Jansson (unpubl.) and Frylestam (unpubl.).

Month	Proportion of Lagomorphs in the diet (%)	Consumed biomass Lagomorphs (kg)	Lagomorph species (%) Rabbit	Hare	Consumed biomass rabbit (kg)	Rabbit mean weight (kg)	Number of rabbits consumed	Consumed biomass hare (kg)	Hare mean weight (kg)	Number of hares consumed
Jan	84	706	100	–	706	1.89	370	–		–
Feb	87	444	73	27	324	1.86	170			
Mar	85	502	91	9	457	1.69	270	216	2.06	110
Apr	77	640	92	8	589	1.70	350			
May	80 52 (juv)	448 478 (juv)	89	11	824	1.75	470			
Jun	89 52 (juv)	490 530 (juv)	95	5	969	0.74	1310	231	2.24	100
Jul	91	1565	95	5	1487	0.67	2220			
Aug	90	1719	100	–	1719	0.95	1810			
Sep	88	1276	100	–	1276	1.53	840	132	2.49	50
Oct	86	1204	89	11	1072	1.59	670			
Nov	90	1125	100	–	1125	1.61	700	–		–
Dec	84	949	100	–	949	1 63	580	–		–
		Σ 12076			Σ 11497		Σ 9760	Σ 579		Σ 260

Tables 6 and 7 show the numbers of *Microtus* and *Arvicola* taken, respectively. The annual amount of consumed biomass were approximately the same for these two species.

Eighty occurrences of large birds could be determined to species, 20 of which were pheasants (*Phasianus colchicus*). In Table 8 it is assumed that 25 percent of all large birds were pheasants, in all periods. Seventy percent of the annual fox predation occurred during April to June, the denning period.

Rabbits constituted nearly 80 percent of annual food demands of the fox population (Fig. 3). *Arvicola, Microtus*, hares, and birds were all equally important as supplementary food and they contributed less than 20 percent to the foxes' annual prey consumption. If foxes, during the prey species'

Table 6. Consumed biomass, and calculated number of *Microtus agrestis* eaten by the fox population. Numbers of *Microtus* eaten are approximated to closest hundred. Mean weight of *Microtus* was assumed to be 30 g from March to September and 21 g from October to February (Hansson unpubl.).

Month	Percentage of weight in the diet	Consumed biomass *Microtus a* (kg)	Number consumed
Jan	10	84	4,000
Feb	8	41	2,000
Mar	8	47	1,600
Apr	8	66	2,200
May	5	28	900
Jun	1	6	200
Jul	2	34	1,100
Aug	2	38	1,300
Sep	2	29	1,000
Oct	6	84	4,000
Nov	4	50	2,400
Dec	10	113	5,400
		Σ 620	Σ 26,100

Table 7. Consumed biomass, and number of *Arvicola terrestris* eaten by the fox population. Numbers of *Arvicola* eaten are approximated to closest ten. Mean weight of *Arvicola* was assumed to be 115 g from March to September and 100 g from October to February (Jeppsson unpubl.).

Month	Percentage of weight in the diet	Consumed biomass Arvicola t, (kg)	Number consumed
Jan	3	25	250
Feb	0	0	0
Mar	3	18	160
Apr	8	66	570
May	15 2 (juv)	84 18	890
Jun	8 2 (juv)	44 20	560
Jul	6	103	900
Aug	4	76	660
Sep	7	102	890
Oct	5	70	700
Nov	2	25	250
Dec	3	34	340
		Σ 685	Σ 6,170

reproduction seasons, preyed to a very high extent on the juveniles, then the calculated number of prey animals taken during the summer could be greater than presented in Tables 5–8. It is obvious that these figures are rough, as young voles, that have not yet left their nests, are not within the calculation of the mean weights in the vole populations. Furthermore, naked juvenile voles will probably not leave any remains in fox scats. However, Englund (pers. comm.) found only few juvenile voles in the fox stomachs he analysed (Englund 1965). Another source of error is to what extent taken large prey are eaten. That is, do foxes leave a large prey once they have filled their stomach or is the prey cached, and later fully exploited? Data presented by Macdonald (1976) suggest that foxes do exploit all their food caches, and that is assumed in this paper.

The annual number of main prey species taken by the fox population, suggest that 10–15 percent of annual prey production was exploited (Table

Table 8. Estimated number of pheasants (*Phasianus colchicus*), approximated to closest ten, eaten by the fox population. The numbers are calculated from the assumption that 25 percent of all occurrences of large birds were pheasants, in all periods. Mean weight of pheasants was assumed to be 1.1 kg from October to June and 0.8 kg from July to September (Göransson unpubl.).

Period	Consumed biomass pheasant (kg)	Number consumed
Dec–Mar	17	20
Apr–Jun	174	160
Jul–Sep	21	30
Oct–Nov	20	20
	Σ 232	Σ 230

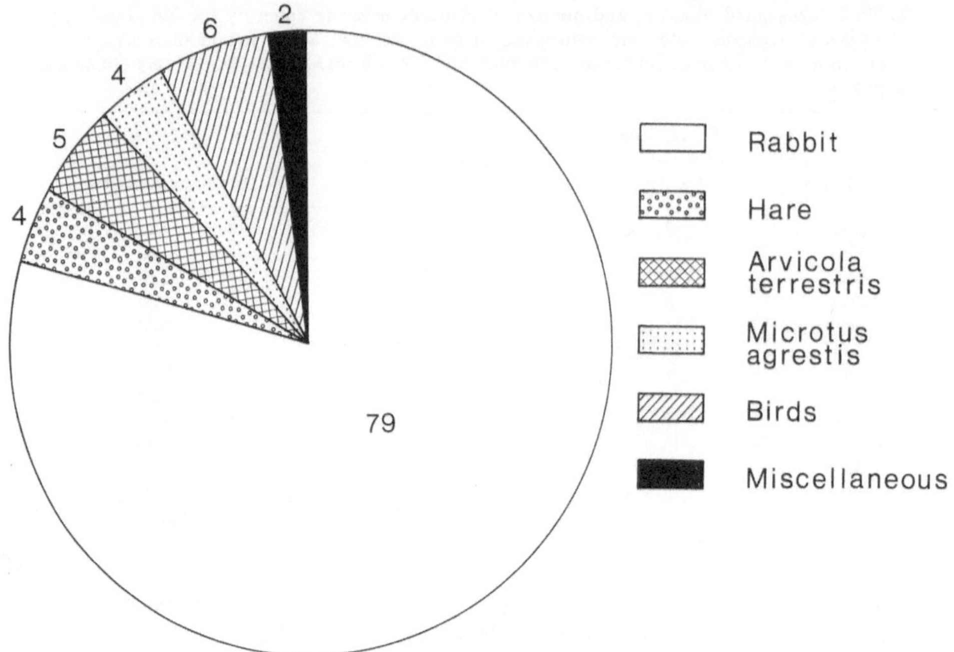

Fig. 3. Percentage of weight of prey items in total annual consumption of the fox population.

9). No data on annual production of *Arvicola* was available. Other small rodents, *Clethrionomys glareolus* and *Apodemus* sp., were exploited to a very small extent (Fig. 2), i.e. 1–2 percent of annual production. This agrees well with Macdonald (1977) who showed that foxes do not prefer these species in a comparison with *Microtus*. In spite of the dominance of rabbits in the foxes' diet they were affected to a smaller extent by the foxes than *Microtus*, which only contributed 4 percent to the foxes' annual prey consumption (Fig. 3).

Preliminary analysis of total predation of all the predators in the study area, suggest that foxes were responsible for 30 percent and 20 percent, respectively, of the total predation on rabbit and *Microtus*. Only 35 percent

Table 9. Percentages of annual prey production of newborn young, or hatched chickens, eaten by the fox population in 1976. Estimated prey production figures were kindly supplied by; Jansson (unpubl.) for rabbits, Frylestam (unpubl.) for hares, Hansson (unpubl.) for Microtus, and Göransson (unpubl.) for pheasants.

Prey species	Number consumed per year	Annual production	Percentage consumed of annual production
Rabbit	9,670	90,000	11
Hare	260	2,100	12
Microtus agresttis	26,100	171,000	15
Pheasant	230	2,400	10

of the annual rabbit production was preyed upon by the predator community, whereas almost the total annual production of *Microtus* was taken. During the winter 1976–1977, with very severe snow conditions, the rabbit population decreased sharply from 820 animals per km^2 in autumn to 120 animals per km^2 in spring (Jansson unpubl.). In 1978 they had still not recovered, which indicates that the predators were able to delay population increase at this low density. The consequence of the rabbit crash was a decrease in fox reproduction to half the number of cubs, compared with the year before. This indicates that fox reproduction was limited by food supply. However, the constant number of foxes in autumn 1975 and 1976, in spite of a 60 percent increase of the rabbit population (Jansson unpubl.), suggests that in 1976 the fox population was limited by other factors than food.

Jensen (1970) showed that the hunting bag of hare, partridge, and pheasant increased by 50 to 100 percent during a rabies campaign against foxes in Denmark. In the Revinge area, hunters each year shoot 5–7 percent of annual production of hares and pheasants, which is a smaller amount than the foxes eat (Table 9). Consequently, if the fox population would be kept on a lower level by increased hunting pressure it is likely that the hunting bag of field game would increase. But this in itself does not imply that fox predation is a limiting factor for these populations.

Acknowledgements

I wish to thank Dr S. Erlinge and my colleagues in the Wildlife Research Group for valuable help and criticism, Mr P. Berglund and Mrs M. Andersson for help with the scat analysis, Dr B. Frylestam, Dr G. Göransson, Prof. L. Hansson, Mr. G. Jansson and Miss B. Jeppsson for supplying data on the prey populations. The study was supported by grants from the National Swedish Environment Protection Board, and from the Swedish Natural Science Research Council, both to S. Erlinge.

REFERENCES

Brunner, H. and Coman, B. J. 1974. The identification of mammalian hair. 176 pp. Inkata Press, Melbourne.

Englund, J. 1965. Studies on food ecology of the red fox (*Vulpes v.*) in Sweden. *Viltrevy* 3 (4): 377–485.

Englund, J. 1969. The diet of fox cubs (*Vulpes vulpes*) in Sweden. *Viltrevy* 6 (1): 1–39.

Goszczyński, J. 1974. Studies on the food of foxes. *Acta Theriol.* 19: 1–18.

Goszczyński, J. 1977. Connections between predatory birds and mammals and their prey. *Acta Theriol.* 22 (30): 399–430.

Goszczyński, J., Ryszkowski, L. and Truszovski, J. 1976. The role of the European hare in the diet of predators in cultivated field systems. Ecology and management of European hare populations. Warszawa.

Jensen, B. 1970. Effect of a fox control programme on the bag of some game species. Trans. IX Int. Congr. Game Biol., p. 480.

Jensen, B. and Sequeira, D. M. 1978. The diet of the red fox (*Vulpes vulpes* L.) in Denmark. *Dan. Rev. Game Biol.* 10 (8): 1–16.

Korschgen, L. J. 1959. Food habits of the red fox in Missouri. *J. Wildl. Manage.* 23 (2): 168–176.

Lockie, J. D. 1959. The estimation of the food of foxes. *J. Wildl. Manage.* 23 (2): 224–227.

Lutz, W. 1978. Beitrag zur Nahrung des Rotfuches (*Vulpes vulpes* L.) in *Nationalpark Bayerischer Wald. Z. Jagdwiss.* 24: 1–9.
Macdonald, D. W. 1976. Food caching by the red fox and other carnivores. *Z. Tierpsychol.* 42: 170–185.
Macdonald, D. W. 1977. On food preference in the red fox. *Mammal. Rev.* 7 (1): 7–23.
Pils, C. M. and Martin, M. A. 1978. Population dynamics, predator-prey relationships and management of the red fox in Wisconsin. Technical Bulletin No. 105, Department of Natural Resources Madison Wisconsin, USA.
Ryszkowski, L., Wagner, C. K., Goszczyński, J. and Truszkowski, J. 1971. Operation of predators in a forest and cultivated fields. *Ann. Zool. Fennici* 8: 160–168.
Sargeant, A. B. 1978. Red fox prey demands and implications to prairie duck production. *J. Wildl. Manage.* 42 (3): 520–527.
Schoener, T. W. 1971. Theory of feeding strategies. *Ann. Rev. Ecol. Syst.* 2: 369–404.

6 THE DIET OF THE RED FOX
Questions about method

Hugo Witt*

Very many investigations have been made into the diet of the red fox. They all have very different results. This is caused on the one hand by the differing methods, which I shall deal with later on, and on the other by the often very different ecological conditions in the examined biotopes. The fox is able to adapt to considerably different living conditions. According to Heptner and Naumov (1974) it has the greatest geographic and individual versatility of all carnivores, for example it is very much greater than that of the wolf which does not have a smaller range of habitat. General details concerning the diet of the red fox are misleading as they make standards which are only applicable in limited areas.

The composition of the diet is influenced by endogenetic and exogenous factors. Amongst the endogenetic factors there are both anatomic and physiological specialities as well as age and sex of the fox. Exogenous factors are all environmental factors such as food supply and availability, structure and climate of the landscape, season and time of day. Most of these factors are variable and difficult to determine. Their importance and reciprocal interaction are hardly known. For this reason it has not as yet been possible to theoretically determine the diet of the fox from the factors stated. For this reason it appears to be all the more important to give as far as possible a precise description of the examined area and the above mentioned factors, where known, when stating the type of diet. Unfortunately this is often omitted.

Investigations have shown that the composition of the diet is also subject to changes over a number of years. It would therefore be useful to investigate the long term feeding habits, so as to verify annual fluctuations. In addition it would of course be preferable to investigate at the same time other carnivores, e.g. mustelidae, cats, etc., so as to note possible competitors for food.

I mentioned already that the method can also be the cause of differing results. Results obtained by differing methods cannot normally be compared. Below I intend to deal with the most common investigation methods, their advantages and weaknesses.

The importance of direct observation made in the natural environment should not be underestimated. Nearly all comments made in the sporting press are based on direct observation, however these take special account of spectacular events. This has caused the public to obtain a fairly distorted

* Institut für Haustierkunde, Neue Universität, D-2300 Kiel, FRG.

impression of the dietary range of the fox. For a number of reasons this method is insufficient (randomness of the individual observations, mainly nocturnal activity periods, relatively poor visibility, etc.).

It is more interesting and more informative for the stated target, namely the composition of the diet of the fox, to carry out indirect observation by following a track. For a trained person it is possible not only to note the type of prey, but also to note the effort required for seizure, the path taken for each quarry, as well as noting unsuccessful attempts at catching prey. For a very good reason, investigations that use this method stem from Scandinavia, Canada and the north of Russia, as snow is a virtual necessity for this method.

A further standard method is to analyse the dietary remains at the foxes den. At the den, however, one only will find parts of prey animals which are too big to be eaten completely. Small mammals, insects, and plants are not inventoried. Further on dietary remains at the dens are often remains of the prey which has been brought by the adults to the cubs. The diet of fox cubs may considerably differ from the diet of the adults.

The most common methods are analyses of scats and stomach contents. Naturally the method of analysing the dietary remains in the foxes excrement is more difficult as many dietary substances can no longer be determined because of the digestive process. Quantitative statements are seldom possible. I shall point out an additional limitation below.

Analysis of the stomach contents have the advantage that the food is easy to determine as the fox hardly chews. In addition it can also be analysed quantitatively. It has been found however that the stomach content is not the same as the food intake. This is because of the different staying time of the different food components.

Behrendt (1955) states: 8 hours for mice, 6 hours for other meat foods, 4 hours for insects and 2 hours for plants. The staying times in the individual sections of the intestine also differ. As substances which often pass through the stomach in a relatively short space of time can stay in the intestine for a very long period, it appears that a combined content analysis of stomach and intestine would be closest to the composition of the dietary intake.

If one compares the food composition in stomach and rectum one can find some considerable differences. As these are the same foxes it is obvious that the results from stomach content and excrement investigations are not comparable.

One method that I cannot leave unmentioned is that used in some recent investigations (e.g. Spittler, 1972, Jensen, 1979). This is the one using statistical correlation of the estimated population of carnivores and prey. The basis of these estimations is however so unsatisfactory that I view this method as being insufficient.

Summing up one can state the following: The best method up to now for finding out the composition of the diet of the fox is an analysis of stomach and intestine content. It can be supplemented, but not replaced by additional methods such as examination of food remains at the foxes den and an analysis of excrement.

Below I wish to mention a factor that is often forgotten: This is the difference between food and prey. In the majority of investigations these terms are used synonymously. Below all the individual "prey" killed by the fox is listed even where it does not form a part of the diet. Under the term "diet" one understands the food content found in the stomach and intestine, which need not necessarily have been caught by the fox itself.

It appears that the fox catches more than it can eat during times of increased food availability. Thus Murie (1936) found on a single day 10 killed but otherwise untouched rabbits along a foxes' trail. Goethe (1955) reported 60 silver gulls killed by one fox in one night. The proportion of prey that does not appear in the diet can therefore be quite considerable. Similarly this could also explain why small mammals, especially insectivora and muridae are noticeably missing in the dietary spectrum of the fox in the majority of areas. That the fox is able to catch these animals is obvious as they are the major part of its diet in some biotopes. Murie (1936) thinks that the fox catches shrews to the same extent as other small mammals. He states: "The fox probably pounces on any available source of a smell or a sound and examines what he has captured later; it is likely that every pounce has an element of sport in it". A selective choice of prey is therefore, by comparison to the choice of diet, not or only partly assumable.

Lund (1962) notes that while following a total of 274 km of fox trails, there was no indication that during these journeys larger mammals were killed. Schofield (1960) also notes that whilst following 1109 miles of fox trails, no larger animals were caught.

The foxes' prey, therefore, often is not apparent either at the den, in the stomach nor in the excrement. It can only be determined by following tracks and by direct observation.

On the other hand, the diet not only consists of the prey ingested. During my own investigations (Witt 1976) I tried to determine the proportion of carrion in the stomach-intestine contents. Carrion was considered to be:

1. Where the prey could not have been caught by the fox because of the size of the animal, e.g. stags, deer, cows, pigs, sheep, etc.
2. When shop pellets and/or fly maggots or cocoons were found in the stomach of the fox which had evidently been taken in during ingestion.

Rzebik-Kowalska (1972) adds as an additional criteria: "heads and paws of poultry or leporids which must have been taken from refuse heaps because carnivores never eat these parts if they have whole carcasses of prey at their disposal".

The exact proportion of carrion cannot be exactly determined with the method used, as when eating shot wild animals pellets are not necessarily ingested. Fly maggots require a longer period for development, and cannot be used to indicate carrion in the winter. Using 1+2 as the criteria is certainly insufficient for determining the carrion proportion.

Rzebik-Kowalska (1972) presumes that the method that she uses also leads to an underestimation of the carrion proportion.

When using these criteria in Schleswig Holstein (FRG) I found that 82% of the mammals ingested (excepting mice), and 61% of the birds were carrion (Witt 1976). Here again the considerable difference between prey and diet is noticeable.

According to the dietary formulation, there either appears in the diet a fraction of the actual prey (e.g. insectivores) or a multiple (e.g. deer, hares, pheasants). Only with a few types, e.g. the field mouse, is it probable that diet and prey are quantitatively equal or similar. As long as the relationships are unknown, the ecological importance of the fox can only be insufficiently determined.

One can make a few conclusions from the formulation of the diet. Firstly one can pose the question as to the sites of food intake and catch. Small mammals often have a structurally definable living territory, which also applies for many insects. Thus from the dietary formulation one can form conclusions, with certain limitations, on the preferred hunting grounds of the fox.

In addition one can obtain information regarding the time of day of food ingestion. If one knows the time when the fox was killed, one will find that the stomach content weight will decline from morning to evening. The ingestion of food is therefore mainly at night. One can verify this by observing the movement activity phases of the ingested carabidae. During the course of the year the ingestion of carabidae is limited to the months May to September (Europe), in other words to their active period. Most probably this will also be during the day. One would therefore have to find nocturnally active carabidae.

Finally I should like to put the following points up for discussion: The fox shows taste preferences in the food it ingests. The choice of food, by comparison to the choice of prey, is selective. For example, if alternative prey is abundantly available the fox will seldom eat or will not eat insectivores. If there are sufficient rabbits available it will hardly catch hares as they are more difficult to catch. Taste preferences and varying degrees of difficulty in catching are important factors in the choice of diet. If however the population of a preferred type of prey should suddenly disappear, e.g. rabbits after an outbreak of myxomatosis, the fox will change to a different type of prey. In biotopes with little in the way of food, an otherwise completely despised type of food can even become the main part of the diet. Thus, e.g. Schueler (1951) states in his investigated area that the shrew is the main diet. This shows on the one hand the flexibility of the fox in the choice of diet, whilst in my opinion permitting the following conclusion: If in the biotope of the fox potential prey is available which does not appear in the diet, then the availability of food cannot be a limiting factor for the size of the fox population. In central Europe the available food does not seem to have a controlling function on the size of the fox population. In other biotopes, e.g. North Scandinavia and areas of Russia, by comparison, it is evidently a limiting factor. From the dietary formulation one can see whether the fox population has reached the biologically bearable level, provided one is informed of the available, potential dietary range.

REFERENCES

Behrendt, G. 1955. Beiträge zur Ökologie des Rotfuchses. Z. Jagdwiss. 1 (1955): 113–145, 161–183.

Goethe, F. 1956. Fuchs (vulpes vulpes linné 1758) reibt Schlafgemeinschaft von etwa sechzig jugendlichen Silbermöven (Larus argentatus Pontopp) auf. Säugetierk. Mitt. 4 (1956) 58–60.

Heptner, V. G. und N. P. Naumov 1974. Die Säugetiere der Sowjetunion. Bd. II Seekühe und Raubtiere. Jena.

Jensen, B. 1979. Population Ecology of the fox in Denmark. Lecture Biogeographisches Kolloquium Saarbrücken 24–27.1.1979

Lund, H. 1962. The red fox in Norway, II: The feeding habits. Z. Jagdwiss. 9 (1962): 156.

Murie, A. 1936. Following fox trails. Univ. Michigan, Misc. Publ. 32 (1936): 1–45.

Rzebik-Kowalska, B. 1972. Studies on the diet of the carnivores in Poland. Acta. Zool. Cracov. 17 (1972): 415–506.

Schofield, R. D. 1960. A thousand miles of fox trails in Michigans ruffed grouse range J. Wildl. Mgmt. 24 (1960): 432–434.

Schueler, R. L. 1951. Red fox food habits in a wilderness area. J. Mamm. 32 (1951): 462–464.

Spittler, H. 1972. Über die Auswirkung der durch die Tollwut hervorgerufenen Reduzierung der Fuchspopulation auf den Niederwildbesatz in Nordrhein-Westfalen. Z. f. Jagdwiss. 18 (1972): 76–95.

Witt, H. 1976. Untersuchungen zur Nahrungswahl von Füchsen (vulpes vulpes linné 1758) in Schleswig Holstein. Zool. Anzeiger 197 (1976): 377–400.

7 POPULATION ECOLOGY OF THE RED FOX (Vulpes vulpes L., 1758) IN THE G.D.R.

Michael Stubbe*

INTRODUCTION

The Red Fox is distributed throughout the GDR. From 1968 to 1977 an average of 56,773 specimens were shot annually. In spite of the great ecological importance of the Red Fox only a very little research work has been undertaken to clarify the population ecology of the predator in Northern Central Europe.

In the foreground of all efforts to make an effective contribution to the control of rabies was a reduction of the population of the Red Fox. In spite of considerable financial expenditures in this regard, only partial success was attained (Sinnecker et al., 1975). The rabies epizooty in the GDR has not diminished for more than 25 years, in spite of large scale control measures. From 1953 to 1970 approximately 120,000 people exposed to rabies were registered, of which one third of the cases concerned children and young persons up to the age of 15 years. In the same period 34 people of our country died of this virus epidemic. The rabies expositions were caused by a total of 77,261 animals which endangered persons by bites or intensive contact. Foxes were among the primary causes with 10.9%, cats with 23.5% and dogs with 38.2% (comp. with Figs. 1 and 2).

The research into predatory animals, still on a small scale, will be of particular importance in the future because of the demand for an effective reduction and, as a result, economic use of furs. In spite of intensive efforts to reduce the population density of the Red Fox, the annual estimate increased from 1973 to 1975, from 50,580 to 61,607. Although in 1974 6% more dens were gassed than in 1973, and in 1974 7% more foxes were shot. Since this trend persisted throughout the year 1975 the causes should be analysed thoroughly. The main reason could be that because of the gassing many dens are destroyed and the surviving animals are forced to build new dens. These are mostly small hidden dens which remain unknown to the hunter and that is why they can be used as litter sites for the following years.

In order to considerably reduce the population density of foxes, gassing during littering at the right time, the keeping of known dens, intensive hunting and above all large scale and in depth research into foxes are necessary.

The most important results for the population ecology of foxes in the GDR are to be found in Pitzsche (1972), Stubbe (1965, 1967, 1974, 1977), Stubbe and Stubbe (1977), Ulbrich (1974, 1977). Goretzki and Paustian

* Wissenschaftsbereich Zoologie der Sektion Biowissenschaften Martin-Luther-Universität Halle-Wittenberg, DDR-402 Halle, Domplatz 4.

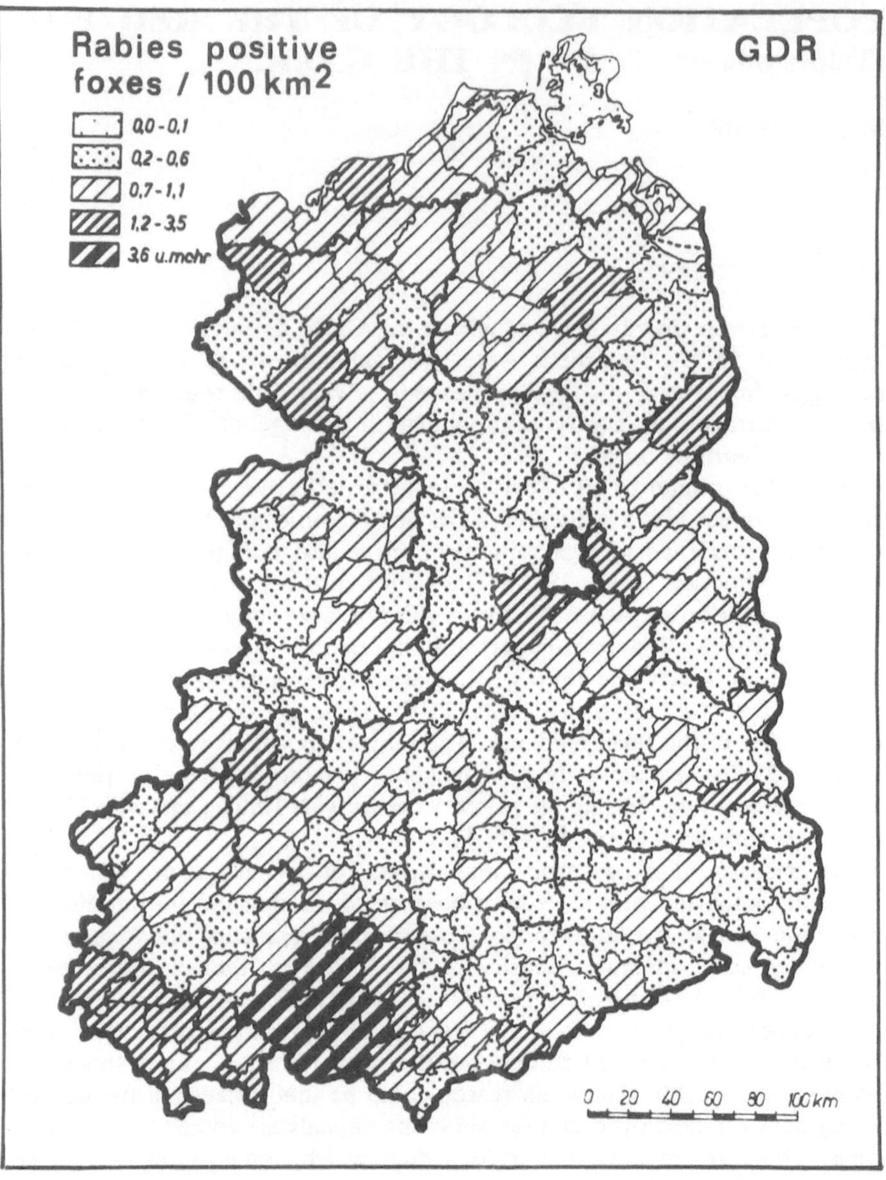

Fig. 1. Rabies positive foxes on 100 km^2 in the districts of the GDR from 1953 to 1970 (after Sinnecker et al. 1975).

(1977) worked on the efficacy of gassing the dens. First investigations into food ecology were made by Proft et al. (1975). The analyses by Creutz (1978) are more comprehensive, although numerous details have not yet been solved. Losert (1975) gave some information about economic calculations. For the epidemilogical importance of the fox Sinnecker et al. (1975) is once again referred to.

In this paper are summarized the most important results of our own

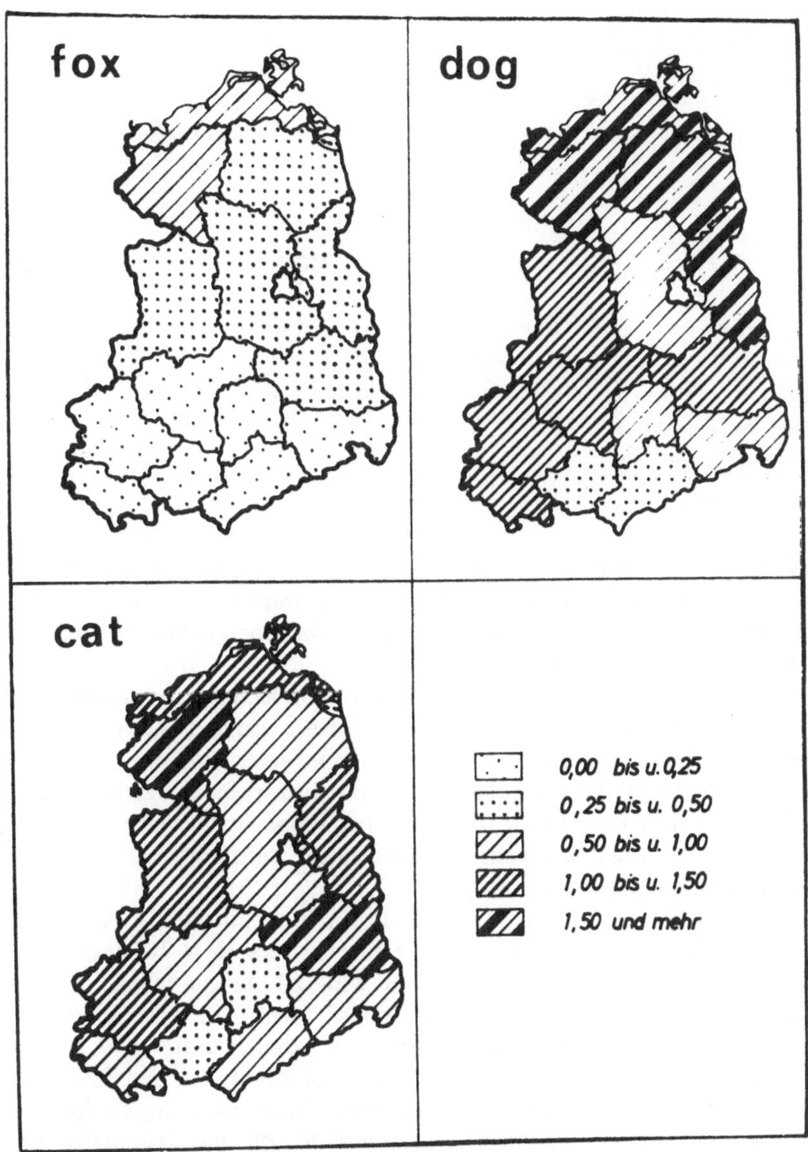

Fig. 2. Number of inocculations/10,000 inhabitants and year after being exposed by fox, dog and cat from 1965 to 1970 (after Sinnecker et al. 1975).

investigations in the Hakel wildlife research area, where we began research into the population ecology of the Red Fox in 1962. The Hakel forest is in the district of Aschersleben, in the county of Halle and has an area of 1,300 ha. As an isolated forest complex of the Hercynic dry space, at the edge of the Magdeburg Börde (Fig. 3), it has been intensively used since the foundation of the wildlife research area Hakel, in the year 1956, for biological investigation into the population of wildlife and birds of prey

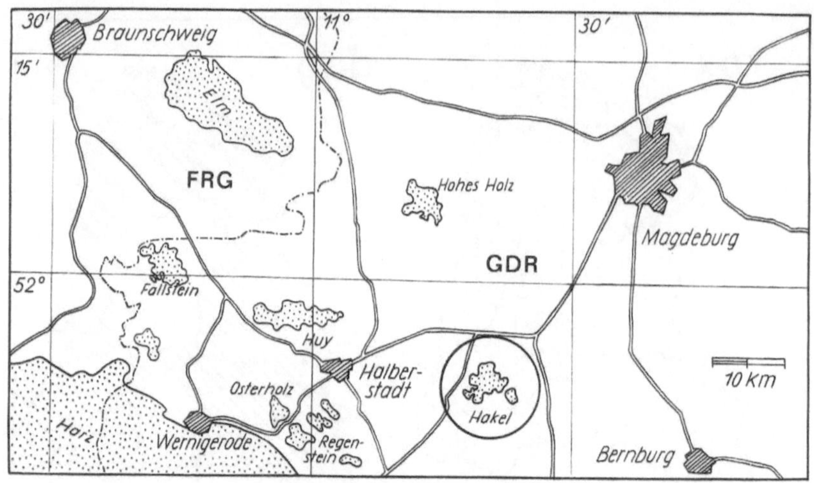

Fig. 3. Location of the wildlife research area Hakel in the GDR.

found there. The Hakel is a mixed deciduous wood, stocked relatively naturally (Fig. 4), and lying in the middle of the most fertile arable land. For more details concerning the Hakel area see the documentation by Stubbe (1971).

Apart from ecological data about population investigations into taxonomy and morphology of foxes in the "Hakel" were carried out (Stubbe and Stubbe, 1977). For the exact treatment of the ecological phenomena it is neccessary to clarify the taxonomy of Central European foxes. In the above mentioned work it was shown that the term *Vulpes vulpes crucigera* (1789) dating back to Bechstein belongs to the typical form whose area stretches across almost the whole of Europe.

The status of *V. v. silacea* (Miller, 1907) in Spain and *V. v. ichnusae* (Miller, 1907) on Corsica and Sardinia merely remain to be verified. The subspecies *crucigera* is to be classified with all its synonyms (Ellerman and Morrison-Scott, 1966) under the typical form. The Irish foxes which obviously have a shorter tail and longer ears than Central and Northern European foxes (Fairley, 1970) also require a clarifying taxonomical classification. The same is true with Bulgarian foxes which, according to Atanassov (1958), have a clearly smaller condylobasal length and zygomatic width. Since Greek foxes also have very small skull sizes, the incidence of a different subspecies from Central Europe for the Balkans or South East Europe is probable. Therefore it is not out of the question that a name synonymous with *crucigera* will be used again. For this purpose thorough taxonomical investigations are required.

THE LOCATING OF DENS AS A BASIS FOR ECOLOGICAL INVESTIGATIONS OF THE FOX POPULATION

The structure of the landscape and soil conditions are the decisive factors for the building of badgers' and foxes' dens which serve as a place of maximum security for birth, rearing and living.

In the Hakel area we find transitions from "Löß" free locations to thin layers of "Löß" up to a metre thick which overlay mesozoic rock. "Löß" and "Löß clay" meet, because of their soil structure, the optimal demands of subterrestrial living mammals, especially of foxes and badgers, keeping in mind the maintenance of dens over the years. On the central plateau of the Hakel area and the North West side of the forest there is almost no "Löß" at all. Only a few small dens are distributed in this area. The geological map shows that the erosion valleys in the North East of this area represent the preferred natural habitation. On the slopes of these valleys one den follows another. Apart from the thick "Löß" layers, as we find them in the South West and in the "Small Hakel", the heat of the sun on the valley slopes play a certain role. Ninety five percent of these dens are on the South East slopes which have an average angle of inclination of 20–30° and offer the carnivores a large colony density.

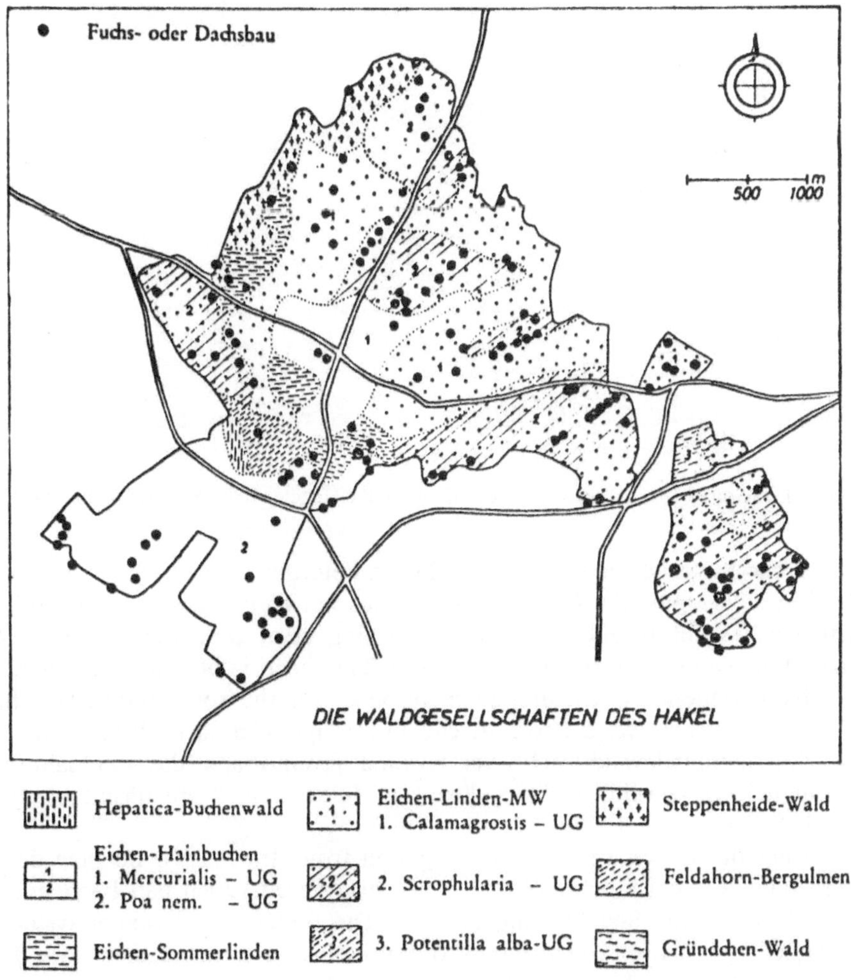

Fig. 4. Location of dens and wood societies of the Hakel area (changed from Stubbe 1971).

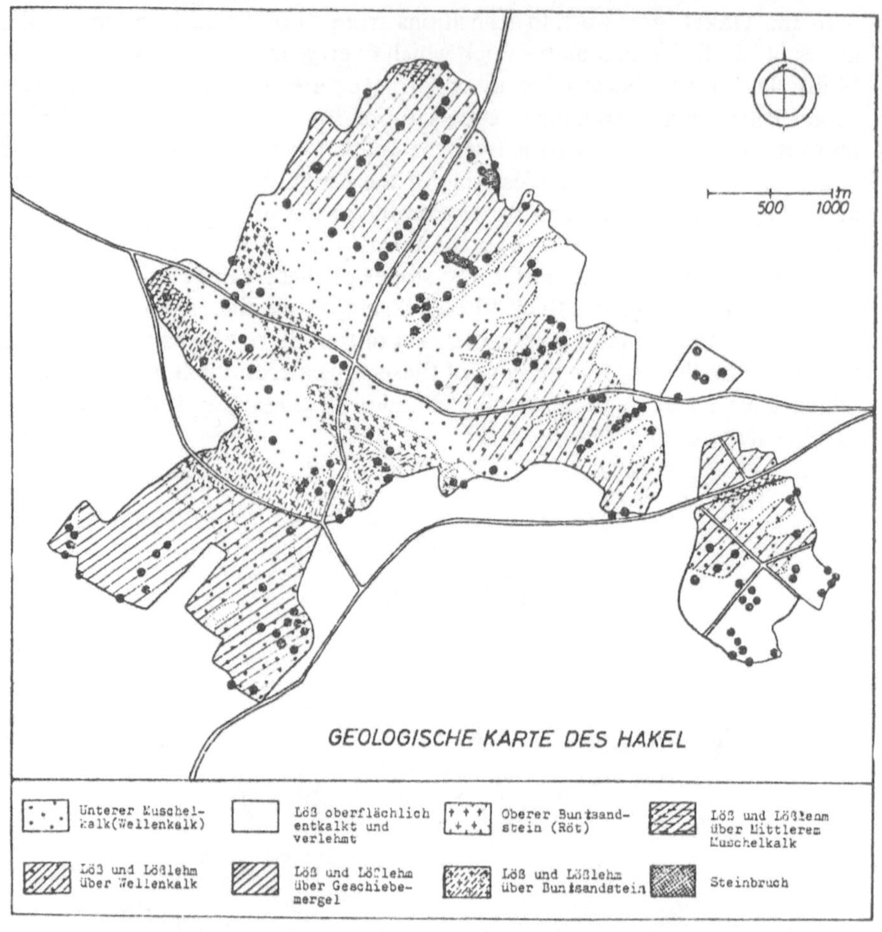

Fig. 5. Location of dens and geological conditions in the Hakel (completed after Stubbe 1965).

The varying thicknesses of "Löß" layers determine a varying stocking of the forest. Weinitschke (1954) distinguishes between eight forest societies with various subsocieties. In spite of having been used for decades the natural image of the forest has been kept. The location of dens in the forest is typical, depending, as already mentioned, on the soil conditions. The "Hepatica" beech wood (from 40 cm on strongly clefted shell-lime stone) the "Steppenheidewald" with its levelled ground and the "Feldahorn-Bergulmen-Mischwald" with its distinct "Mullrendzina" are unsuitable for den habitation.

During the fifteen years of investigation from 1962–1976 no new dens were built. Only a few small dens were discovered which had been overlooked during the first locating of dens. The underground tunnel systems remained because of the firmness of the soil and because there was no gassing of dens which therefore were inhabited for decades. The majority of dens are certainly built by badgers and taken over by foxes.

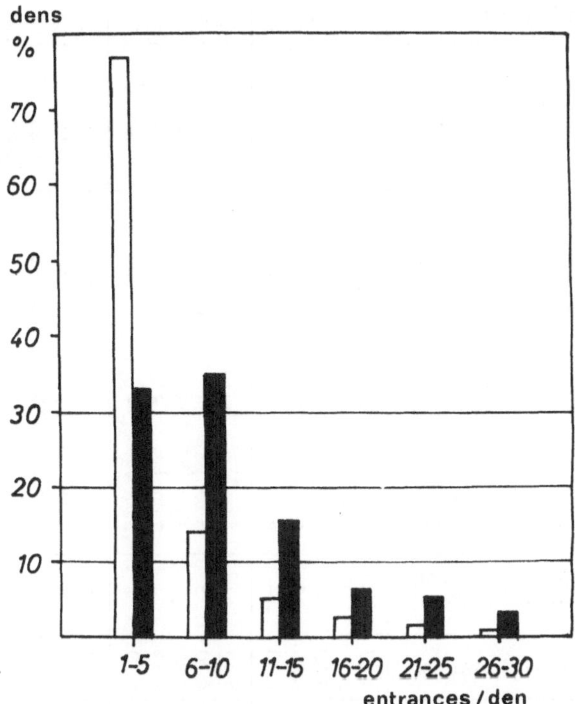

Fig. 6. Percentage of dens ($n = 146$, white columns) in the different size groups and the proportional distribution of fox litters from 1962 to 1976 ($n = 108$, black columns).

The total number of dens located in the Hakel amounts to 146 of which 76% have no more than five tunnels (Fig. 6). To every 100 ha there were 11.2 fox- or badger's dens. This alone shows how this area is predestined for carnivores. Knowledge about all dens is a fundamental condition for the investigation of reproductive potentials and numerous other ecological parameters. In the GDR all hunting societies are obliged to keep a census of dens. For the literature on the subject of the density of dens see Stubbe (1973).

REPRODUCTION RATE

Over a 15-year period the annual reproduction rate for the "Hakel" population amounts to 190% (119–265%) on the average referring to the parent population on the 1st of April of each year under the precondition of a sex ratio of 1.5♂♂: 1♀ (Table 1). The average over the last five years amounts to 208% and therefore lies above the long-term results. Only the year 1976 reached a clear low point again with 153% for the first time since 1969 (161%). The reason for this low reproduction rate was probably the raw April weather in 1976, to which a high percentage of the new born pups fell victim. The minimum losses can be calculated as 30% assuming a minimum number of five pups per litter up to the age of four to six weeks. This is the period of life from the time of birth up to the appearance of the pups on the

Table 1. Reproduction, losses, population density of the Red Fox between 1962 and 1976 in the wildlife research area Hakel.

Years	Litter size											Total litters	Observed pups	Observed pups/litter	Losses of pups in the first 4–6 weeks	Total number of pups/litter	Losses of the first 4–6 weeks as % of the total number of pups	Reproduction rate of the observed pups in relation to the parent population in %	Population density per 100 ha hunting area on 1st April (field: wood = 1 : 1)	Number of litters per 100 ha of wood
	0	1	2	3	4	5	6	7	8	9	10									
1962	—	—	—	4	—	1	—	—	1	—	—	6	25	4.17	8	5.50	24	139	0.69	0.46
1963	—	—	—	1	1	—	2	—	—	—	—	4	19	4.75	3	5.50	14	158	0.46	0.31
1964	—	—	1	1	1	—	—	1	—	—	1	5	26	5.20	6	6.40	19	200	0.50	0.38
1965	—	—	1	2	2	1	1	—	—	—	—	7	27	3.86	9	5.14	25	159	0.65	0.54
1966	—	—	—	—	—	6	1	2	2	—	—	11	66	6.00	—	6.00	—	244	1.04	0.85
1967	2	—	1	4	3	1	—	—	—	—	—	11	31	2.82	24	5.00	43	119	1.00	0.85
1968	—	—	—	2	1	1	1	—	—	—	—	5	21	4.20	5	5.20	19	162	0.50	0.38
1969	—	—	2	3	2	3	1	—	—	—	—	11	42	3.82	14	5.09	25	161	1.00	0.85
1970	—	1	—	—	—	—	—	4	2	—	—	7	45	6.43	4	7.00	8	265	0.65	0.54
1971	—	—	—	1	—	4	2	—	—	1	—	8	44	5.50	2	5.75	4	244	0.69	0.54
1972	—	—	2	—	1	—	1	—	2	—	—	6	30	5.00	7	6.16	19	200	0.58	0.46
1973	—	—	—	—	—	3	1	1	1	—	—	6	36	6.00	—	6.00	—	240	0.58	0.46
1974	—	—	1	—	1	3	2	—	—	—	—	7	33	4.71	4	5.29	11	194	0.65	0.54
1975	—	—	—	1	—	2	1	1	1	1	—	7	43	6.14	2	6.43	4	253	0.65	0.54
1976	—	1	—	3	1	1	—	1	—	—	—	7	26	3.71	11	5.29	30	153	0.65	0.54
total	2	2	8	22	13	26	13	10	9	2	1	108	514	4.76	99	5.68	16	190	0.69	0.55

dens. This mortality rate now amounts to an average of 16% (0–43%) over 15 years.

So the reproduction rate can diverge by approximately 150%. More precisely expressed it is not (or only rarely) the birth rate which underlies the considerable deviations but the annually varying survival chances of the embryos or new born pups, which depend on climatic factors, diseases and food supply. The supply of prey is fully subordinate to these factors in Central Europe as a regulation mechanisms, as the fox is very adaptable in its feeding habits and its young can be reared on variable prey. The supply of prey has no primary importance for the annual growth in Central Europe and Southern Sweden (Englund, 1970). However in Northern Sweden there is supposed to be a positive relation between both factors.

The observed litter size of 108 fox litters in the Hakel area amounts to 4.76 pups per litter (0–10). Taking into account the losses during the first four to six weeks of life the rate increases to 5.68 (see Table 1).

For the 108 litters calculated from 1962 to 1976 54 dens were necessary. One den was used six times as a litter site one other den five times, 6 dens four times, 8 dens three times, 11 dens twice and 27 dens once (Figs. 7 and 8). Two female foxes reared their pups together in one den twice, whereby the maximum distance between two litters is nought. Once a large den was inhabited simultaneously by a fox and badger litter. On another four occasions female foxes moved with their pups after being disturbed in their litter den to dens inhabited by badgers.

We found, in contrast to the observations of Behrendt (1955), that up to 67% of litter dens were large dens in the Hakel area and could be left after a few weeks, especially after disturbances. Also small litter dens have often been known for decades. The average number of litters in the Hakel area amounted to 7.2 litters per year (4–11). Under litter size 0 in Table 1 two litters are mentioned which, presumably, were totally lost in the first four weeks of life because of rabies.

The number of non-pregnant female foxes is obviously very small. According to Ulbrich (1975) the number in the county of Dresden was about 3.5%. In the Hakel area only two such animals could be shot. Pitzschke (1972) investigated 301 female foxes in the county of Gera in the GDR between February and June and established that 67 animals (22.3%) were not reproductively active. Among the fertile females the fetuses of 30

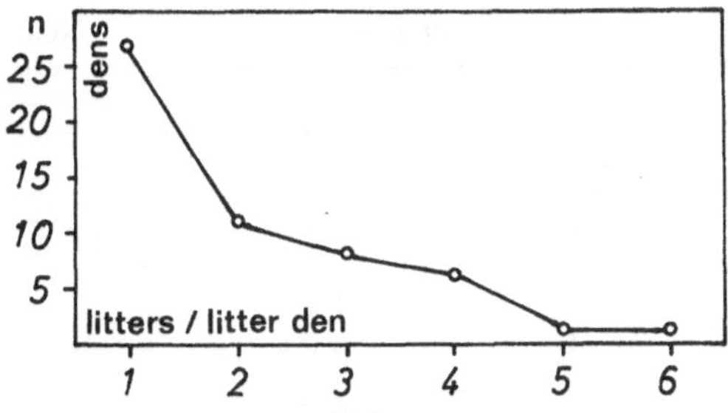

Fig. 7. Number of litters per litter den.

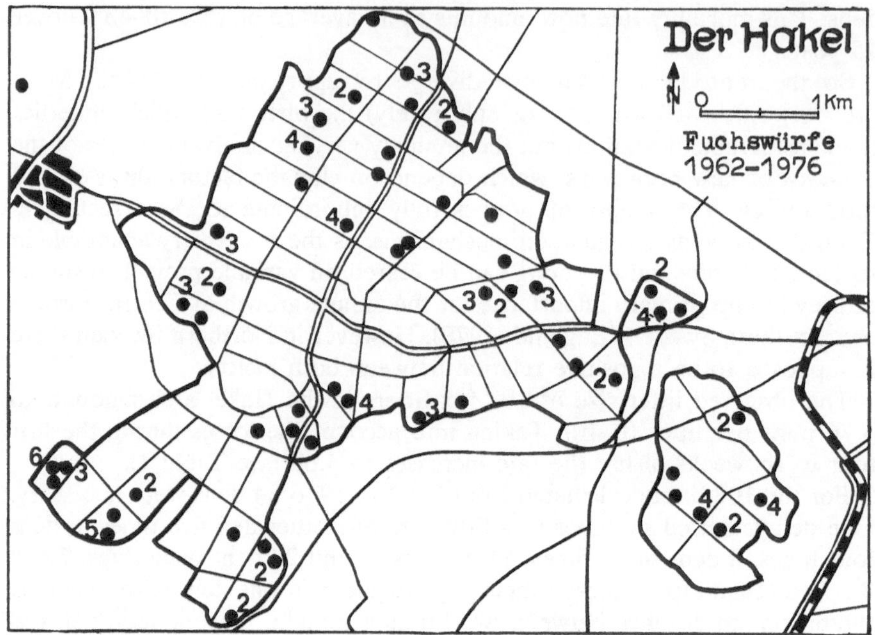

Fig. 8. Distribution of litter dens and number of litters ($n = 108$) in the wildlife research area Hakel.

animals (12.8%) died of embryonic miscarriage. So altogether 97 females (32.3%) were eliminated from reproduction. This percentage is unusually high and shows how important are parallel investigations in several areas. Further comparison data are noted by Stubbe (1973).

A comparison between the number of fetuses (Table 3) and the rate for observed pups per litter (Table 4) shows that one has to take into account a loss of 1.54 fetuses or pups between the time of the embryonic development and the fourth postnatal week (see Fig. 9). In this the losses of ova and young embryos are not included. An extraordinarily high reproduction rate of 6.3 observed pups per female fox ($n = 32$) is given by Pielowski (1976) for Poland.

Table 2. Number of dens ($n = 146$) and of litters ($n = 108$) per den size from 1962–1976 in the Hakel area.

Number of tunnels/den	Number of dens		Number of litters	
	n	%	*n*	%
1 to 5	112	76.7	36	33.3
6 to 10	20	13.7	38	35.2
11 to 15	7	4.8	17	15.7
16 to 20	4	2.7	7	6.5
21 to 25	2	1.4	6	5.6
26 to 30	1	0.7	4	3.7
Total	146	100.0	108	100.0

Table 3. Number of fetuses or placental scars* for female foxes in the GDR and their distribution in percent.

Author (year)	Place of origin	n	Vixens fetuses or placental scars												Fetuses/ vixen
			1	2	3	4	5	6	7	8	9	10	11	12	
Pitzschke (1972)	county of Gera	67	2	1	1	7	3	25	8	13	5	1	1	—	6.36
Ulbrich* (1977)	county of Dresden	132	1	3	5	16	28	27	22	20	4	4	—	2	6.08
Stubbe (1974)	Tharandt forest	2	—	—	—	—	—	1	—	—	1	—	—	—	7.50
Stubbe	Hakel area	7	—	—	—	—	—	—	1	—	2	2	1	1	9.17
total	GDR n	208	3	4	6	23	31	53	31	33	12	7	2	3	6.30
	%	100	1.4	1.9	2.9	11.1	14.9	25.5	14.9	15.9	5.8	3.4	0.9	1.4	—

Table 4. Number of observed pups per litter size and their distribution in percent in the Hakel area.

	Litter size												
Litters	0	1	2	3	4	5	6	7	8	9	10	Total	Pups/Vixen
n	2	2	8	22	13	26	13	10	9	2	1	108	4.76
%	1.9	1.9	7.4	20.4	12.0	24.0	12.0	9.3	8.3	1.9	0.9	100	—

The maximum number of reared pups in general correlates with the number of existing milk teats. In contrast to the information given by Usinger (1950), van den Brink (1957), and Gaffrey (1961) as well as Haltenorth and Roth (1968) female foxes have not three, but four or five pairs of teats. This is definitely proved by findings with foxes in the Hakel, where out of seven vixens four had eight, two females had nine, and one specimen had ten teats.

In the Hakel six pregnant females from 1962 to 1972 were examined. Since, according to Naaktgeboren (1965) head presentation is supposed to occur more often than pelvic presentation during birth, special attention was paid to the position of the fetuses, their sex and weight during dissection (Table 6). Because of the relatively small amount of material the findings of Naaktgeboren could not be confirmed. Out of 57 fetuses 30 were in front position, but when the 15 fetuses from the Tharandt forest are also taken into account, the relation becomes almost entirely balanced (see Table 5). A conformity in presentation and sequence of presentation and sex could not be found (see Table 6).

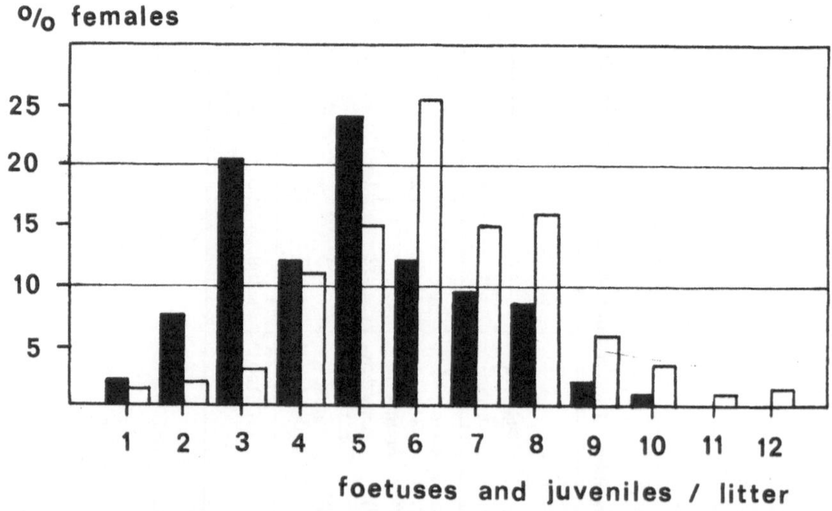

Fig. 9. Proportional distribution of vixens per litter size four weeks after birth ($n = 108$, black columns) in comparison to the number of fetuses or uterus scars ($n = 208$, white columns) with a clear phase displacement of both reproduction data.

Table 5. Position of fetuses in the vixens.

Origin	Number of vixens	Head presentation ♂♂	Head presentation ♀♀	Pelvic presentation ♂♂	Pelvic presentation ♀♀	Not clear presentation ♂♂	Not clear presentation ♀♀
Hakel	6	19	11	16	11	2	—
Tharandt forest	2	5		10		—	
total	8	35		37		2	

SEX RATIO

A preponderance of male foxes in the sex ratio has been established by numerous authors (see Stubbe, 1973, Ulbrich, 1974). This is also true of the Hakel population (see Tables 7–9). In calculating the sex ratio three age groups were taken into consideration; fetuses, pups shot on the dens and young or old foxes shot separately in the preserve. In the county of Dresden Ulbrich (1974) was only able to establish a surplus of males in the second and third group of 1.45 or 1.25 : 1 following the sequence given. In the cases of 142 fetuses the ratio of 0.95 : 1 had moved slightly in favour of the vixens. Altogether the ratio in the cases of 2.569 investigated individuals was 1.31♂♂ : 1♀. Pitzschke (1972) came to varying results in the county of Gera. From 177 fetuses 90 males were able to be registered (1.03♂♂ : 1♀). Out of 875 pups up the age of six months the females predominated with 51.7% (0.94♂♂ : 1♀), among 1.421 older foxes the males were predominant with 57.2% (1.34♂♂ : 1♀). Altogether 2,473 foxes with a sex ratio of 1.16♂♂ : 1♀ were examined by Pitzschke. In the Hakel area the proportion of males in the evaluation of all three age groups is clearly greater than that of females with 1.58 : 1 ($n = 642$) (Table 10).

The variance of the sex ratio of foxes shot separately in the Hakel area is especially high (2.01 : 1, $n = 325$). The ratio is subject to a certain rhythm in the annual cycle, which has its origins partly in defined behaviour norms of the fox and partly in the shooting habits of the hunters (see Table 7, Fig. 10).

While from May to August the sex ratio is to, a great extent balanced, from September to April many more males than females are shot. Since from May to June whole litters with adult females are eliminated, the proportion of females is much higher than in the other months. From July to August it is mostly the still inexperienced pups which are shot in the preserves and harvested fields, and the relation corresponds, apart from a slight predomination of males, almost exactly to the results found with young foxes from April to June (Table 8).

In the other months the males are obviously more active, have a greater radius of action and are more easily shot. During the main breeding time the females are mainly in the dens and the searching males are therefore more easily shot. An intensification of den hunting which, because of the danger of rabies, is at the moment still prohibited in the GDR, would especially at

Table 6. Sex ratio, presentation and mass of fetuses in six vixens from wildlife research area Hakel. The direction of arrow marks the head presentation.

No.	Date	Age	Teats	Fetuses left 7	6	5	4	3	2	1	Fetuses right 1	2	3	4	5	6	7	♂♂:♀♀	Head-rump-length mm
1967/7	11.3	1	4/4			♂ →8	♂ ←8	♂ ←9	♀ →8	♂ →8	♂ ←8	♂ ←8	♀ →8	♂ →8	♀ ←8	♂ ←8	♂ ←8	9:3	50
1967/6	11.3	1	4/4					♂ →48	♀ →49	♂ ←50	♂ ←53	♀ →48	♀ →42	♀ →45	♀ →45	♀ ←48		3:6	100
1972/5	17.3	very old	5/4		♂ ←102	♀ ←54	♂ ←(†)	♂ ←70	♀ ←72	♀ →83	♀ ←94	♂ ←74	♀ →81	♂ →82				5:5	100 to 120
1972/6	19.3	1	4/4					♀ →71	♂ →74	♂ →72	♂ ←71	♂ →78	♂ ←68	♀ →76				5:2	110
1970/2	26.3	1	4/4		♀ →32	♂ →30	♂ →34	♂ →31	♂ →34	♀ →35	♂ ←30	♂ →35	♂ →30	♂ →34	♀ →32			8:3	80
1970/7	9.3	1	?			(3♂♂+1♀) 3♂♂ ← 1♀ → 29.5–35 g								(4♂♂+2♀♀) 2♀♀ ← 1♂ ← 1♂ → 2♂♂ ? 30–35.5 g				7:3	95

Table 7. Number and sex ratio of older than 3 months old foxes shot in the wildlife area Hakel ($n = 411$).

Months	Foxes shot ♂♂	♀♀	Sex?	Total n	%	Sex ratio ♂♂:♀♀
January	23	9	3	35	8.51	2.56:1
February	9	1	10	20	4.87	9.00:1
March	19	11	2	32	7.78	1.72:1
April	14	0	4	18	4.38	14.00:0
May	10	12	5	27	6.57	0.83:1
June	6	7	10	23	5.60	0.86:1
July	11	9	6	26	6.33	1.22:1
August	16	14	8	38	9.24	1.14:1
September	8	3	3	14	3.41	2.67:1
October	9	5	5	19	4.62	1.80:1
November	49	19	16	84	20.44	2.58:1
December	43	18	14	75	18.25	2.39:1
total	217	108	86	411	100.00	2.01:1

this time mean a greater number of females would be shot and contribute to the reduction of the fox. Already in March the proportion of males considerably decreases in comparisons with February. At the beginning of April most of the females give birth, they are longer in the den and are hidden. The hunter also spares the suckling females at this time and mostly males are shot. The April data from Table 7 are, however, not representative of a larger area, since in a larger area a certain percentage of females is also shot. Moreover the natural mortality of females seems to increase with age compared with the males.

Lund (1959) in Norway found very similar results in the monthly analysis of the sex ratio, Pielowski (1976) in Poland and Ulbrich (1974) in the county

Table 8. Number and sex ratio per month of young foxes shot on the dens in the wildlife research area Hakel ($n = 305$).

Months	Foxes shot ♂♂	♀♀	Sex?	Total n	%	Sex ratio ♂♂:♀♀
April	11	9	9	29	9.51	1.22:1
May	119	100	47	266	87.22	1.19:1
June	6	4	—	10	3.27	1.50:1
total	136	113	56	305	100.00	1.20:1

Table 9. Sex ratio of the foxes in the Hakel area.

Age	n	♂♂	♀♀	♂♂:♀♀
Fetuses (of 7 vixens)	68	40	28	1.43:1
Young foxes (shot on the dens)	249	136	113	1.20:1
Foxes (shot seperately in the preserve)	325	217	108	2.01:1

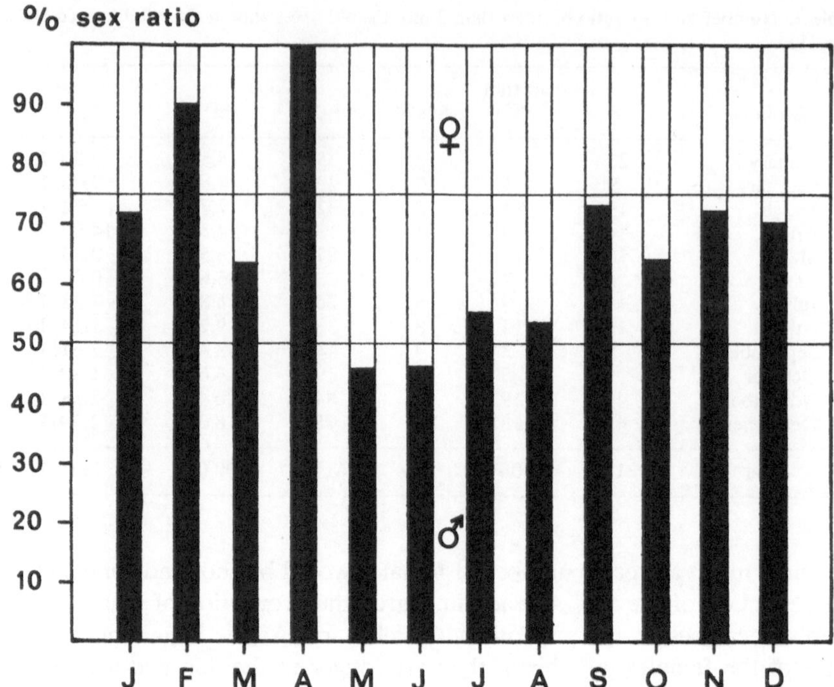

Fig. 10. Proportional distribution of sex ratio of foxes shot separately in the Hakel area ($n = 325$) in the annual cycle.

of Dresden also found the same. The total average for the GDR, according to present proportional figures (Table 10), shows a sex ratio of 1.35 ♂♂ : 1 ♀, the foxes being six months and older. Taking into account that a certain proportion of females is not pregnant, it seems to be safe to base one's reproductive calculations on the ratio 1.5 : 1. After determining the number of litters (n) in a certain area, the size of the parent population can be relatively accurately calculated ($n \times 2.5$).

POPULATION STRUCTURE

Growth and mortality, sex ratio and life expectancy determine the turnover of a population. As with many other mammals several methods for determining the age of the fox are used: second dentition (complete at the age of 5 or 6 months) and the teeth wear, fusion of the epiphysis of certain bones of the extremities (ossified at the age of one year), ossification of the skull sutures, mass of the eye lenses (12 month old = 212 mg, 197 to 223 mg), mass and length of the baculum (juv <50 mm, ad >5 g), counting of the cementum or dentin layers (literature, see Stubbe, 1973).

What is important is the immediate recognition of animals up to the age of one year and therewith the distinction from older animals. For this purpose the best means of determining the age of the Hakel population proved to be the wear of teeth (Stubbe, 1965, 1973) which comprehensive

Table 10. Sex ratio of foxes of different age groups from the GDR.

Author year	n	Fetuses ♂♂	♀♀	Sex ratio	n	Young foxes ♂♂	♀♀	Sex ratio	n	Foxes >6 mths ♂♂	♀♀	Sex ratio	n	Total ♂♂	♀♀	Sex ratio
Pitzschke (1972)	177	90	87	1.03:1	875	423	452	0.94:1	1,421	813	608	1.34:1	2,473	1,326	1,147	1.16:1
Ulbrich (1974)	142	69	73	0.95:1	1,063	630	433	1.45:1	1,364	758	606	1.25:1	2,569	1,457	1,112	1.31:1
Stubbe & Stubbe (1977)	68	40	28	1.43:1	249	136	113	1.20:1	325	217	108	2.01:1	642	393	249	1.58:1
total	387	199	188	1.06:1	2,187	1,189	998	1.20:1	3,110	1,788	1,322	1.35:1	5,684	3,176	2,508	1.27:1

follow up investigations also confirm in comparison with teeth profiles up to the age of two. Brömel and Zettl (1974) as well as Geiger et al. (1977) came to the same results. In contrast, v. Bree et al. (1974) found that evaluation of cementum layers in teeth profiles, in comparison with that of teeth wear, only corresponded in 35% of the cases of foxes from Holland and 47% from France, whereby they especially refer to the third year of life of older individuals after tooth grinding as well. This should be avoided in order to prevent greater mistakes in the evaluation in the ages of animals more than two years old.

Animals up to the age of 12 months do not generally have any tooth grinding worth mentioning on the tips of the lower incisors. The first upper incisor (I^1) is only worn a little. The triple pointed crown is usually still well recognisable. By this method a quick determination of age of trapped live animals is also possible.

In the case of an average reproduction rate of 190% (without counting the losses of the first four to six weeks of life), related to the parent generation before the new reproduction cycle in spring, the annual increase or decrease of the fox population amounts to approximately two thirds. The proportion of pups in the population after the birth period and the proportion of one year old animals during the reproduction period are about the same, whereby similar mortality rates among young animals under one year and adults can be concluded (also see Lloyd et al., 1976). From the findings a turnover model can be easily deduced, according to which the age groups, nought to five, give the following percentages: 65.52%, 22.76%, 7.58%, 2.76%, 1.03%, 0.35%. In the case of a mortality rate being evenly distributed over the whole year, the average life expectancy of a fox would then be exactly 12 months, theoretically speaking. Since the main losses take place during the fur maturing period as a result of human interference and as the losses occur on average between November and January, the life expectancy of a fox in hunted populations may theoretically be higher, about 15 months.

How the age pyramid looks for populations not hunted is largely unknown. In the Hakel area where the population has been repeatedly hunted to below the "zero level" (see Fig. 12) in recent years and where immigration from outside has occurred, the age distribution has shifted much more

Table 11. Theoretical life table of fox population in the Hakel area.

Age class	1	d	1_x	d_x	q_x	e_x
0	290	190	1,000	655	655	1.02
1	100	66	345	228	661	1.00
2	34	22	117	76	650	0.99
3	12	8	41	28	683	0.89
4	4	3	13	10	769	0.73
5	1	1	3	3	1,000	0.50
total	441	290	1,519	1,000	659	—

strongly in favour of age group nought than as calculated in the theoretical model above. This will be investigated in a special paper.

POPULATION DENSITY

For years much effort has been made to decrease the population density of the fox on a large scale. In order to eliminate rabies in the GDR, legislation demands a Spring stock (1st of April) of two foxes on an area of 1,000 ha. This means, with a sex ratio of 1.5 : 1, one litter on 1,250 ha. Other authors see in a relative fox density, hunting indicator of population density (HIPD, after Bögel et al., 1974), of 0.2 to 0.4 foxes shot per 100 ha per year an effective bar against the spreading of rabies. In spite of the intensive gassing of dens, we remain, with an HIPD of about 0.6, still far away from this goal, as shown by the annual shooting rate in the GDR.

The spring stock can not be exactly estimated for the territory of the GDR, since the annual shooting rate reflects the shooting of the calender year and not of the hunting year which is from 1st April to 31st March. Moreover the foxes killed during the gassing of dens are not entered in the statistics. One can certainly assume that in this way 50% of the juveniles are eliminated. Theoretically one could expect 8,664 litters and a parent population of 21,660 individuals calculated in the GDR annually based on 2 foxes per 1,000 ha or a litter per 1,250 ha. With an average reproduction rate of 190%, related to the parent population, this would be an annual growth of 41,154 young foxes which would immediately have to be eliminated again as an annual shooting rate in order to have the same number of original stock. This would correspond to an HIPD of 0.38 individuals per 100 ha.

If gassing is continued and keeping in mind the assumption that 50% of the young foxes are eliminated by this means, the theoretical figure for the annual shooting rate and all other calculations based on this could still be fixed 50% lower. The annual spring stock in the GDR still probably lies between 50,000 and 60,000 foxes.

The average of the HIPD values in the years 1972 and 1973 for the several districts of the GDR is to be found in Fig. 11. Analyses referring to this have to be continued and to be thoroughly evaluated.

In the Hakel area the average population density for Spring (March) amounted to 0.69 (0.46–1.04) individuals per 100 ha. (see Table 1) over 15 years of investigation. There were 0.28 litters per 100 ha. of preserve area (field : wood = 1 : 1). According to the research principles no den was gassed but an intensive shooting of litters was carried out instead. It was possible to reduce the fox population of the Hakel area in most of the years down to the level of the preceeding Spring stock, and in 1972/73 as well as in 1976/77 and the following years down to zero. This vacuum was always filled again from surrounding preserve areas until the next reproduction phase (see Fig. 12). Pielowski (1976) came to similar results in Poland. This shows clearly that neighbouring preserves have to take still more intensive measures to drastically decrease the population density of the Spring stock.

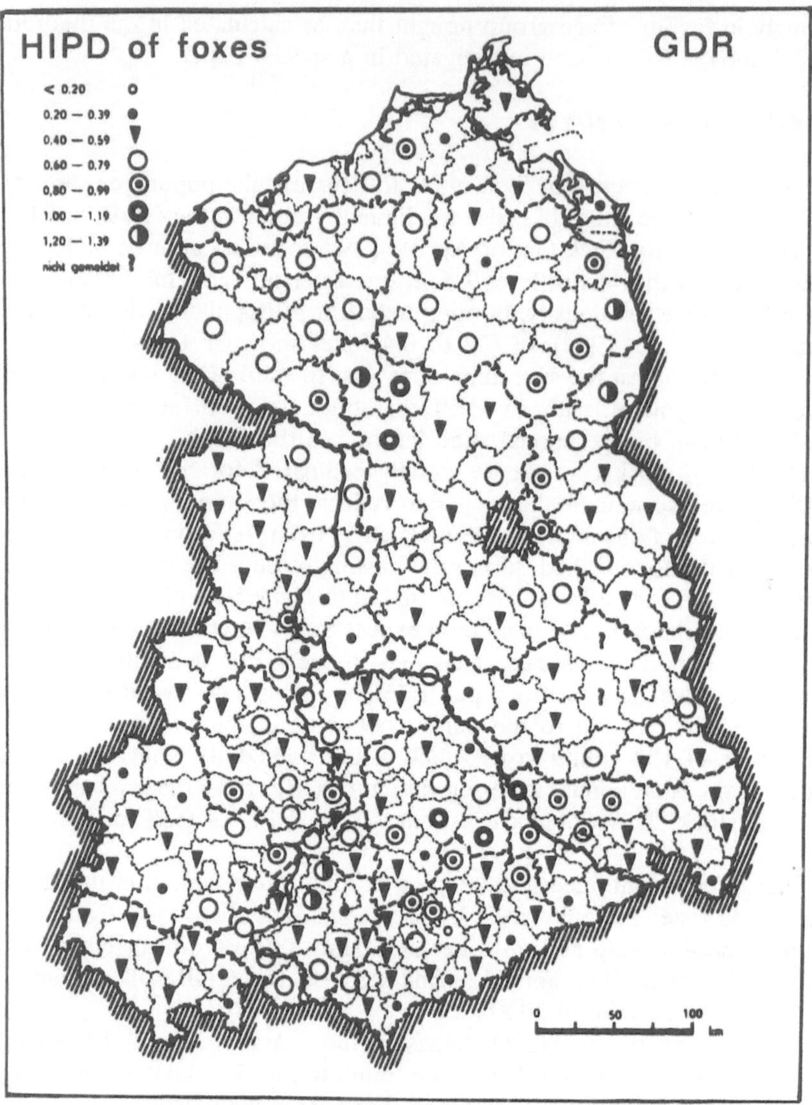

Fig. 11. Number of foxes shot in the districts of the GDR on 100 ha preserve area in the average of the years 1972 and 1973 (after Stubbe 1977).

It was demonstrated in the Hakel area, that the shooting of young foxes represents an effective method of reduction which is of the same value as the gassing of dens at random. It has, as already mentioned, the essential advantage of keeping all known dens, so that the foxes are not forced to build small hidden dens in order to breed their pups unnoticed during the next reproduction phase.

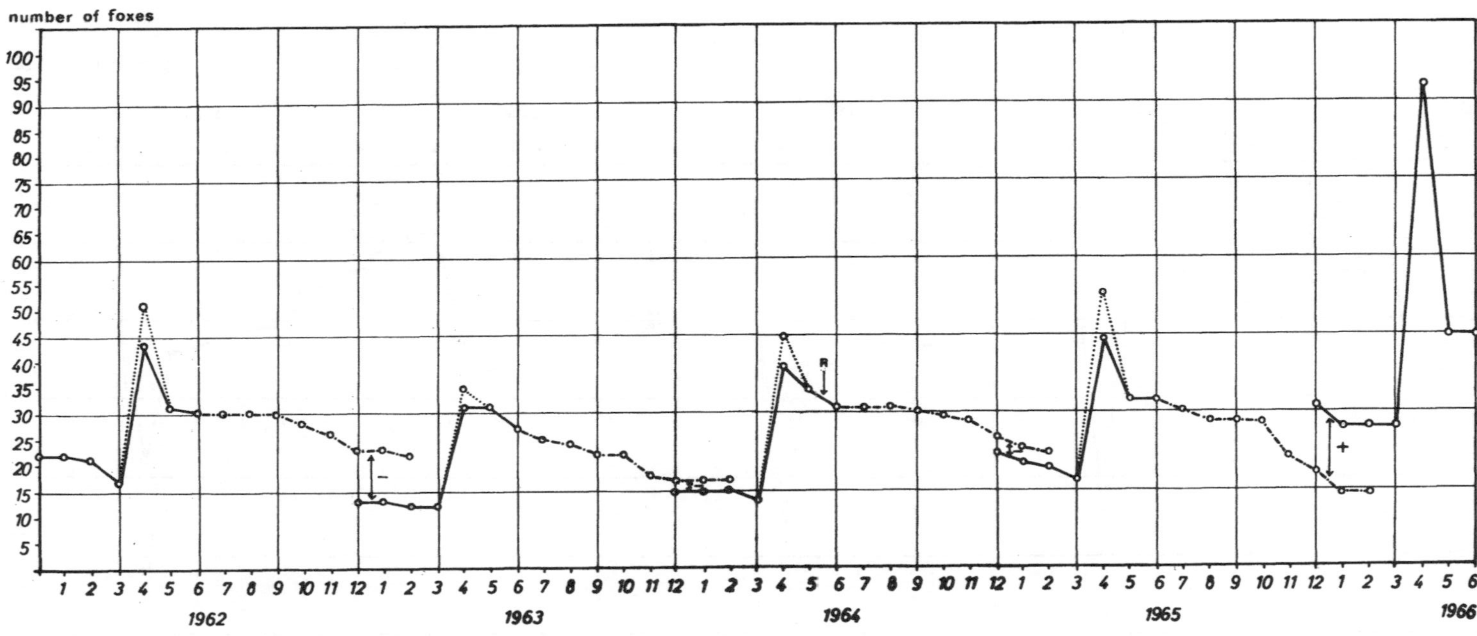

Fig. 12. Annual population dynamic of the Red Fox in the wildlife research area Hakel from 1962 to 1976. The dotted curves from March to May show the estimated losses of young foxes. The arrows with the marks – or + point to a population decrease or population increase (imigration). The R means rabies which did not cause any essential losses apart from the year 1967, however.

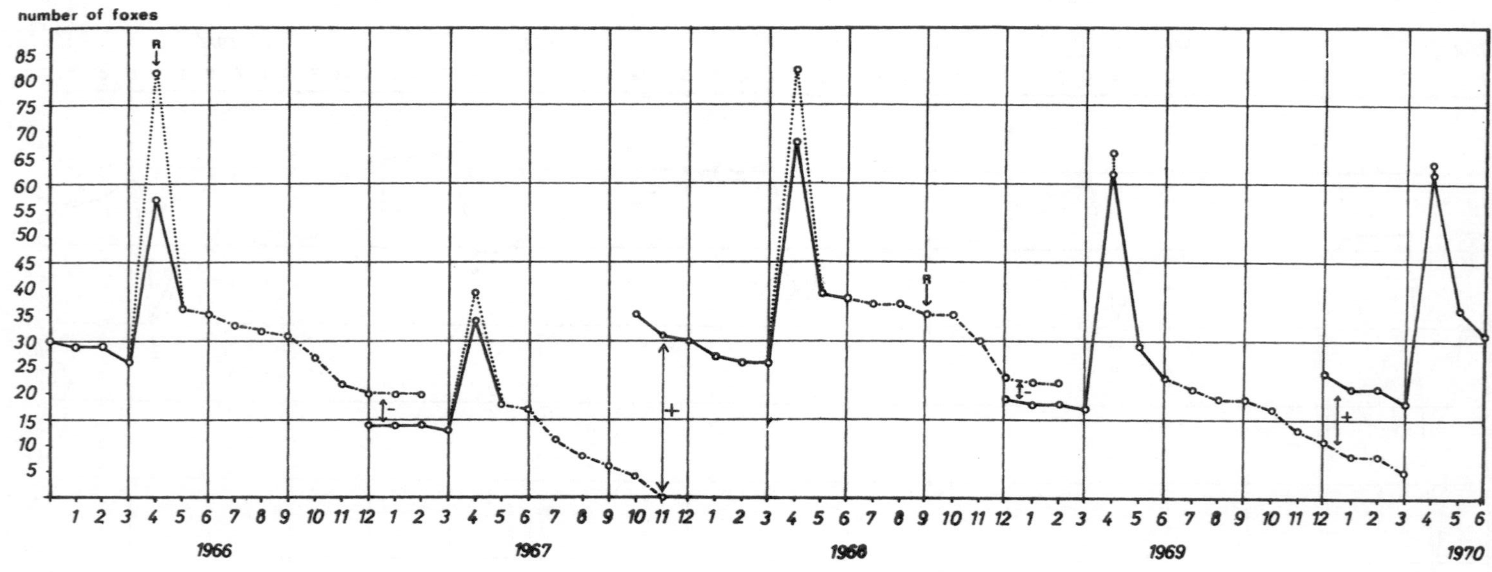

Fig. 12 continued.

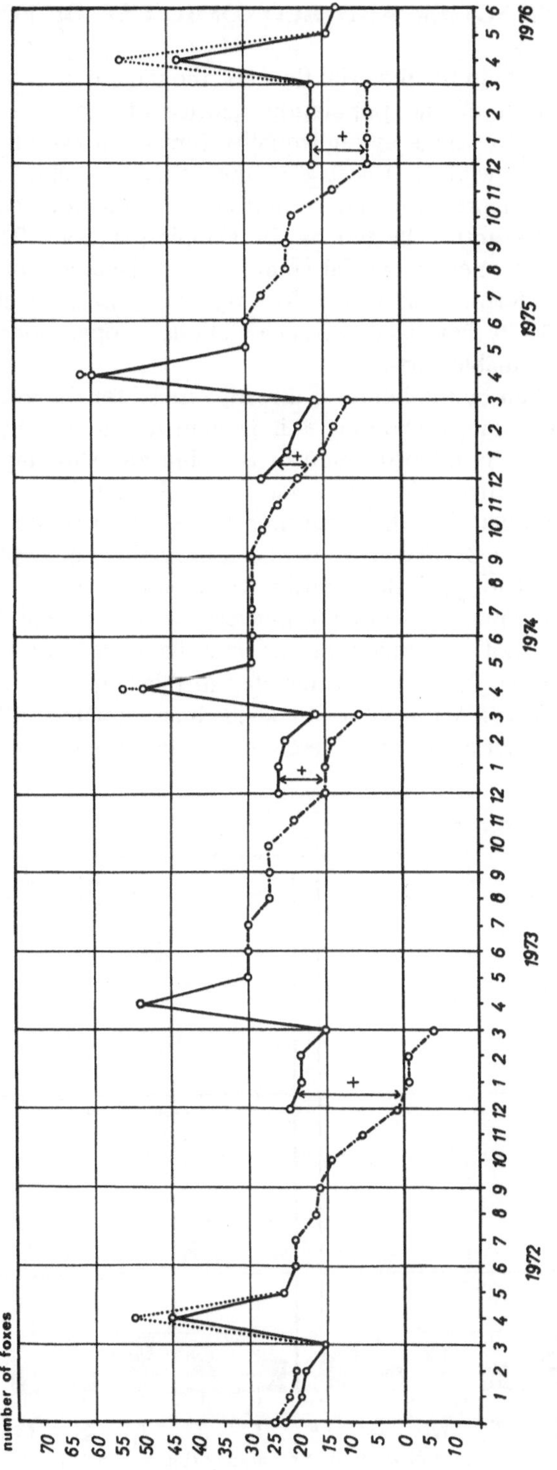

Fig. 12 continued.

HUNTING PRACTICE AND ECONOMIC USE OF FOXES

As already shown in the material thus far, the most urgent task is to be seen in the reduction of the population density of the fox. Moreover, the exploitation of fox skins as high quality furs is an essential factor in the economic use of wildlife. Because of the danger of rabies in the GDR at present, the shot predator and vermin are put into foil sacks and kept in lockable barrels until collected by the skinning stations. The accumulated predators are skinned in special skinning establishments of several animal exploitation enterprises and some forestry enterprises as well as by hunting societies. In 1976 there were 29 such stations in operation in the GDR to preserve the valuable furs.

The care taken by the hunter is already crucial for the quality of the furs. First of all the time of shooting is important for the condition of the furs. For this reason the intensification of trapping and shooting during the fur maturing period, from November 1st to March 31st, is of vital importance. The yield of predators in the given time period varies in the individual GDR counties, between 39 and 67% of the total yield. Fur exploitation during the October 1st to April 30th period is now aimed at.

In the Hakel area, as the long term average shows (Table 7, Fig. 13), 60% of the foxes shot are killed in the months from November to March (October to April 68.9%). Not included in this sum were the young foxes killed on the dens from April to June, which are equated with the quota lost by gassing in other areas. In November and December the main proportion,

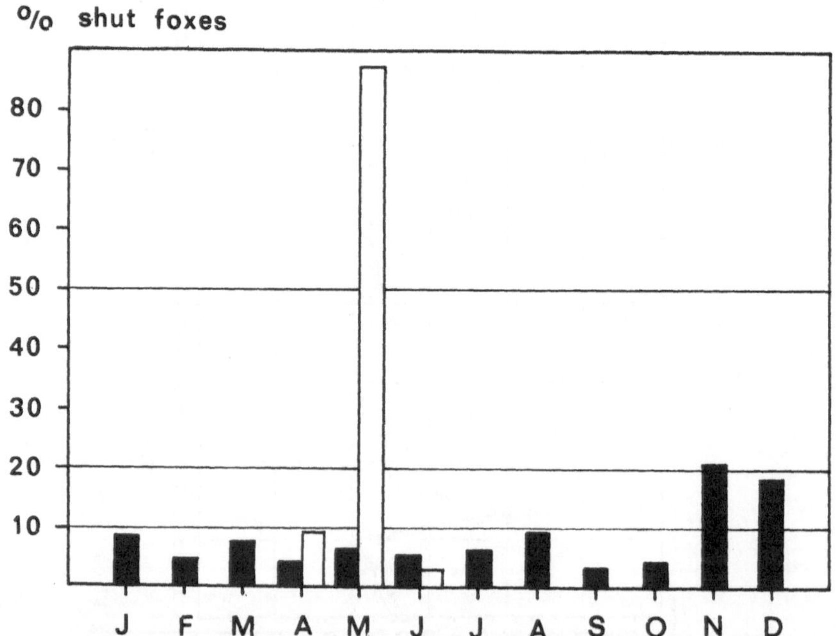

Fig. 13. Monthly percentage of foxes shot separately (black, $n = 411$) or young foxes shot on the dens (white, $n = 305$) in the Hakel area.

39%, is hunted. From April to October the shootings per month vary between 3.4 and 9.2% of the total yield. The maximum summer shootings coincide with the harvesting of the fields in August. Stubble-fields are preferred feeding grounds of the fox.

SUMMARY

In this paper the most important findings about the population ecology of the Red Fox in the GDR are summarized. In the Hakel, a large isolated wood of 1,300 ha on the edge of the Magdeburg Börde the fox population was investigated from 1962 to 1976. Information about the taxonomy of the Central European foxes is given. They belong to the typical form; the subspecies *Vulpes vulpes crucigera* (Bechstein, 1789) is a synonym for *Vulpes vulpes vulpes* (L., 1758). The population ecology of the Hakel foxes is characterized by investigations into the reproduction rate and birth biology, into the sex ratio and its population density as well as the population structure. The annual reproduction rate over a fifteen year average amounted to 190% on April, 1st, related to the parent population. On average 4.76 pups per litter survived the first 4 to 6 weeks of the post natal period. The post natal mortality rate amounted, at this time, to at least 16%. The male foxes predominate in the sex ratio. One can base one's reproduction calculations on the ratio 1.5♂♂:1♀. In the case of an average reproduction rate 190% the theoretical population structure of the age groups 0 to 5 amount to 65.5%, 22.8%, 7.6%, 2.8%, 1.0%, 0.3%. Average life expectancy is approximately 12 to 15 months. The absolute age seldom exceeds 6 years. The "hunting indicator of population density" was about 0.6 shot animals/100 ha. in the GDR taken as an average over the years 1966 to 1975. The GDR annual spring stock is assumed to be 50,000 to 60,000 foxes. For an effective suppression of rabies all conceivable measures must also be taken in the future, in order to reduce the population density effectively on a wide scale. To this end knowledge in the areas of wildlife biology, hunting and epidemic hygene must be efficiently coordinated. Instruction for the economic use of foxes orientates around an approved harvest and usage of the mature furs from October to April. In the Hakel area 68.9% of the annual yield are sent for skin producing in this period.

REFERENCES

Atanassov, N. 1958. Der Fuchs (Vulpes vulpes crucigera Bechstein) in Bulgarien. Morphologie, Biologie und wirtschaftliche Bedeutung. Sofia.

Behrendt, G. 1955. Beiträge zur Ökologie des Rotfuchses (Vulpes vulpes L.). *Z. Jagdwiss.* 1: 133–144, 162–183.

Bögel, K., Arata, A. A., Moegle, H. and Knorpp, F. 1974. Recovery of Reduced Fox Populations in Rabies Control. *Zbl. Vet. Med.* B 21: 401–412.

Bree, P. J. H. v. et al. 1974. Tooth wear as an indication of age in Badgers (Meles meles L.) and Red Foxes (Vulpes vulpes L.). *Z. Säugetierk.* 39: 243–248.

Brink, F. H. van den 1957. Die Säugetiere Europas westlich des 30. Längengrades. Berlin und Hamburg.

Brömel, J. and Zettl, K. 1974. Beitrag zur Altersbestimmung beim Rotfuchs (Vulpes vulpes L., 1758). *Z. Jagdwiss.* 20: 96–104.

Creutz, G. 1978. Zur Ernährung des Rotfuchses, *Vulpes vulpes* (L. 1758), in der DDR. *Zool. Garten* N.F. 48: 401–417.
Ellerman, J. R. and Morrison-Scott, T. C. S. 1966. Checklist of Palaearctic and Indian Mammals 1758 to 1946. London.
Englund, J. 1970. Some aspects of reproduction and mortality rates in Swedish Foxes (*Vulpes vulpes vulpes*), 1961–63 and 1966–69. *Viltrevy* 8 (1): 1–82.
Fairley, J. S. 1970. The food, reproduction, form, growth and development of the fox Vulpes vulpes (L.) in Northeast Ireland. *Proc. Royal Irish Acad.* 69 B, No. 5: 103–137.
Gaffrey, G. 1961. Merkmale der wildlebenden Säugetiere Mitteleuropas. Leipzig.
Geiger, G. et al. 1977. Konkordanz verschiedener Methoden der Altersbestimmung beim Rotfuchs (*Vulpes vulpes* L., 1758). *Z. Jagdwiss.* 23: 57–64.
Goretzki, J. and Paustian, K.-H. 1977. Zur Effektivität der Fuchsbaubegasung. *Beitr. Jagd- und Wildforsch.* 10: 327–331.
Haltenorth, Th. and Roth, H. H. 1968: Short review of the biology and ecology of the red fox. *Säugetierk. Mitt.* 16: 339–352.
Lloyd, H. G. et al. 1976. Annual Turnover of Fox Populations in Europe. *Zbl. Vet. Med.* B 23: 580–589.
Losert, J. 1975. Möglichkeiten zur Steigerung des Aufkommens an Raubwildbälgen und Erlösen. *Unsere Jagd* 25: 112–113.
Lund, H. M. 1959. The red fox in Norway. I. *Papers Norwegian State Game Res.*, 2. Ser. (5): 1–57.
Naaktgeboren, C. 1965. Die Fortpflanzung des Rotfuchses, *Vulpes vulpes* (L.), mit besonderer Berücksichtigung von Schwangerschaft und Geburt. *Zool. Anz.* 175: 235–263.
Pielowski, Z. 1976. The Role of Foxes in the Reduction of the European Hare Population. Ecology and management of European hare populations (Warszawa) 135–148 (Proceedings of an international symposium held in Poznań, December 23–24, 1974).
Pitzschke, H. 1972. Untersuchungen über die Fuchspopulation – ein Beitrag zur Erforschung von Grundlagen für eine wirksame Tollwutbekämpfung. *Monatshefte Veterinärmed.* 27: 926–932.
Proft, G. et al. 1975. Untersuchungen über die Nahrung des Fuchses im Bezirk Gera. *Landschaftspflege und Naturschutz in Thüringen* 12 (3): 50–56.
Sinnecker, H. et al. 1975. Die Entwicklung der Tollwut in der DDR 1953 bis 1970. *Z. gesamte Hygiene* 21: 39–45.
Stubbe, M. 1965. Zur Biologie der Raubtiere eines abgeschlossenen Waldgebietes. *Z. Jagdwiss.* 11: 73–102.
Stubbe, M. 1967. Zur Populationsbiologie des Rotfuchses vulpes vulpes (L.). *Hercynia* (N.F.) 4: 1–10.
Stubbe, M. 1971. Wald-, Wild- und Jagdgeschichte des Hakel. *Arch. Forstwesen* 20: 115–204.
Stubbe, M. 1973. Der Fuchs (*Vulpes vulpes* L.), in Stubbe, H.: Buch der Hege, Bd. 1, Berlin.
Stubbe, M. 1974. Zur Populationsbiologie des Rotfuchses *Vulpes vulpes* L. II. *Beitr. Jagd- und Wilforsch.* 8: 385–395.
Stubbe, M. 1977. Raubwild, Raubzeug, Krähenvögel, Grundlagen der Bewirtschaftung. Berlin.
Stubbe, M. and Stubbe, W. 1977. Zur Populationsbiologie des Rotfuchses *Vulpes vulpes* (L.), III. *Hercynia* (N.F.) 14: 160–177.
Ulbrich, F. 1974. Zu einigen Fragen den Fortpflanzungsbiologie des Rotfuchses (*Vulpes vulpes* L.). *Beitr. Jagd- und Wildforsch.* 8: 397–405.
Ulbrich, F. 1977. Weitere Angaben zur Fortpflanzungsbiologie des Rotfuchses (*Vulpes vulpes* L.). *Beitr. Jagd- und Wildforsch.* 10: 322–326.
Usinger, A. 1950. Der Fuchs. – Merkblätter des Niederwildausschusses des DJV, Nr. 5, München.
Weinitschke, H. 1954. Die Waldgesellschaften des Hakels. *Wiss. Z. Univ. Halle* 3: 947–978.

8 EIN MODELL FÜR DIE FUCHSPOPULATIONSDYNAMIK IN DER BUNDESREPUBLIK DEUTSCHLAND

A. V. Braunschweig*

ZUSAMMENFASSUNG

Der Fuchsbestand ist in der freien Wildbahn nicht zählbar. Im Vortrag wird dargelegt, wie man aus den bekannten Daten wie Abschuß, Vermehrungsrate, ermittelte Tollwutzahlen und Frühverluste annähernd den Fuchsbestand errechnen kann. Dabei ergibt sich, daß etwa nur 2% der tatsächlichen Tollwutfälle bekannt werden. Ferner wird gezeigt, daß Fuchsbestände von mehr als 1.3 Füchsen je 100 ha Jagdfläche am 15.II. d. Jahres kaum erreicht werden können, weil die Tollwut als Hauptregulator des Fuchsbestandes dann eine drastische Fuchsminderung auf etwa 0,8 Füchse je 100/ha am 15.II.d. Jahres herbeiführt. Die Jagd spielt für den Fuchsbestand keine entscheidende Rolle.

SUMMARY

Fox numbers cannot be accurately counted in the wilds. In this paper it is shown how the fox population can be estimated from such known parameters as hunting mortality, increment, the number of rabid foxes and spring mortality. Whereby it is known that only 2% of all rabid foxes are found. In addition, it is shown that fox densities greater than 1, 3 foxes per 100 ha of huntable ground are hardly ever reached on the 15th February of each year since rabies, as the major regulator, decimates the fox numbers to roughly 0.8 foxes per 100 ha. Hunting mortality does not play an important role in regulating fox numbers.

Als Jagdwissenschaftler bin ich nach jahrelanger Beobachtung und Auswertung der einschlägigen Literatur zu folgenden Leitsätzen für die Fuchspopulationsdynamik gekommen.

(1) Der Fuchsbestand und sein wichtigster biologischer Regulator, die Tollwut, dürfen zur Zeit unabhängig voneinander nicht betrachtet werden.
(2) Der Jagd auf den Fuchs kommt zweitrangige Bedeutung zu.
(3) Fuchszählungen führen zu irreführenden Ergebnissen, ebenso wie Zählmethoden bei anderen freilebenden Tieren. Das ist unbestritten.
(4) Rückschlüsse aus bekannten Zahlen wie Vermehrungsrate, Jungendsterblichkeit, Abschußzahlen usw. auf den Bestand – am besten über

* Institut für Wildbiologie und Jagdkunde, Göttingen, FGR.

mehrere Jahre hinweg-ergeben die besten Anhaltspunkte für die wahren Bestandeshöhen. Das gilt auch für den Fuchs.

(5) Der derzeitige Tollwutseuchenzug läßt eine permanente Fuchsdichte von erheblich mehr als 1 Stück je 100 ha am 15.II. eines jeden Jahren nicht zu. Das gilt es an dieser Stelle glaubhaft zu machen und ist gleichzeitig als Maßstab für die tatsächliche Höhe des Fuchsbestandes anzusehen.

(6) Die amtlichen Tollwutzahlen geben tatsächlich nur etwa 2% der gesamten Tollwutverluste beim Fuchs wieder.

Mit diesen Leitsätzen habe ich versucht klarzumachen, daß der Wirt (Fuchs) in engem Abhängigkeitsverhältnis zu seinem Hauptparasiten (dem Tollwutwirus) steht.

Gleichzeitig messe ich den Jagderfolgen zweitrangige Bedeutung zu.

Es entsteht bei meinen Betrachtungen ein ganz deutliches Dreicksverhältnis zwischen Fuchsbestand, Tollwut und Bekämpfung. Dabei spielt biologisch die entscheidende Rolle ein ausgewogenes Wirt-Parasit verhältnis. Das, was dabei den Parasiten zurückdrängen kann, ist der Geist oder Ungeist des Menschen in Form von Schaffung eines zur Zeit optimalen Lebensraumes für den Fuchs einerseits und die Jagd u.a. Bekämpfungsmethoden auf der anderen Seite. Leider findet man heute noch Angaben über Fuchsbestandeshöhen ohne Nennung des Stichtages. Das ist nach meinem Dafürhalten bei Tieren mit einer Jungenzahl bis zu 11 von einem Weibchen absolut unzulässig.

Die jeweilige Fuchsbestandeshöhe läßt sich auf keinen Fall auszählen. Ich glaube, darüber dürfte es keinen Zweifel geben.

Die Zählmethoden bei freilebenden Tieren, besonders auf dem Lande, haben bislang zu keinen brauchbaren Ergebnissen geführt. Aber aus Streckenergebnissen und biologischen Daten wie Anzahl der Jungen, Jugendsterblichkeit, Krankheitszahlen usw. läßt sich am ehesten eine annähernd richtige Bestandeshöhe ermitteln.

Auf diesen Daten habe ich meine Aussagen aufgebaut und jagdliche Erfahrung mit eingebracht. Folgende bekannte Zahlen sind verwendet worden:

Jahr	Fuchsstrecke	Tollwutfälle	Bekämpfung
1971/72	122,703	1,494	Begasung
1974/75	177,248	3,103	keine
1976/77	186,814	6,296	keine
1977/78	194,030	2,639	keine

Die Größe aller Reviere in der Bundesrepublik beträgt 23,688,716 ha. (DJV-Handbuch)

Das bedeutet rechnerisch folgende Fuchsbestandeszahlen:

Alt-Füchse/ 100 ha	Fuchsbestand ♂, ♀ am 15.II. d.J.	Anzahl der Jungen 4	5
1,5	355,330	710,660	888,325
1,2	284,264	568,528	710,660
1,0	236,887	473,770	592,215
0,8	189,509	379,018	473,700
0,6	142,132	284,264	355,330

Diese Daten müssen jetzt mit Sachverstand geprüft werden, ob sie in freier Wildbahn möglich oder sogar wahrscheinlich der tatsächlichen Besiedlungsdichte unseres Landes durch den Fuchs entsprechen.

Dazu müssen wir noch wissen, daß in den Fuchsstrecken das Verhältnis von Jung- zu Altfuchs etwa 2–(3): 1 beträgt und daß die Füchse im Durchschnitt nicht älter als 2 Jahre werden. Die Verluste sind also sehr hoch. Die Regenerationsrate gleicht das aber leicht aus.

Die Jugendverluste der Welpen liegen nach Stubbe bei 21% und nach Behrend zwischen 25–32%. Ich setze also 20% als Frühverluste in meine Berechnungen ein. Damit liegt man sicher nicht zu hoch. Es erscheinen also von 5 Jungen nur noch 4 am Bau. Das entspricht den Beobachtungen, die wir alle haben. 3–5 Junge zählen die meisten Gehecke, die man zu sehen bekommt. Pitzschke, Leßmann, Stubbe, Ulbrich liegen alle mit ihren Ergebnissen von trächtigen Fuchsfähen zwischen 5,8 und 6,2 (7,0) Plazentationen. Ulbrich gibt 3,7% fetalen Frühtod an. Deshalb muß die Zahl der geborenen Jungfüchse durchschnittlich mit Sicherheit unter 6 mit hoher Wahrscheinlichkeit bei 5 Welpen liegen. Nach Ulbrich waren noch 3,9% der Fähen nicht trächtig. Meine eigenen Erfahrungen haben mich gelehrt, daß 4 Jungfüchse am häufigsten am Bau beobachtet werden können. Deshalb muß auch die Geheckstärke mit 5 Welpen angesetzt werden.

Mit diesen Zahlen darf ich Ihnen ein Beispiel des Jahres 1971/72 darstellen, wie s.Zt. der Fuchsbestand augesehen haben müßte.

Die Abschußzahlen von 1970–1972 lagen zwischen 114,398 und 122,703 pro Jahr d.h. sie waren etwa gleichbleibend. Gleiches gilt für die Tollwutfälle die bei 1500 bis 1900 lagen. Zu jener Zeit vor allem 1970–71 wurden die Baue begast. Es muß sich damals der Fuchsbestand in Zuwachs und Verlusten etwa die Waage gehalten haben. Bei Jungfüchsen kommen Krankheiten außer Tollwut und Unfälle häufiger vor als bei Altfüchsen, weil die Immunitätsverhältnisse noch nicht so ausgebildet sind und den Jungfüchsen Lebenserfahrungen fehlen.

Derjenige der weiß, wie schwer ein Altfuchs zu erbeuten ist, wird mir zustimmen, daß 21% Abschuß von Altfüchsen schon schwierig ist.

Dagegen dürften 16% der Jungfüchse leicht zu erbeuten sein, auch wenn die doppelte Anzahl von Jungfüchsen gegenüber Altfüchsen erreicht werden muß.

Rechnet man 10% Verluste durch Tollwut bei Altfüchsen (20,000) und

Tabelle 1. Fuchsbesatz ~0,8/100 ha am 15.II.1972. Tollwut im Besatz, Bekämpfung findet statt, Zuwachs und Verluste ~ ausgeglichen. Jahres 1971/72 122,703 Füchse. Tollwut beim Fuchs 1971 = 1,494.

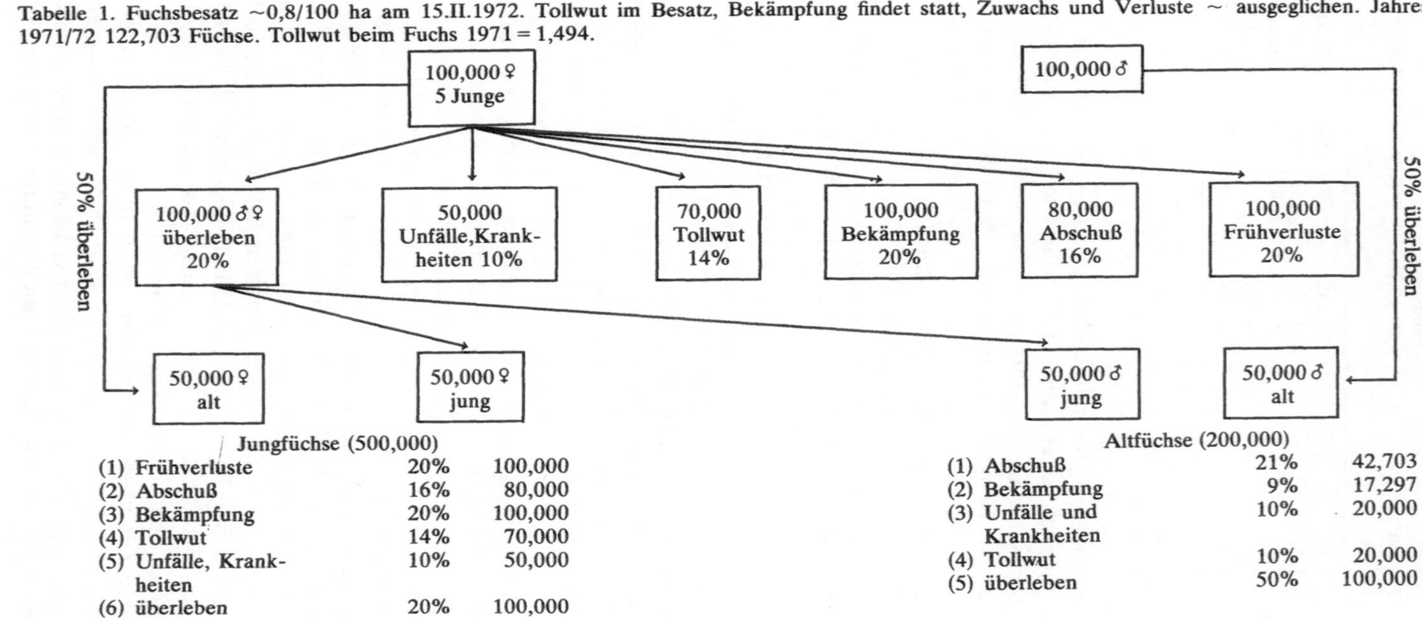

Jungfüchse (500,000)		
(1) Frühverluste	20%	100,000
(2) Abschuß	16%	80,000
(3) Bekämpfung	20%	100,000
(4) Tollwut	14%	70,000
(5) Unfälle, Krankheiten	10%	50,000
(6) überleben	20%	100,000

Altfüchse (200,000)		
(1) Abschuß	21%	42,703
(2) Bekämpfung	9%	17,297
(3) Unfälle und Krankheiten	10%	20,000
(4) Tollwut	10%	20,000
(5) überleben	50%	100,000

etwa 14% bei Jungfüchsen (70,000) so ergibt sich ein Jahresverlust von 90,000 Füchsen alleine durch Tollwut. So steht es hier in der ausgeglichenen Rechnung. Festgestellt wurden aber nur 1,494 Tollwutfälle bei Füchsen, das sind 1,66% der vermutlichen Tollwutverluste.

Einige Autoren (Lund, Richards u. Shelton, Johannson, Lessmann, Stubbe) geben das Geschlechterverhältnis ausgewachsener Füchse mit 1,2–1,5 : 1 an. Diese Feststellung stammt aus Streckenuntersuchungen. Das fetale Geschlechterverhältnis wird von Ulbrich 1 : 1 angegeben. Dem vermag ich bei Altfüchsen nicht zu folgen und rechne mit 1 : 1, weil beim Raubwild die Strecken der Männchen höher liegen als die der Weibchen, das hängt mit geschlechtsgebundener größerer Vorsicht der Weibchen zusammen.

Die durchschnittliche Wurfgröße beträgt nach Ulbrich an 132 Fähen gemessen 6,07 Feten pro Fähe. Intrauterines Fruchtabsterben wurde mit 13,2% für die erste Hälfte und mit 3,5% für die 2. Hälfte der Trächtigkeit nachgewiesen. Wandeler gibt die fetale Sterblichkeit mit 5% und Fairley mit 8% an. 3,9% der Fähen waren nicht trächtig (Ulbrich).

Die Autoren Zwingenberg, Richards, Hine, Schofield, Lichatshev, Stscherbina, Sovis, Stubbe, Wandeler beobachteten am Bau bei 2,354 Gehecken 3,9–5,54 Welpem mit Schwerpunkt bei 5 Jungen. Darum habe ich diese Zahl in meine Rechnung eingesetzt. Stubbe mißt der Bekämpfung durch den Menschen und durch Krankheiten eine große Bedeutung für die Bestandesreduzierung bei. Ich glaube, mein Rechenansatz mit 20–25% Frühverluste liegt damit auch etwa richtig. Die Frühverluste der Welpen betragen nach Stubbe minimal 14% und nach Behrend 25–32%.

Aus 23-jähriger eigener Erfahrung in mehreren Instituten mit dem Fuchs weiß ich, daß Krankheiten außer Tollwut bei über 2 Monate alten Füchsen eine geringe Rolle spielen. Unfälle sind da schon etwas bedeutender.

Damit habe ich die bekannten und die annähernd bekannten Daten der Zuwachs- und Verlustrate des Fuchses erwähnt und in der Tabelle 1 mosaikartig zusammengetragen und unbekannte Größen nach ihrer Wahrscheinlichkeit eingefügt.

Da in den Jahren 1970–1972, während der Begasungsaktionen die Strecken und die Tollwutfälle etwa gleichbleibend waren, darf man zu Recht folgern, daß auch der Fuchsbesatz in diesen Jahren etwa gleich hoch war.

Die Tabelle 2 zeigt den Fuchsbestand, wie er etwa 1974/75 bestanden haben dürfte.

Die Strecke hatte sich wesentlich auf 177,248 Füchse erhöht. Die Tollwutfälle waren stark gestiegen. Sie betrugen 3103 Fälle beim Fuchs. Die Bekämpfung durch Baubegasung war ausgesetzt worden. Deshalb muß man mit mehr als 0,8 Füchsen/100 ha rechnen. Ich bin bei den erhöhten Abschuß-und Tollwutzahlen durch Variieren der Fuchsdichte darauf gekommen, daß s.Zt. etwa 1,0 Füchse/100 ha vorhanden gewesen sein müssen.

Geht man mit der Rechnung höher, so kommt man zu Zahlen, die irreal sind. Die Frühverluste sind ebenfalls mit 20% veranschlagt. Der Abschuß ist bekannt und wurde auf Alt- und Jungfüchse wiederum mit ~ 1 : 2 verteilt. Auch die Unfälle und Krankheiten außer Tollwut bleiben mit 10% in der

Tabelle 2. Fuchsbesatz ~ 1,0/100 ha am 15.II.1975: Tollwut im Besatz weit verbreitet, Bekämpfung entfällt. Zuwachs und Verluste nicht ausgeglichen. Jahresstrecke 1974/75 = 177,248 Füchse. Tollwut beim Fuchs 1974 = 3,103.

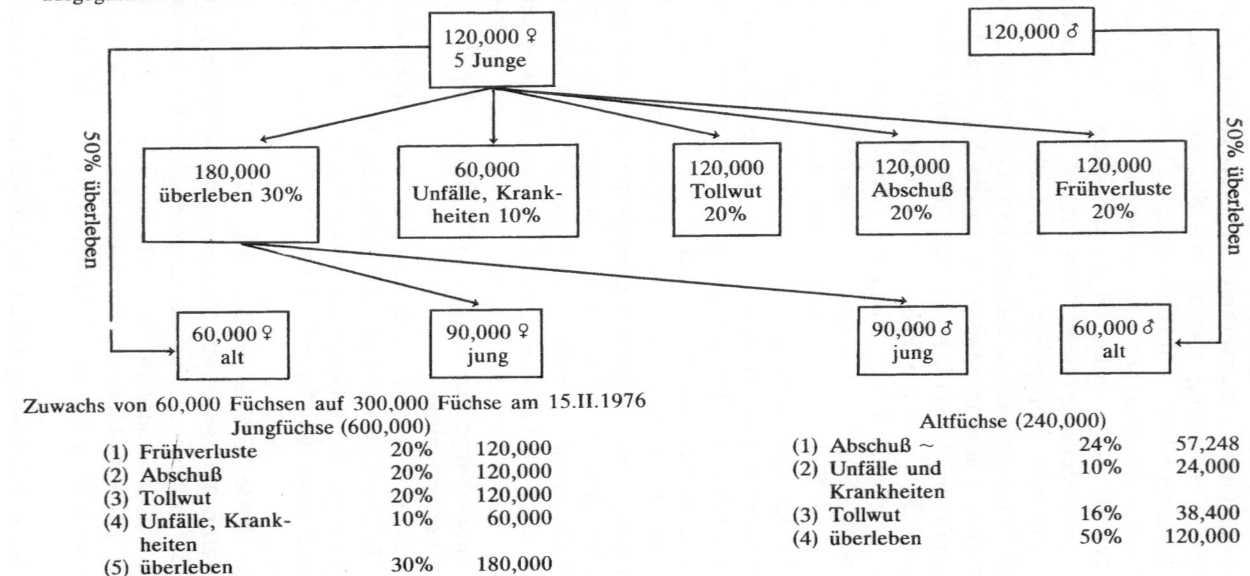

Zuwachs von 60,000 Füchsen auf 300,000 Füchse am 15.II.1976

Jungfüchse (600,000)

(1) Frühverluste	20%	120,000
(2) Abschuß	20%	120,000
(3) Tollwut	20%	120,000
(4) Unfälle, Krankheiten	10%	60,000
(5) überleben	30%	180,000

Altfüchse (240,000)

(1) Abschuß ~	24%	57,248
(2) Unfälle und Krankheiten	10%	24,000
(3) Tollwut	16%	38,400
(4) überleben	50%	120,000

Rechnung. Aber die Tollwut muß zunehmen, wenn man nicht zu unglaublichen Fuchsbeständen kommen will. Trotzdem steigt der Bestand um 60,000 Füchse im nächsten Jahr zum 15.II.

Das ist die Situation, die nun auf den kommenden Zusammenbruch des Bestandes hintreibt. Kauker und Zettl haben schon 1963 beschrieben, daß der Fuchsbestand etwa alle 3 Jahre durch die Tollwut stehr stark dezimiert wird. Wir sehen also, daß oberhalb von 1,0 Füchse/100 ha eine scharf greifende Regulation notwendig wird. Bis 1975/76 steigt die Strecke dann auf ihren Höchststand mit 219,550 Füchsen. Die Tollwutzahlen verdoppeln sich und erreichen 6296 Tollwutfälle beim Fuchs. Nach Tabelle 3 muß man jetzt schon mit 1,3 Füchsen je 100 ha rechnen. Dabei wäre ein Bestand von 312,000 Füchsen am 15.II.76 vorhanden. Das Jagdjahr 1976/77 bringt seit Jahren erstmalig einen kleinen Rückgang der Strecke. Im Jagdjahr 1976 erleidet der Bestand einen deutlichen Rückschlag. 1977/78 werden trotzdem 194,030 Füchse erlegt. Es gab 1977 aber nur noch 2,639 Tollwutfälle beim Fuchs. Damit sollten wir mit dem Fuchsbestand unter den Bestand des Jahres 1974/75 gelangt sein (3,103 Tollwutfälle). Die Fuchsstrecke liegt aber höher als damals. Dies bedarf der Interpretation. Bayern erhöhte seine Strecke von 1974/75 von 29,425 auf 47,756 und das Saarland von 2,035 auf 3,105 im Jagdjahr 1977/78. Ich glaube, das liegt an intensivierter Bejagung. Daher stammt die Gesamtstreckenerhöhung um rund 20,000 Füchse. Die Situation eines in dieser Form abnehmenden Bestandes gibt die Tabelle 3 wieder. – Der Ausgangsbestand betrug etwa 312,000 am 15.II. 76 mit ca. 156,000 Fähen.

Für hohe Wildtierbestände gilt, daß die Frühverluste, Krankheiten und Unfälle zumindest etwas ansteigen. Darum habe ich die Frühverluste mit 25% und die Unfälle und Krankheiten mit 13% in Ansatz gebracht. Der Abschuß liegt fest und wurde wieder auf Alt- und Jungfüchse etwa im Verhältnis 1:2 verteilt. Die Tollwutzahl ist dann der rechnerische Restbetrag mit 195,000 bei den Jungfüchsen.

Das erscheint auf den ersten Blick fast nicht glaubhaft. Es kämen außerdem bei den Altfüchsen noch einmal 125,000 Stück dazu. Es müßten also 320,000 Füchse durch Tollwut umkommen. 6,296 Tollwutfälle wurden aber nur amtlich registriert. Das würde darauf hinweisen, daß die Aufklärungsquote etwa 2,0% beträgt.

Diese Zahl liegt damit im gleichen Bereich wie 1974.

Man muß natürlich daran denken, daß bei hohen Besiedlungsdichten die Vermehrungsrate sinken kann. Ich persönlich glaube beim Fuchs nicht recht daran, daß bereits bei 1,3 Füchsen/100 ha am 15.II.d.J. Nahrungsmangel oder Streßsituationen im Bestand auftreten. Dennoch habe ich die gleiche Rechnung wie in Tabelle 3 noch einmal gemacht und dabei aber nur 4 Welpen pro Geheck angenommen. Es müssen bei so geringem Nachwuchs die Frühverluste und Unfälle und Krankheiten wieder etwas geringer um insgesamt 4% angesetzt werden. Außerdem muß die Zahl der Überlebensrate mit 250,000 gleich bleiben.

Da ja die begründete Voraussetzung besteht, daß am 15.II.1977 etwa die Bestandeszahlen von 1974–75 erreicht worden sein müssen. Die Abschuß-

Tabelle 3. Fuchsbesatz ~1,3/100 ha am 15.II.1978. Tollwut sehr stark verbreitet, Bekämpfung entfällt. Zuwachs durch Verluste übertroffen. Jahresstrecke 1975/76 = 219,550 und 1976/77 = 186,814 Füchse. Tollwutfälle = 6,296 u. 1977 = 2,639 Füchse.

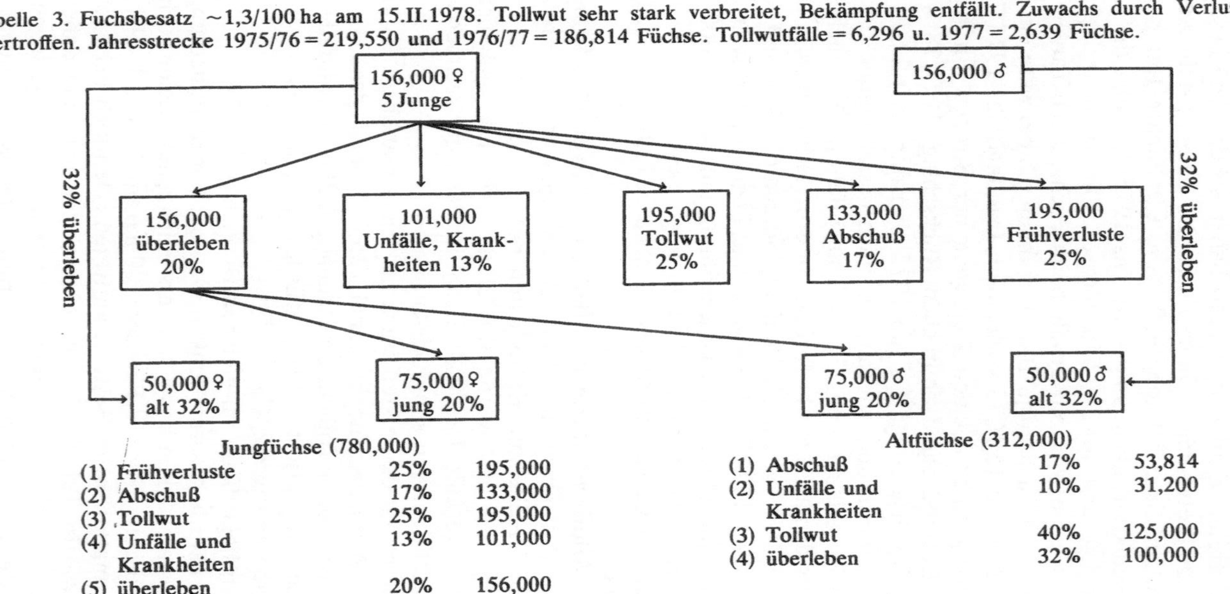

Jungfüchse (780,000)

(1)	Frühverluste	25%	195,000
(2)	Abschuß	17%	133,000
(3)	Tollwut	25%	195,000
(4)	Unfälle und Krankheiten	13%	101,000
(5)	überleben	20%	156,000

Altfüchse (312,000)

(1)	Abschuß	17%	53,814
(2)	Unfälle und Krankheiten	10%	31,200
(3)	Tollwut	40%	125,000
(4)	überleben	32%	100,000

Tabelle 4. Fuchsbesatz 1,3/100 ha am 15.II.1976. Tollwut sehr stark verbreitet, Bekämpfung entfällt. Zuwachs durch Verluste übertroffen. Jahresstrecke 1975/76 = 219,550 und 1976/77 = 186,814 Füchse, Tollwutfälle = 6,296 u. 1977 = 2,639 Füchse.

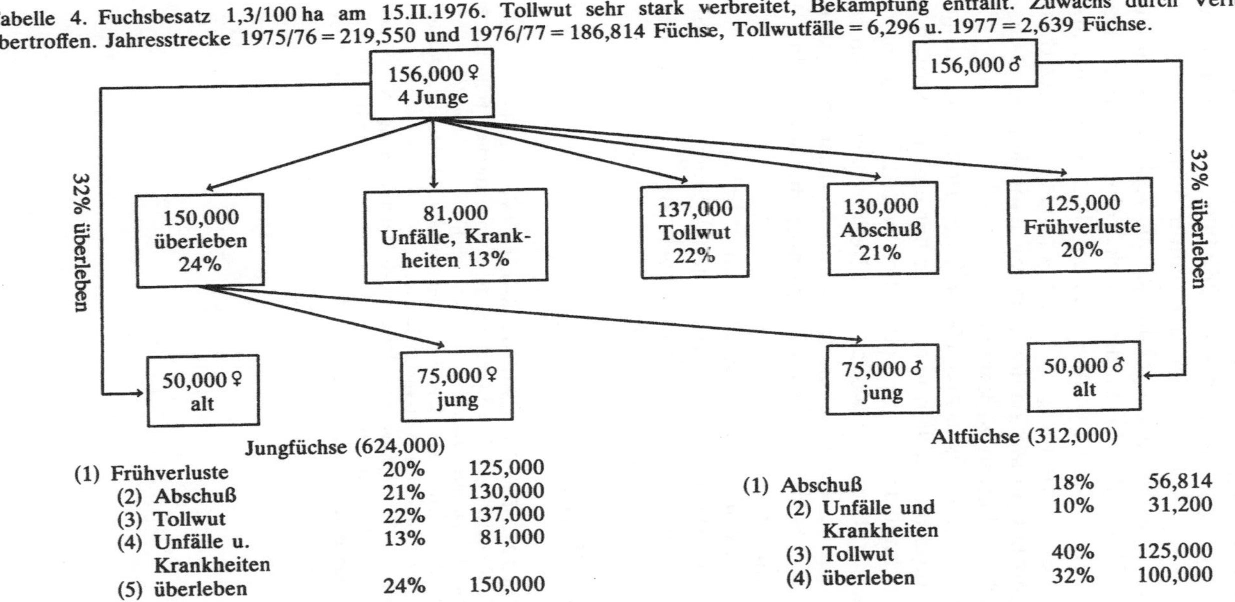

Jungfüchse (624,000)

(1) Frühverluste	20%	125,000
(2) Abschuß	21%	130,000
(3) Tollwut	22%	137,000
(4) Unfälle u. Krankheiten	13%	81,000
(5) überleben	24%	150,000

Altfüchse (312,000)

(1) Abschuß	18%	56,814
(2) Unfälle und Krankheiten	10%	31,200
(3) Tollwut	40%	125,000
(4) überleben	32%	100,000

zahlen liegen auch fest, so daß die vornehmlich veränderte Zahl im Bereich der Tollwut bei den Jungfüchsen liegen muß. Die Altfüchse sind mit gleichen Zahlen wie in Tabelle 3 aufgeführt. Die Überlebensrate sinkt wegen des Bestandzusammenbruchs weit unter 50% (32%). Es bleiben in diesem Fall bei Jungfüchsen 137,000 und bei Altfüchsen 125,000 Tollwutopfer übrig, das sind insgesamt 262,000. 6,296 erkannte Tollwutfälle ergäben ~2,4% Bekanntwerden der wahren Verluste durch Tollwut.

Es kommen bei diesen Berechnungen als Aufklärungsquote etwa immer ~2% der Verluste des Bestandes durch Tollwut heraus. Das könnte ein Weiser für den wahren Fuchsbestand werden. Man muß dies weiter verfolgen, um diese Vermutung zu bestätigen oder zu verwerfen.

Die übrigen bekannten Zahlen gehören natürlich dazu, um den Gesamtfuchsbestand abschätzen zu können.

Die Strecke könnte zwischen 16–24% des Bestandes ausmachen.

Die Abschußzahlen stellen ganz sicher nur einen Faktor in der Rechnung dar. Deshalb halte ich es für bedenklich von daher auf den tatsächlichen Fuchsbestand schließen zu wollen. Bei freilebenden Tieren sollte man *alle* bekennten und in etwa bekannten Größen zusammenfassen und daraus Schlüsse ziehen. Selbst dann werden wir noch mit beträchtlichen Fehlerquellen rechnen müssen.

Mir wären unwiderlegbare Zahlen auch lieber. Aber wir haben leider nicht mehr als Abschußzahlen, festgestellte Tollwutfälle, Wurfgrößen und vielleicht etwas mehr oder weniger Sachverstand, um damit der Wirklichkeit des Fuchsbestandes nahe zu kommen.

LITERATURVERZEICHNIS

Burrows/Matzen 1971. Der Fuchs, BLV-Jagdbiologie München, 1–196.

Kauker, E. und K. Zettl 1963. Zur Epidemiologie der sylvatischen Tollwut in Mitteleuropa und zu den Möglichkeiten ihrer Bekämpfung. *Vet. Med. Nachrichten* 2/3: 1963, 181–204.

Stubbe, H. Ed. 1973. Buch der Hege Bd. I. Haarwild v. M. Stubbe, Der Rotfuchs, 181–212 VEB Deutscher Land-wirtschaftsverlag Berlin.

Stubbe, H. Ed. 1977. Beiträge zur Jagd- und Wildforschung X, Weitere Angaben zur Fortpflanzungsbiologie des Rotfuches v. F. Ulbrich, 322–331. VEB Deutscher Landwirtschaftsverlag Berlin.

9 POPULATION DYNAMICS OF THE RED FOX (Vulpes vulpes L., 1758) IN SWEDEN

Jan Englund*

INTRODUCTION

The population dynamics of the red fox has been studied in two different habitats, the northern and the southern coniferous belts in Sweden. In the northern area the amount and quality of food varies greatly from year to year. Litter sizes, proportion of barren vixens, and mortality rates of cubs vary depending on the abundance of rodents. When rodents are rather few, young males more often than others disperse long distances. In the southern coniferous belt rodents also fluctuate in numbers but to a lesser extent. Alternative food is also more abundant, which results in a better and more stable food supply than in the north. The fox populations are also at higher densities here probably with smaller variations in number. Litter sizes vary a little but they are always large and the proportion of barren vixens is high. The mortality rate among cubs is also high with some variation between years. There is no correlation between the abundance of rodents and productivity or mortality rate. Population density *per se* is therefore assumed to be a factor of importance in regulating the number of foxes in the southern coniferous belt.

The red fox is found throughout Sweden from the mountainous areas in the north to the arable land of the south. In Sweden, the red fox is regarded as a game animal which hunters believe is harmful to other game species. The study reported here is part of a broad study on the ecology and behaviour of the fox aimed to provide an understanding on the way in which the fox might impinge on man's interests. The population dynamics of the fox is crucial to this aim and studies were undertaken in two ecologically different areas from 1966 to 1970 inclusive. Most data refer to autopsy examinations of foxes killed by hunters. The techniques used have been described earlier (Englund, 1970). Neither rabies, which is absent from Sweden, nor mange, absent at least during the period of study, have any effect upon the population dynamics of foxes as reported here.

STUDY AREAS

The provinces of Jämtland–Härjedalen lie in the central part of Sweden at 62–64° north in the northern coniferous forest. Less than 10% of the area is

* Swedish Museum of Natural History, Section for Vertebrate Zoology, S-104 05 Stockholm, Sweden.

cultivated and the forest that covers most of the land consists mainly of pine (*Pinus silvestris*) and spruce (*Picea excelsa*).

In these northern areas rodent numbers fluctuate dramatically, usually with a periodicity of 4 years. During an 8 year study in the province of Västerbotten, the number of rodents caught per 100 trap nights varied between 0.3 (2,825 trap nights) and 50.6 (2,900) (Hörnfeldt, 1978).

During peak years rodents constitute the main food of foxes. In other years foxes are to a large extent, from necessity, forced to scavenge on carcasses, and garbage found at rubbish dumps. In these northern areas garbage is of very poor nutritional quality (Englund, 1965).

Uppland and Södermanland (58.5–60° north) and parts of adjoining provinces belong to the southern coniferous belt. This is a more productive area where agricultural land constitutes about 30% of the land area.

Rodent numbers fluctuate here also, but less so than in the northern areas of Sweden, and the amount of alternative food is also greater. When rodents are least abundant, birds occur more frequently in the diet of foxes. Though foxes will also scavenge, the quality of the food at rubbish dumps is better than in the north. Food availability and quality in the agricultural areas is never as poor as in northern Sweden (Englund, 1965).

PRODUCTIVITY AND MORTALITY

Jämtland–Härjedalen

From 1966 to 1970 inclusive the abundance of rodents varied greatly. They were very common in 1966–67 and extremely rare in the summer of 1967 and the winter of 1967–68. They increased somewhat in numbers in the winter of 1968–69 and were very common again in 1969–70. This is reflected in the numbers of rodents found in fox stomachs shown in Fig. 1.

After the very good rodent year of 1966–67 the mean litter sizes of foxes were large, being about 5.4 for yearlings and 6.3 for older vixens. About 54% of the yearlings were barren compared with about 13% in older vixens. The productivity was therefore high in the spring of 1967, or about 2.5 cubs per yearling vixen, 5.5 for older ones and about 3.9 per vixen for the total population (Fig. 1). The proportion of cubs in the spring population therefore was about 66% if the spring sex ratio among adults was about 1:1 (3.9 out of 5.9).

Food was scarce in the summer of 1967 since rodent numbers had crashed, and of the foxes killed in the following hunting season (1967/68) the population was composed of about 44 and 46% subadult females and males respectively. The large discrepancy between the proportion of cubs in the spring and of subadults in the following hunting season indicate a high mortality rate during the spring–summer of 1967.

During the winter of 1967–68 rodents were very scarce and no more than a mean of 0.1 to 0.2 rodents were found per fox stomach containing food. Litter sizes in 1968 were reduced to about half the size of those observed in 1967 and barrenness increased to about 88% in yearlings and 67% in older ones. This had a drastic effect on productivity which fell to 0.8 cubs per vixen for the entire population. Thus the proportion of cubs in the spring population was about 28% and subadults represented in the 1968–69 samples were about 26–27%. These data indicate a lower mortality rate of juveniles in 1968 than in the preceding year.

The proportion of subadults, one and two year old foxes were about the same in the 1968–69 sample. In fact the cohort from 1966 was somewhat greater than that of 1968 in spite of being hunted for two seasons (furthermore, juveniles are over-represented; see below).

The moderate rodent increase of the summer of 1968 resulted in a slightly improved food situation in the winter of 1968–69. The stomach rodent index increased to 0.5–0.6 and in the spring of 1969 the litters of old vixens increased in size. The proportion of barren vixens of all age also decreased with an increase of productivity as a result. The proportion of cubs in the spring of 1969 was about 56% and the proportion of subadults in the sample obtained from the next hunting season was 68–69%. Thus there was a higher proportion than in the spring (Table 1). How is that possible?

It is improbable that the adults died to a higher extent than the cubs did in 1969. Nor is the situation caused by chance since the samples are large and any bias would probably operate to the same degree each year. The expected proportion of cubs in the spring at $\pm 2s/\sqrt{n}$, (95% confidence limits) was 47–61 and the higher proportion of subadult males and females is calculated from 483 males and 345 females respectively. Some of the sample therefore must give false information. The productivity might have been larger than calculated (in which case the proportion of barrenness has probably been overestimated). Another possibility is that juveniles represented by the samples obtained by shooting are over-represented either because they are more easily hunted or because they disperse, to a higher extent than adults, from areas where hunting pressure is slight into areas where hunting is more intense and pressure on space and food less.

This bias towards juveniles, by shooting, is revealed by the recovery rates of foxes tagged in hunting areas. Of 100 juveniles (mostly 1.7 to 3.0 kg) and 173 adults tagged in the south of the northern coniferous zone, 54 juveniles and 58 adults were recovered during the first hunting season after tagging. Thus juveniles were represented by 36.6% in the tagged population and by 48.2% in the hunting sample. The proportion of juveniles in the samples used in this study has to be corrected by a factor of 0.76 to get the true proportion of 36.6 (0.76×48.2). A comparison of the proportion of juveniles in a shot sample with the proportion of 1 year old foxes out of all adults shot next hunting season also shows a large difference. A correction factor of about 0.80 reduced the discrepancy to a minimum, for several areas and years (Englund, 1970), and is proposed as a method of calculating the true proportion of juveniles from killed foxes. The excessively high frequency of juveniles discussed here is clearly due to their greater probability and risk of being killed than other animals.

If the proportion of juveniles in the autumn–winter population is estimated by the correction factor of 0.80, there is no mortality among adults in

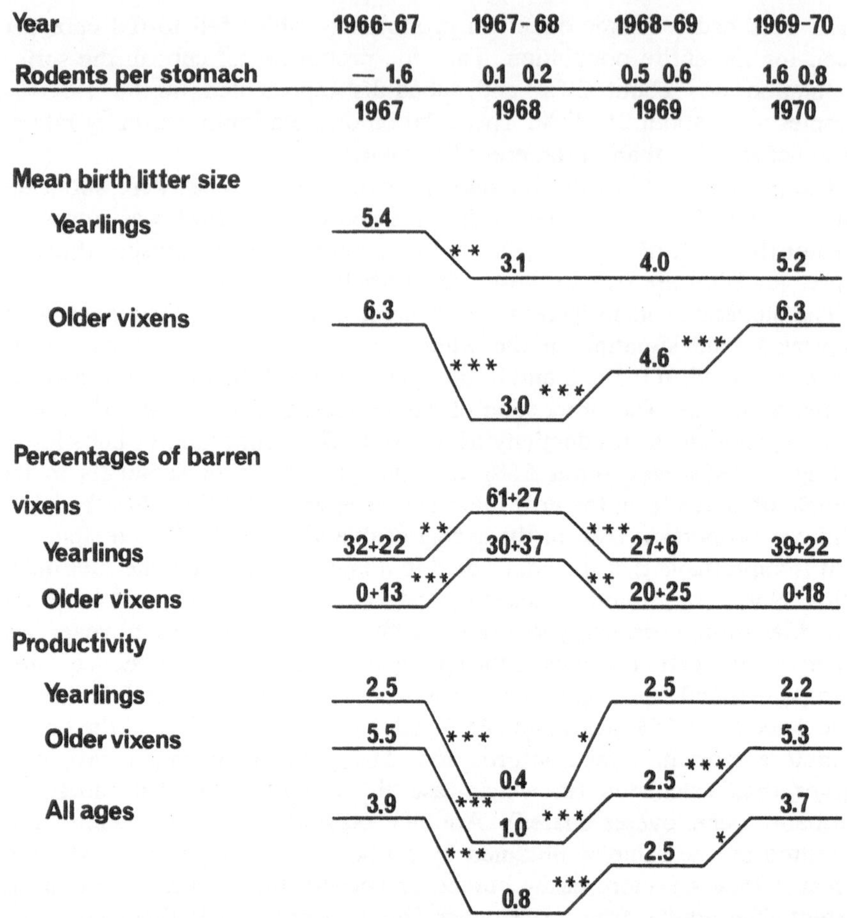

Year
1966-67
1967-68
1968-69
1969-70
Rodents per stomach
— 1.6
0.1 0.2
0.5 0.6
1.6 0.8
1967
1968
1969
1970
Mean birth litter size
Yearlings
5.4
3.1
4.0
5.2
Older vixens
6.3
4.6
3.0
6.3
Percentages of barren
vixens
61+27
Yearlings
32+22
30+37
27+6
39+22
Older vixens
0+13
20+25
0+18
Productivity
Yearlings
2.5
2.5
2.2
Older vixens
5.5
5.3
0.4
2.5
All ages
3.9
3.7
1.0
2.5
0.8

Figs. 1 and 2. Mean number of rodents per stomach containing food from foxes killed in January–February and March–April respectively, mean birth litter-sizes, percentages of barren vixens (with *no* signs of placental scars + vixens with gray scars only) and productivity. When there are significant differences in yearly data these are shown by a bend in the curves. The significant level is shown by the number of asterisks (* = $0.05 \geq P > 0.01$, ** = $0.01 \geq P > 0.001$, *** = $0.001 \geq P$).

the spring or summer and the risk of dying by natural causes is the same for juveniles and adults during the autumn–winter, the mortality rate during the spring–summer period can be assessed more precisely (Englund, 1970). For the four years 1967–1970 the mortality rate of young foxes from birth to autumn is estimated to be 70, 34, 7 and 23% in Jämtland–Härjedalen. The decrease in mortality in the first three years is well correlated with the abundance of rodents which were extremely few in 1967, increased slowly in 1968 and were very common in the summer of 1969.

Uppland–Södermanland

In the winter of 1965–66, rodents were very common (according to observations of many hunters and farmers) and in the following spring litter sizes

Year	1965-66	1966-67	1967-68	1968-69	1969-70
Rodents per stomach	High	1.0 1.1	0.5 0.5	0.7 0.7	0.6 0.5
	1966	1967	1968	1969	1970
Mean birth litter size					
Yearlings	5.0	4.6	4.4	4.2	4.1
Older vixens	6.6	** 4.6	4.3 **	5.4	5.0
Percentages of barren vixens					
Yearlings	40+0 ?	50+10	38+26	31+13 ?	33+25
Older vixens	8+15 ?	27+9	12+22	9+9 ?	0+25
Productivity					
Yearlings	3.2	1.8	1.8	2.1	1.5
Older vixens	5.0	* 2.8	3.1	4.0 ?	1.8
All ages	4.5	* 2.3	2.6	3.2	* 1.6

(1) *Jämtland–Härjedalen.* Yearly number of vixens examined for litter sizes: yearlings 15, 7, 11, 9 and older vixens 25, 33, 37, 21 and for barrenness and productivity: yearlings 28, 49, 15, 18 and older vixens 24, 99, 56, 17.

(2) *Uppland–Södermanland.* Yearly number of vixens examined was for litter sizes: yearlings 14, 17, 26, 21, 7 and older vixens 18, 29, 67, 44, 8 and for barrenness and productivity: yearlings 5, 20, 34, 16, 12 and older vixens 13, 22, 50, 23, 4.

were large with a mean of 5.0 for yearlings and 6.6 for older vixens (Fig. 2). Inadequate material gives very uncertain data on barrenness as well as on productivity. According to the material however, the latter was very high or 4.5 per vixen, which gives a spring population of 69% cubs. During the summer rodents were plentiful and the sample of foxes from the following hunting season contained 54 and 63% subadult females and males respectively.

In the winter of 1966–67 rodents were still very plentiful and the stomach rodent index was 1.0 to 1.1. Nevertheless the mean litter size of old vixens fell to 4.6, productivity decreased to 2.3 and a high proprotion of vixens were barren.

In 1967–68, rodent numbers declined (index 0.4–0.5) but this was not followed by any change in litter size, barrenness or productivity in foxes. In 1968–69, rodents were again common and litter sizes of old foxes increased,

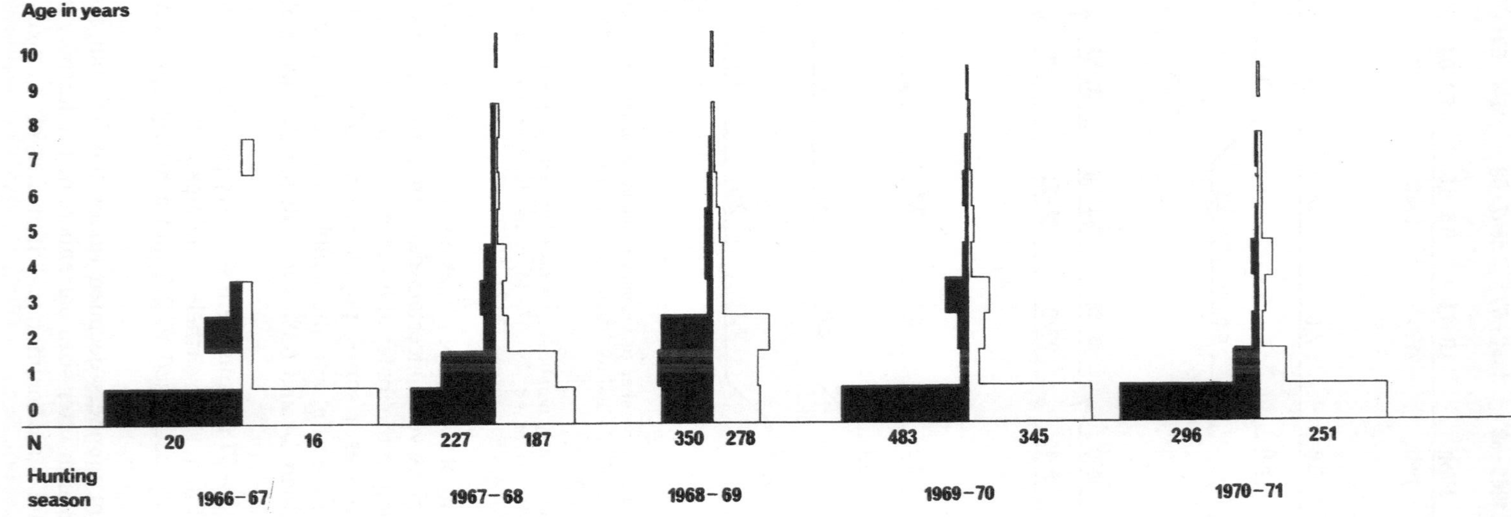

Fig. 3. Jåmtland–Härjedalen. Proportion of foxes of different ages observed in the samples of foxes killed in the hunting season. (Correction factors not applied.) Males to the left and females to the right.

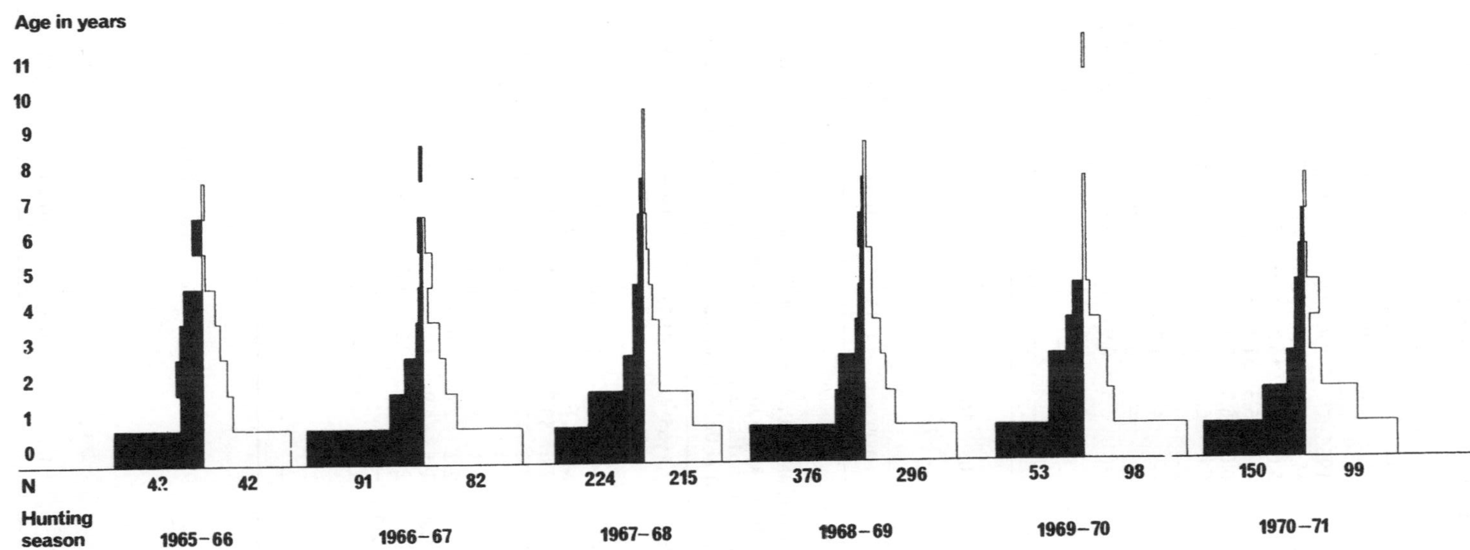

Fig. 4. Uppland–Södermanland. Proportion of foxes of different ages observed in the samples of foxes killed in the hunting season. (Correction factors not applied.) Males to the left and females to the right.

Table 1. Proportion of cubs in the spring calculated from data on productivity (the sex ratio of adults in the spring is assumed to be 1:1) and age structure of foxes killed in Jämtland–Härjedalen.

Spring of			1967		1968		1969		1970	
% of cubs	—		66.1		28.6		55.6		64.9	
Foxes killed in	1966–67		1967–68		1968–69		1969–70		1970–71	
Age in years	M	F	M	F	M	F	M	F	M	F
0	75.0	75.0	46.3	43.9	27.1	25.5	68.9	67.8	76.4	69.7
1	—	6.3	31.3	33.7	30.0	25.2	5.6	5.8	14.5	15.1
2	20.0	6.3	5.7	7.5	28.3	30.6	7.2	9.3	2.4	2.0
3	5.0	6.3	7.5	3.7	3.1	5.0	12.0	11.6	2.0	2.4
4	—	—	6.2	5.9	4.0	5.4	2.7	0.6	3.7	7.2
5	—	—	1.3	1.1	4.3	4.0	1.0	2.3	0.7	0.8
6	—	—	0.4	2.1	2.0	2.2	1.7	1.7	—	0.8
7	—	6.3	0.9	0.5	1.0	1.4	0.2	0.6	0.3	1.6
8	—	—	0.4	1.1	—	0.4	—	0.3	—	—
9	—	—	—	—	—	—	0.6	—	—	0.4
10	—	—	—	0.5	—	0.4	—	—	—	—
Number of foxes examined	20	16	227	187	350	278	483	345	296	251
Number killed per sq km according to hunting records	0.09		0.15		0.08		0.08		0.10	

Table 2. Uppland–Södermanland. Proportion of cubs in the spring calculated from data on productivity (the sex ratio of adults in the spring is assumed to be 1:1); and age structure of foxes killed. The sex ratios in the last two years are not strictly comparable because some hunters were asked to send males (M) in the autumn only and females (F) in late winter.

Spring of	1965		1966		1967		1968		1969		1970	
% of cubs	—		69		53		57		62		44	
Foxes killed in	1965–66		1966–67		1967–68		1968–69		1969–70		1970–71	
Age in years	M	F	M	F	M	F	M	F	M	F	M	F
0	47.6	47.6	62.6	53.7	48.2	43.3	62.0	50.3	47.2	53.1	54.7	49.5
1	11.9	16.7	17.6	17.1	29.0	27.4	15.4	16.6	18.9	16.3	22.7	28.3
2	14.3	14.3	9.9	12.2	10.7	8.8	12.8	12.2	18.9	12.2	9.3	9.1
3	11.9	9.5	3.3	8.5	4.5	8.8	4.0	9.1	9.4	9.2	4.7	3.0
4	9.5	7.1	2.2	2.4	4.9	5.1	1.9	3.4	5.7	3.1	4.7	8.1
5	—	2.4	1.1	4.9	1.3	3.3	1.6	3.7	—	2.0	3.3	1.0
6	4.8	—	2.2	1.2	0.9	1.4	2.1	2.0	—	2.0	0.7	—
7	—	2.4	—	—	0.4	0.9	0.3	2.0	—	1.0	—	1.0
8	—	—	1.1	—	—	0.5	—	0.7	—	—	—	—
9	—	—	—	—	—	0.5	—	—	—	—	—	—
10	—	—	—	—	—	—	—	—	—	—	—	—
11	—	—	—	—	—	—	—	—	—	1.0	—	—
Number of foxes examined	42	42	91	82	224	215	376	296	53	98	150	99
Number killed per sq km according to hunting records	0.29		0.35		0.32		0.48		0.40		0.34	

Table 3. Mortality rate of cubs in the spring-summer periods in Jämtland–Härjedalen (calculated figures).

Year	1967	1968	1969	1970
Proportion of cubs in the spring (%)	66	29	56	65
Abundance of rodents in spring-summer	very few	moderate	abundant	?
Mortality rate of juvenile foxes (%)	70	34	7	23

Table 4. Mortality rate of cubs in the spring-summer periods in Uppland–Södermanland (calculated figures).

Year	1966	1967	1968	1969
Proportion of cubs in the spring (%)	69	53	57	62
Abundance of rodents in spring-summer	abundant	?	few	?
Mortality rate of juvenile foxes (%)	62	47	35	63

but there were no detectable changes in litter size, barrenness or productivity.

The mortality rate of juveniles, calculated, with a correction factor, in the same way as for the other material, was 62, 47, 35 and 63% for the four years 1966–1969. In the summer of 1966, rodents were extremely abundant, and in 1968, when the lowest mortality rate of the cubs was observed, rodents were few.

As can be seen in Table 2 and Fig. 4 there were only small variations in the age structure from year to year. A chi-square test (where all foxes older than two years (3+) are lumped together) showed no significant difference between years for females ($x^2 = 19.031$; $f = 15$; $0.25 > P > 0.10$). For males the situation was somewhat different, however, since the frequency of old males in 1965–66 and of one year olds in 1967–68 were higher than normal ($x^2 = 35.497$; $f = 15$; $P < 0.005$).

DISPERSAL

A study of the dispersal movements, as revealed by tagging began in 1974 in the provinces of Hälsingland and Gästrikland, in the northern coniferous zone. So far about 700 juvenile and adult foxes have been tagged.

The abundance of rodents varied between years and the mean number per stomach containing food, in Nov–Dec, for the four years 1974–77 was about 0.7, 0.3, 0.8 and 1.0. The population densities in the autumn also varied according to the number of foxes killed by hunters and during each of the four hunting seasons from 1974/75 to 1977/78 the numbers killed were estimated to be 3,435, 2,208, 2,800 and 3,300 respectively. These numbers

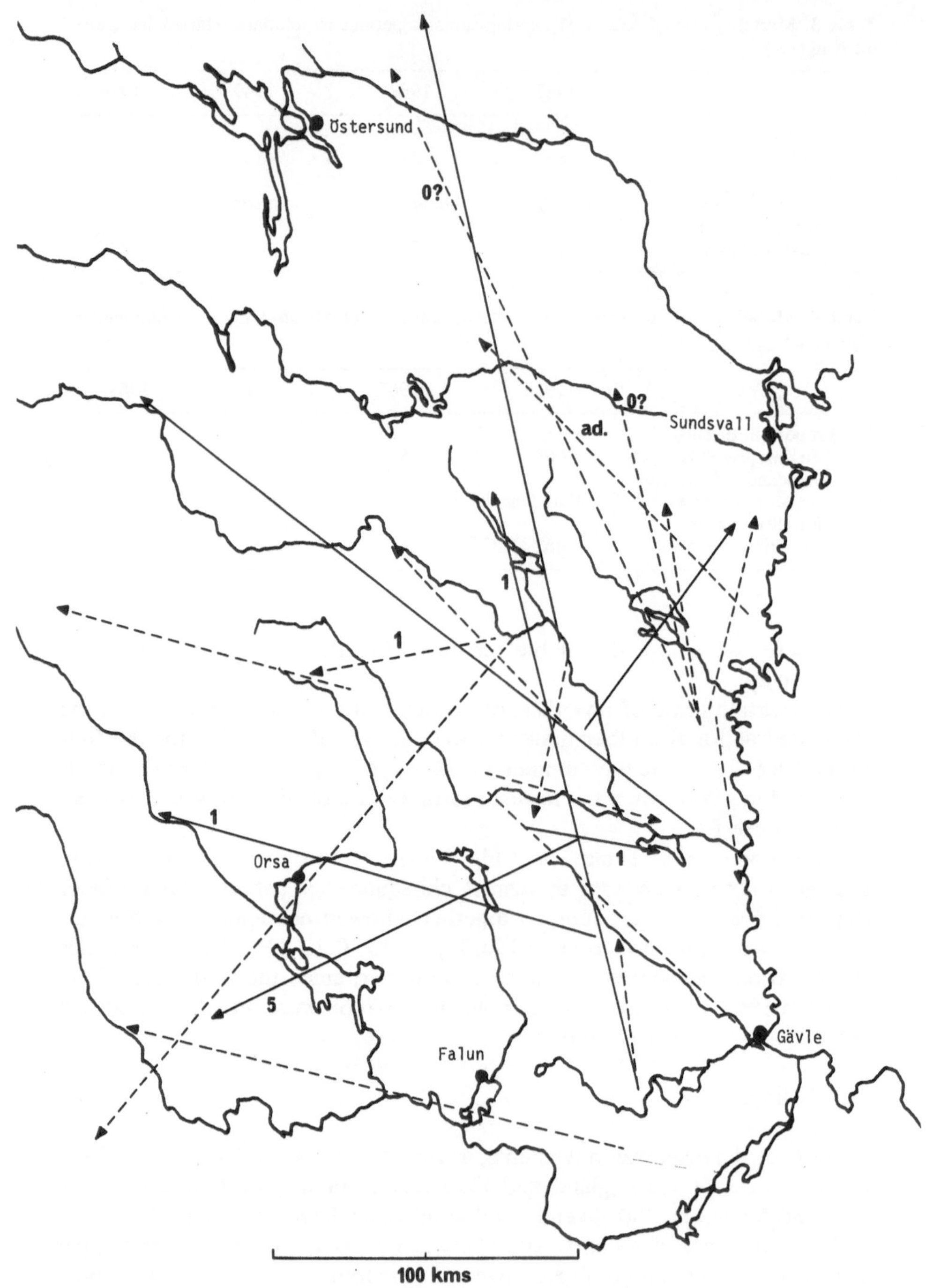

Fig. 5. Recoveries of foxes tagged in the provinces of Hälsingland and Gästrikland during 1974 through 1977. Only foxes that have moved at least 40 kms are included. The age in years at the time of tagging is given for adults (ad = adult with unknown age; - - - - males; —— females).

Table 5. Recoveries of foxes from the 1st of September during the first year after tagging. Foxes tagged in 1974 and 1977 are pooled.

		Number of recoveries	Distance in km from the tagging place in % 0–5.0	5.1–10.0	10.1–20.0	20.1–50.0	>50
Juv. males	74+77	31	52	13	19	16	—
	75	16	31	6	19	19	25
Juv. females	74+77	23	91	4	—	—	4
	75	9	89	11	—	—	—
Ad. males	74+77	10	50	10	—	10	30
	75	14	71	14	14	—	—
Ad. females	74+77	20	95	—	—	—	5
	75	10	70	20	—	10	—

represent a take of 0.19, 0.12, 0.15 and 0.18 foxes per sq. km respectively. Though it is not known how reliable these figures are, they do however, give an indication of the yearly variation in the population density.

The collection of data on foxes from hunting areas introduces some complexities. In the autumn of 1975 there were few foxes and few rodents, but since fox skins fetch high prices many hunters make bait stations with meat and the food situation might not have been so bad in the autumn of 1975. In the autumn of 1974 and 1977 there were according to hunting records more foxes than in 1975, and there were plenty of rodents too.

In all years some foxes of all ages and both sexes disperse (Fig. 5). Three males, probably juveniles, were shot about 150, 200 and 220 km, respectively, from the tagging place and two young females had moved about 200 and 210 km respectively, in less than 120 days. Another female, 5 years old when tagged, was shot two years later 124 km away. But do foxes disperse more often when rodents are few and when the fox populations are dense?

The few recoveries at hand indicate that young males dispersed more often in 1975 than in other years (Table 5) and 63% of these were killed more than 10 km away. All juvenile females and most adult foxes on the other hand were killed within 10 km of the place of tagging. There is no reason therefore to believe that food shortage or starvation *per se* caused the high dispersal rate among the young males. Also, if the number of foxes killed reflect population levels – which are apparently related to abundance of rodents – there should be more unoccupied land at least if the resident foxes did not take up larger territories, when the rodent numbers decreased.

Unfortunately there are extremely few data from the agricultural provinces of Uppland–Södermanland. Four of five tagged young males, however, left their home areas, one moving nearly 250 km away. Three young females on the other hand were all killed within 2 km of the tagging place.

DISCUSSION

As shown there is good correlation in the northern areas between rodent numbers and litter size, barrenness and mortality rate. When rodents are so scarce that the mean number found per stomach containing food is as low as 0.2 to 0.4 the availability of suitable food for foxes then seems to be critical for successful productivity. Foxes, it seems, are not subjected to starvation conditions however, since the mean weights of juvenile vixens at critical times does not vary between years. In 1967/68/69 the mean weights with 95% confidence limits were 5.0 (±0.5), 5.1 (±0.2) and 4.8 (±0.3) kg respectively (Englund, 1970). Foxes probably had to search much harder and longer to get enough food in the winter of 1967–68 which, together with the poorer quality of the food might have affected productivity.

Apart from increased time and effort spent searching for food of low quality, the scarcity of rodents in the northern areas will probably result in an increased mobility (dispersal) of the fox populations. In times of severe food shortage most foxes have to disperse. In the winter of 1967/68 for example when rodents were extremely scarce, hunters reported that fox tracks were seldom seen except where rubbish dumps occurred around villages and cities.

The patchy distribution of food in poor rodent years will result in a locally clumped distribution of foxes, possibly even at higher densities than occur when foxes are more evenly distributed in years of rodent abundance. This enforced mobility of foxes may even prevent stable territories being formed in time for the mating season, and some factor associated with this may have contributed to the low productivity of breeding in 1968.

In the autumn of 1975 rodents were rather scare in the south-east of the northern forest zone (0.3 rodents per stomach containing food). In the autumn of the year more young males than usual dispersed. Females as well as older foxes, on the other hand, dispersed less as usual. The high dispersal rate of the young males in that year was not therefore caused directly by starvation since the other components of the fox population did not behave similarly. They must have dispersed for some other reason.

In the winter of 1975/76 fewer foxes than usual were shot. Since the recovery rate of tagged foxes is roughly the same in all years (Englund, 1973), there is a strong indication that hunting records give a good index of the change in density of the autumn fox populations. Thus, because foxes were fewer in the autumn of 1975 population density alone cannot be responsible for the higher tendency of young males to disperse long distances. Might the reduced abundance of food have resulted in an increase in size of ranges by the older animals thus forcing the least tolerated animals – the juveniles – to move away?

An alternative hypothesis whereby it is postulated that it might be possible to predict which cubs of a litter are most likely to disperse is relevant (Bekoff, 1977). High-ranking individuals, he says, more often than others invite littermates to play but they are less successful in soliciting play than low-ranking individuals which only rarely will invite play but when they do they are usually successful. Subdominant cubs, of intermediate rank, on the other hand often invite play and are often successful in this. The social ties within the group will therefore vary. Since the dispersal of red foxes seems to be a "passive" progress, that is the individuals do not emigrate because they are forcibly ejected, Bekoff suggests that the first to leave should be the ones with the weakest ties within or to the group, that is the highest and lowest-ranking individuals in the litter. That the desire to disperse might be dependent of the strength of the social bonds to group-members is very interesting. Might it be so that the social bonds between littermates was lower in 1975 caused by increased fights for food, when the overflow of food was reduced by the scarcity of rodents? And if that is the case why did it only effect the males?

Summing up for the northern area, it is clearly uncertain that the lower recruitment rates following periods of rodent scarcity and the greater mobility among foxes at such time, are caused by lack of suitable food only. There are clear indications that the social organisation of foxes in the northern coniferous belt may be significant in these respects and need to be investigated.

The situation is different in the agricultural areas. Here the availability of food is better, with less variation. Rodents are never so few as in the north (0.5 per stomach containing food was the lowest found in the southern area during the study period, which for foxes in the north would be a moderately good level of food, see Fig. 1); furthermore the abundance of alternative food of high quality is also better (Englund, 1965). In spite of that, a high proportion of the vixens were barren, even after winters when rodents were extremely numerous. It varied from 40 to 64% for yearlings (31 to 50% with *no* signs of placental scars) and 18 to 36% for older vixens (up to 27% with no signs of scars). In all years, however, nearly all vixens ovulated (Englund, 1970 and Lloyd et al., 1973).

The mortality rates of cubs were also usually high, and the highest recorded, in 1966, occurred when rodent numbers were at a peak. The lowest mortality rate was in 1968 when rodents were scarce.

Thus there is no correlation between the abundance of rodents and productivity and mortality rates. Since the productivity varied during the study period but the age-structure was about the same all years (at least for vixens), the mortality rate among cubs, during the spring–summer period, seems to be inversely correlated with the productivity. When many cubs are born there will be a high mortality rate; the converse obtains when few cubs are born.

About 60% of the spring population are cubs but in the autumn they probably form no more than about 40% (corrected figure). Accepting these calculated figures then, as can be seen from the footnote,* about 55% of the cubs has died before the autumn. Examination of the numbers of cubs found in litters dug out of earths indicates that litter sizes decrease with the age of the (unweaned) cubs, up to 1.8 kg weight, but this loss is too small to account for the calculated mortality of 55% during the spring summer period.

---

* The spring population consists of 60 juvenile and 40 adult foxes that is 100 specimens. Suppose that x juvenile and all adults will survive the summer. Thus $x/(x+40) = 40/100$ (that is the correction factor of 0.80 multiplied with the percentages of juveniles killed in the autumn winter period). The number of cubs that survived will then be 27 and 33 died.

Evidence gained by cub tagging suggests that there is not a high mortality of cubs in late summer because there is a high recovery rate of cubs tagged at 2–3 kg weight. During the first hunting season after tagging in the southern part of the northern area, there is a higher recovery rate of cubs than of adults. If all foxes not recovered during the first year were still alive (and had not lost tags) then the recovery rate of these during the following years will be about the same for foxes tagged as cubs or as adults – 36 and 30% respectively (9 out of 25 tagged as cubs and 25 out of 84 tagged as adults). The main part of the large mortality rate among cubs therefore must be in spring or early summer. And since the size in the litters only slightly decreases with age, then the main part of the mortality must lie in the loss of entire litters.

It is difficult to elucidate the causes of low vixen productivity, high cub mortality and high dispersal rates of young males. Population densities may have an important influence on these features but nothing is known of number of foxes in the study area, except that foxes are more abundant than in the north.

Regarding Macdonald's observations (1977) the over-population itself might cause the formation of groups with such a social organisation that most vixens in one way or another will be either barren (alternatively not mated) or mistreat their cubs in such a way that these will die.

Elsewhere in agricultural areas, in the midwest in the US for example (Storm et al., 1976) and in Ontario (Johnstone pers. comm.) nearly all vixens reproduce. The same situation obtains in at least some areas in Europe both in Britain, the Netherlands, Switzerland and Germany (Lloyd et al., 1976). In some of these countries where data have been collected rabies is present and the hunting pressure is high (1 to 2 foxes shot per sq km; Lloyd et al. 1976). The population densities in these areas therefore may be low well prior to the mating season. These data contrast fundamentally from the agricultural areas in Sweden, where we have no rabies (or mange) and where the number of foxes killed per sq km is estimated to be around 0.3 to 0.5 per year during the study period, but population densities of foxes might be less in Sweden. Furthermore if fox population densities are reduced to low levels compared with the level that that habitat can support, food will not then be a limiting factor, and then reproduction and productivity will probably be high and natural mortality low. Evidence from this study suggests that population density is a key factor in the regulation of fox numbers in Sweden, but habitat requirements and food resources interact with population density and have a modifying influence upon reproduction, productivity, mortality and dispersal. No data so far therefore will upset the theory that the foxes in southern Sweden are regulated by the density itself in one way or another. Many data, however, support the idea.

REFERENCES

Bekoff, M. 1977. Mammalian dispersal and the ontogeny of individual behavioural phenotypes. *Amer. Natur.* 111: 715–732.

Englund, J. 1965. Studies on food ecology of the red fox (*Vulpes v.*) in Sweden. *Viltrevy* 3: 377–485.

Englund, J. 1970. Some aspects of reproduction and mortality rates in Swedish foxes (*Vulpes vulpes*), 1961–63 and 1966–69. *Viltrevy* 8: 1–82.

Englund, J. 1978. Rävens vandringar – en fråga om gnagarbrist eller stress? *Svensk Jakt* 116: 322–325.

Hörnfeldt, B. 1978. Basinventering gnagare. Progress report No. 11 (mimeographed).

Lloyd, H. G. and Englund, J. 1973. The reproductive cycle of the red fox in Europe. *J. Reprod. Fert. Suppl.* 19: 119–130.

Lloyd, H. G., Jensen, B., Van Haaften, J. L., Niewold, F. J. J., Wandeler, A., Bögel, K. and Arata, A. A. 1976. Annual turnover of fox populations in Europe. *Zbl. Vet. Med. B.* 23: 580–589.

Macdonald, D. W. 1977. The behavioural ecology of the red fox, *Vulpes vulpes*; a study of social organisation and resource exploitation. D.Phil. thesis. Oxford University.

Storm, G. L., Andrews, R. D., Phillips, R. L., Bishop, R. A., Siniff, D. B. and Tester, J. R. 1976. Morphology, reproduction, dispersal, and mortality of midwest red fox populations. *Wildlife Monographs* No. 49, 82 pp.

10 SOCIAL FACTORS AFFECTING REPRODUCTION AMONGST RED FOXES (*Vulpes vulpes* L., 1758)

David W. Macdonald*

INTRODUCTION

In some habitats foxes live in social groups comprised of one adult male and several adult vixens. These groups occupy territories from which neighbouring groups are excluded. Elsewhere, foxes live in territorial pairs but recent research suggests that at least small groups are found in many habitats (Ables, 1975, Macdonald, 1980). On Boar's Hill, Oxfordshire, a rural-suburban habitat where foxes live in stable groups of between 4–5 adults in territories of about 40 hectares, radio-tracking, direct observation, handling during trapping and occasional post-mortems all confirmed that many vixens did not rear cubs (Macdonald, 1979a). For instance, in 1974 one group consisted of three vixens, aged 9, 6, and probably 2 years, of which only the eldest reared cubs. Considering the four groups for which I have complete data in 1973–74, 6 out of 15 vixens (40%) reared cubs. Of these four groups a mean of 38.75% (S.D. ±15.5) of vixens in each group bred, that is only one or two vixens bred. More fragmentary information from the other groups on Boar's Hill also suggested that many adult vixens did not rear cubs. There was no evidence of cub mortality, suggesting either that "barren vixens" did not conceive or that the embryos did not reach full term. Lumping data from 1973, 1974 and 1975, 12 of 24 vixens studied on Boar's Hill reared young. In other study areas similar observations were made. A group of four vixens were observed at Ein Gedi (Israel) in 1976 and only one of these bred; within one group on Oxfordshire agricultural land one vixen did not breed in 1974, but did so in 1975 after the disappearance of another vixen who had bred the previous year.

These data may have bearing on the widely different figures for proportions of barren vixens in fox populations studied (post mortem) in different regions. In Sweden alone Englund (1970) reports yearling fecundity to vary between 5–88%. In this paper I will describe observations on both wild and captive foxes which clarify vixen productivity figures from different study areas.

Where several vixens occupy a group territory all the available evidence suggests that they are close relatives. This evidence includes long-term radio-tracking of juvenile vixens as they become integrated into their parental group, and observations on colour similarities between group members.

* Animal Behaviour Research Group, Department of Zoology, South Parks Road, Oxford, OX1 3PS.

In the course of searching for possible explanations for the wide differences in vixen productivity mentioned above I will describe reproductive behaviour of both wild and captive foxes.

MATERIALS AND METHODS

Observations on wild foxes have been made between 1972–79. Foxes were captured and equipped with 102 MHz radio transmitters. Radio-tracking was used predictively to increase the chances of direct observation using infra-red binoculars. To further increase opportunities for watching foxes they were fitted with fluorescent β-lights and reflective markers (Macdonald 1978, Macdonald and Amlaner, 1980).

Radio tracking "fixes" and direct observations were used together to draw by eye the exact borders of fox home-ranges. For quantitative comparison with other data, eliptical estimates of home range size are also given (using Jennrich and Turner's (1969) method) although all indices of home range size and configuration have drawbacks (see Macdonald et al. 1980).

Observations in captivity were made between 1973–79 in two 1000 m^2 enclosures, described below.

REPRODUCTION IN WILD FOXES

Changes in movement pattern

During January, February and March, freshly dug fox earths and abandoned exploratory excavations became a noticable feature of each of my study areas. Radio-tracking of 10 pregnant vixens revealed that as birth approached, they made increasingly frequent visits to favoured earths, which they continued to excavate. After the cubs were born vixens normally remained in or near the natal earths for 7–10 days and nights. Thereafter they made forays from the earth, which initially lasted under an hour but become prolonged as the cubs grow. By the time the cubs are 3 weeks old the vixen may spend most of her time outside the earth, lying nearby or even in a neighbouring earth. When the cubs start eating meat at about 4 weeks old the vixen makes frequent trips back to the earth during the night, sometimes taking food. During the vixen's "confinement" prey are deposited at the mouth of the earth by other foxes. The fox bringing the prey calls the nursing vixen from her earth with a "warble" cry. The cubs may be moved to another earth at any time and are invariably moved after human intrusion. Sometimes the family is split up with cubs being deposited at different earths; Storm et al. (1976) and Niewold (pers. comm.) have also found split litters. These depots may be slightly enlarged rabbit "stops" or caverns under a tree into which an adult fox could not fit. These generalisations are illustrated by observations on one vixen during the spring of 1973.

Before the birth of her cubs the vixen's range was delimited by roads and tracks (Fig. 1). Typically, she covered most of this area (250 ha) every night. Prior to the birth of her cubs she visited several of the earths in her territory. The cubs were born on 15th March at map ref.

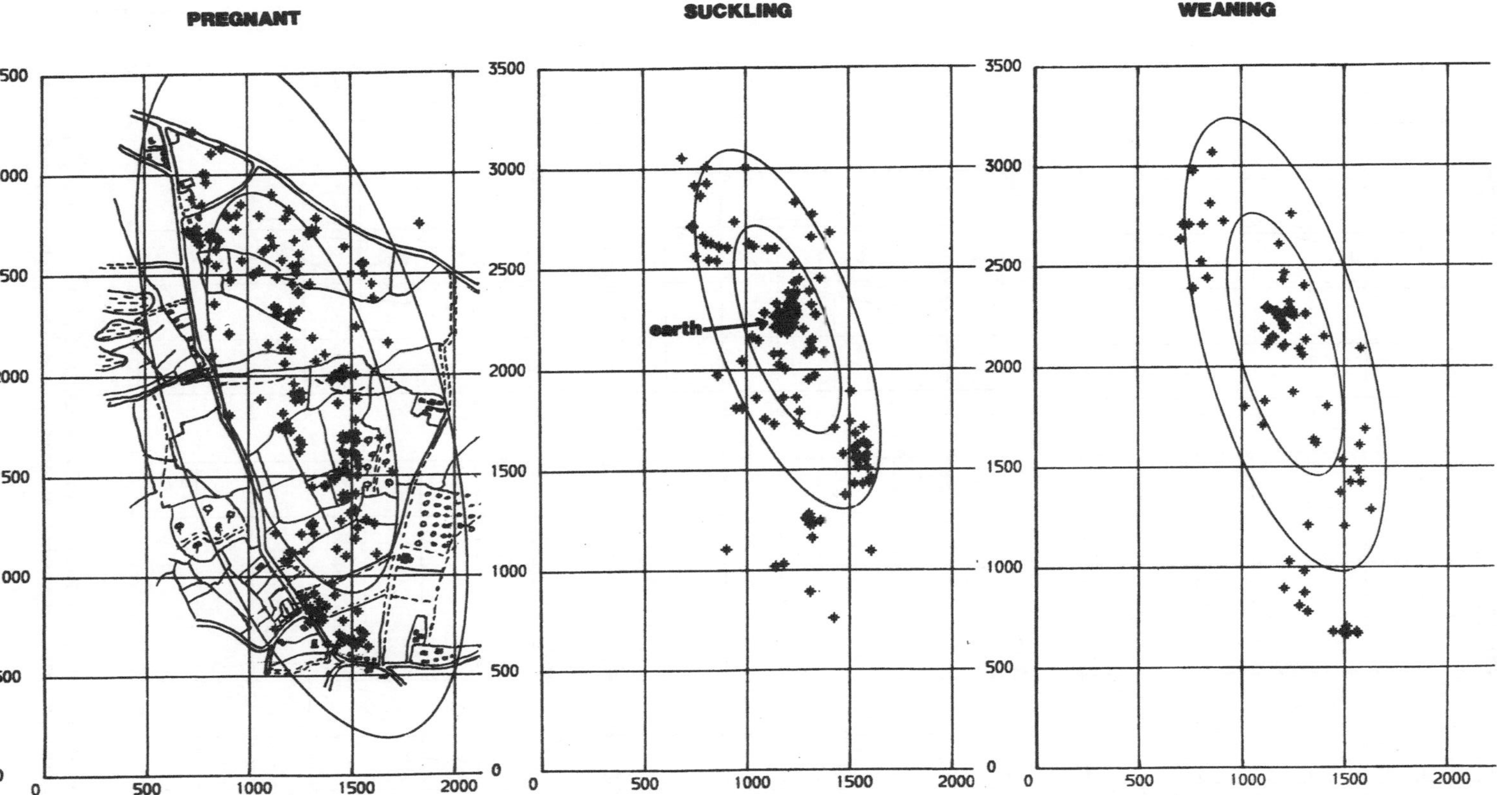

Fig. 1. Radio-fixes on a vixen during different periods on Oxfordshire agricultural land (scales in metres). The borders of her territory were bounded by roads and tracks. 62.3% and 95% probability elipses are drawn around the fixes (see Table 1).

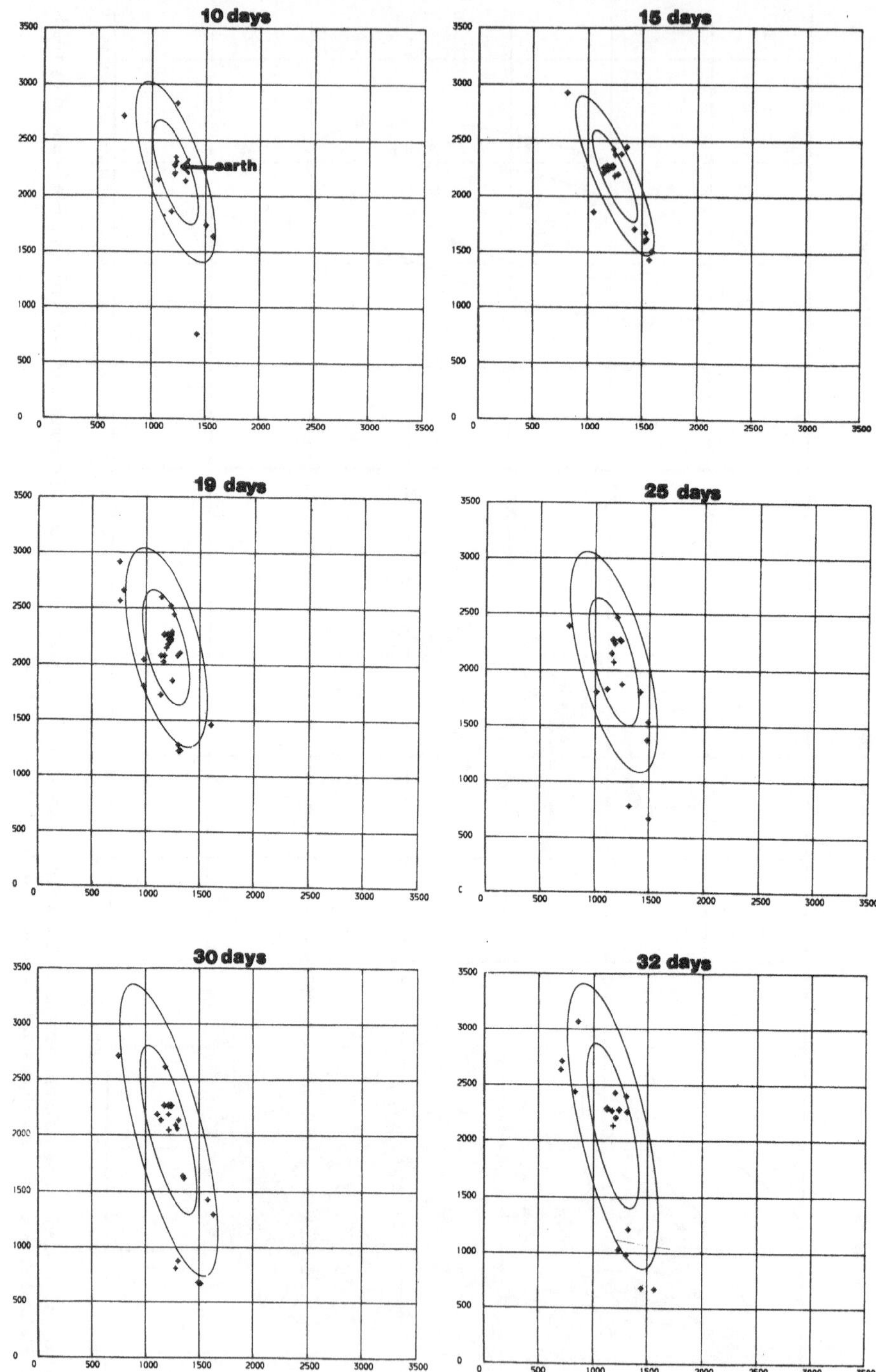

Fig. 2. Records of a vixen's movements for a selection of nights following parturition. With passing time she spent less of the night near the earth. Locations can be compared with the map on Fig. 1.

Table 1

Period	Cubs age (days)	Eliptical estimates (ha) n	62.3%	95%
Pregnant	—	229	123.3	369.9
Always in earth (14–25th march)	0–10	183	—	—
Suckling	10–19	331	33.6	100.8
Starting solid food (11–19th april)	24–32	192	48.1	144.2
28/3/73	10	40	22.9	68.7
2/4/73	15	40	15.3	45.9
6/4/73	19	40	29.9	89.7
12/4/73	25	38	42.2	126.7
17/4/73	30	37	32.9	98.6
19/4/73	32	39	43.8	131.5

1500, 2000 (Fig. 1), an area of rushes and marshland. 5 cubs were born above ground in a tussock of sedge, surrounded by water. Three days later the vixen moved her cubs to the earth at ref. 1400, 2300. During the following nights she remained in the earth until the 21st when she made a brief excursion of 150 m lasting for about 20 minutes and thereafter remained in the earth until dawn. During the next 3 nights she spent progressively more time out of the earth although always in the immediate vicinity until 25th March. Table 1 summarises eliptical estimates of her movements during three periods (see Jennrich and Turner, 1969). During this time the vixen was radio-tracked all night every night (by foot or by car) and point locations recorded every 15 minutes were possible. The vixen's movements throughout the month after the birth of her cubs are illustrated by 6 complete nights' data on Fig. 2 (see Table 1). Her nightly tracks initially concentrate either around alternative earths, favoured feeding sites (such as farmyards and gardens) and, for the majority of the night (>80%), around the breeding earth.

Similar changes accompanied the birth of cubs to all radio-controlled vixens, although the scale of the changes in movement pattern was affected by home range size. For instance, a vixen tracked on the fells of Cumbria normally ranged over 1,000 hectares. When she began taking food to her cubs she remained in the section of her territory nearest to them, although hunting an area greater than the full extent of most Oxfordshire fox territories.

Observations at earths

An earth in a territory occupied by one dog fox and three adult vixens was watched during 3 consecutive all-night vigils which confirmed that barren vixens do visit natal earths. The eldest vixen, T'peg, was the only one to breed:
e.g. 26th May, 1974*

I took up position in the tree at 1900 hours. At 2120 hours a fox moved beneath a bush 30 m ahead of me, and uttered a greeting "warble". Scar Nose ♂ emerged from the bush, closely followed by T'peg ♀ who slithered on the ground behind him in submissive greeting. Scar Nose ♂ walked above the entrance to the earth and T'peg's whimperings developed into a submissive shriek. Scar Nose ♂ turned towards her and she crept forward and licked his muzzle, her screams reaching a crescendo. Scar Nose ♂ moved to the entrance of the earth and token urine marked into it (see Macdonald 1979b). While T'peg ♀ sniffed his mark, Scar Nose ♂ went back into the rhodedendron from whence came greeting screams from another adult fox. T'peg ♀ moved off up the bank as the screams from the bush reached a peak before Scar Nose ♂ and

* All names based on characteristics by which I identified the foxes.

Husky ♀ emerged. Husky trotted up the path after T'peg ♀ and on reaching her, crouched flat to the ground, tail lashed and submitted briefly before continuing up the path. At 2145 hours the third vixen visited the earth and at 2215 hours T'peg ♀ returned. Throughout the night all three vixens continued to visit the earth.

Sargeant and Eberhardt (1975) mention dog foxes carrying food to their vixens.

Multiple litters

Contrary to the conventional description of foxes as "solitary" carnivores (e.g. Fox, 1971) there are reports of vixens sharing a den and rearing cubs together (Pils and Martin, 1974; Storm et al., 1976). This phenomenon is well known amongst gamekeepers and hunt-servants. I have made sufficient observations of multiple litters in England to believe that they are a widespread phenomenon. How do the vixens involved in communal earths behave towards each other and how do they react to each other's cubs?

Two vixens attended a mixed litter during May and June 1973. One litter consisted of 4 cubs estimated at 2 weeks older than another solitary cub. Radio-tracking and other circumstantial evidence strongly suggested that the two vixens were often together with the cubs simultaneously. When they were seen independently at the earth they both suckled all the cubs. e.g. 30th May:

At 2030 hours Blackfringe ♀ appeared near the earth; pausing she made a low 'wowwow-wow" call at which all five cubs tumbled out of the earth. Blackfringe ♀ dropped a vole she was carrying and, amidst the greeting confusion, a cub grabbed it. The cubs fought fiercely over the food. Afterwards they ran back to Blackfringe ♀ and all suckled from her. 2040 hours: Blackfringe ♀ watched the cubs as they played. The younger cub was frequently attacked. Blackfringe ♀ left at 2045 hours but the cubs continued to play. 2130 hours: Downwhite ♀ came into sight and called the cubs. She did not bring food but suckled the cubs. 2137 hours: Downwhite ♀ left and the cubs continued to play.

Detailed observations were made on the group to which these two vixens belonged. One dog fox and four adult vixens comprised the group and a summary of interactions between them is given in Fig. 3 (for more details see Macdonald (1977a)). These observations were made around a feeding site where initially Blackfringe ♀ dominated Downwhite ♀, e.g. on 12th May, 1973 (during a 1 hour observation) three separate bouts of serious aggression involving "sideways barging" and intense gekkering occurred between Blackfringe ♀ and Downwhite ♀, each of which was conclusively won by Blackfringe ♀.

Subsequently, however Blackfringe ♀ lost encounters, both around the feeding site and elsewhere. In fact, the reversal of status when Downwhite ♀ became dominant to Blackfringe ♀ could be tied down to the night of 20th May, roughly the date when the cubs were weaned. Seven of the 13 interactions (46.2%) observed before 20th May involved serious attacks of Blackfringe ♀ on Downwhite ♀, while afterwards, 7 of the 8 interactions (88.9%) observed, were in the opposite direction, with Downwhite ♀ seriously attacking Blackfringe ♀ (Fig. 4). Subsequent observations in captivity (below) document similar changes in status within a family group.

These flow diagrams record only the course of observed interactions and

Serious Aggression

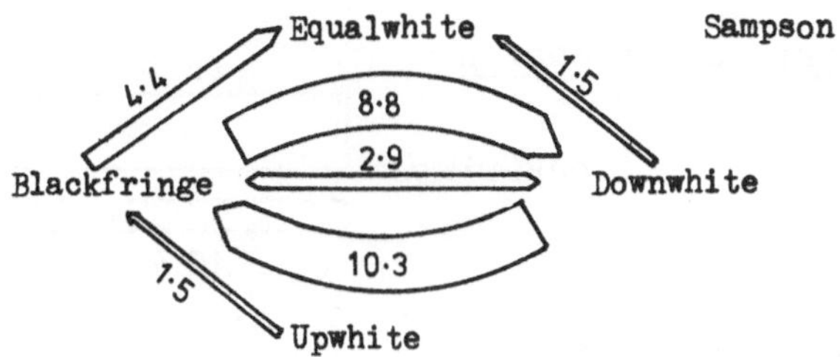

Mild Aggression

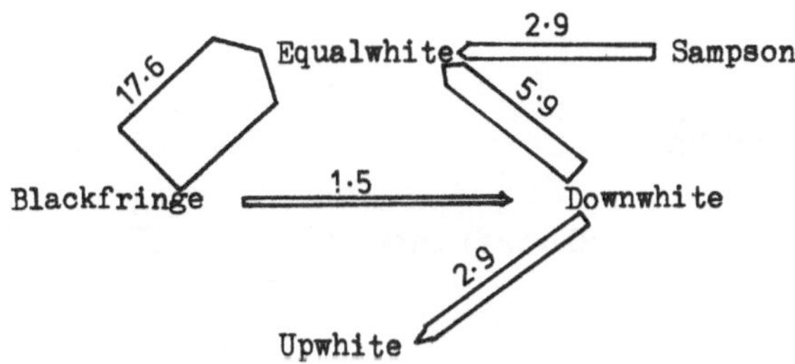

Amicable

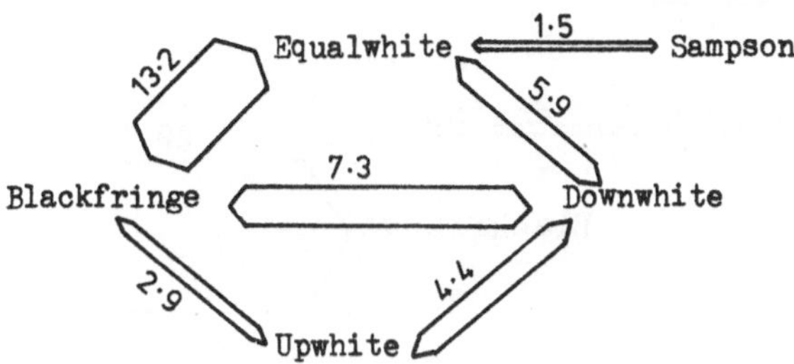

Play

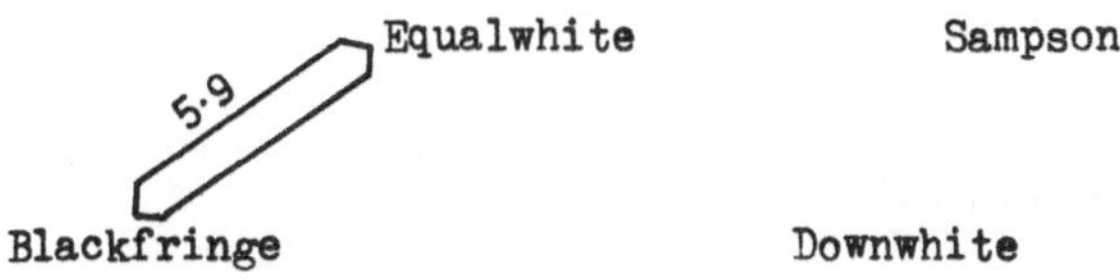

Upwhite

Fig. 3. Summary of 68 interactions between group I foxes around the feeding site.

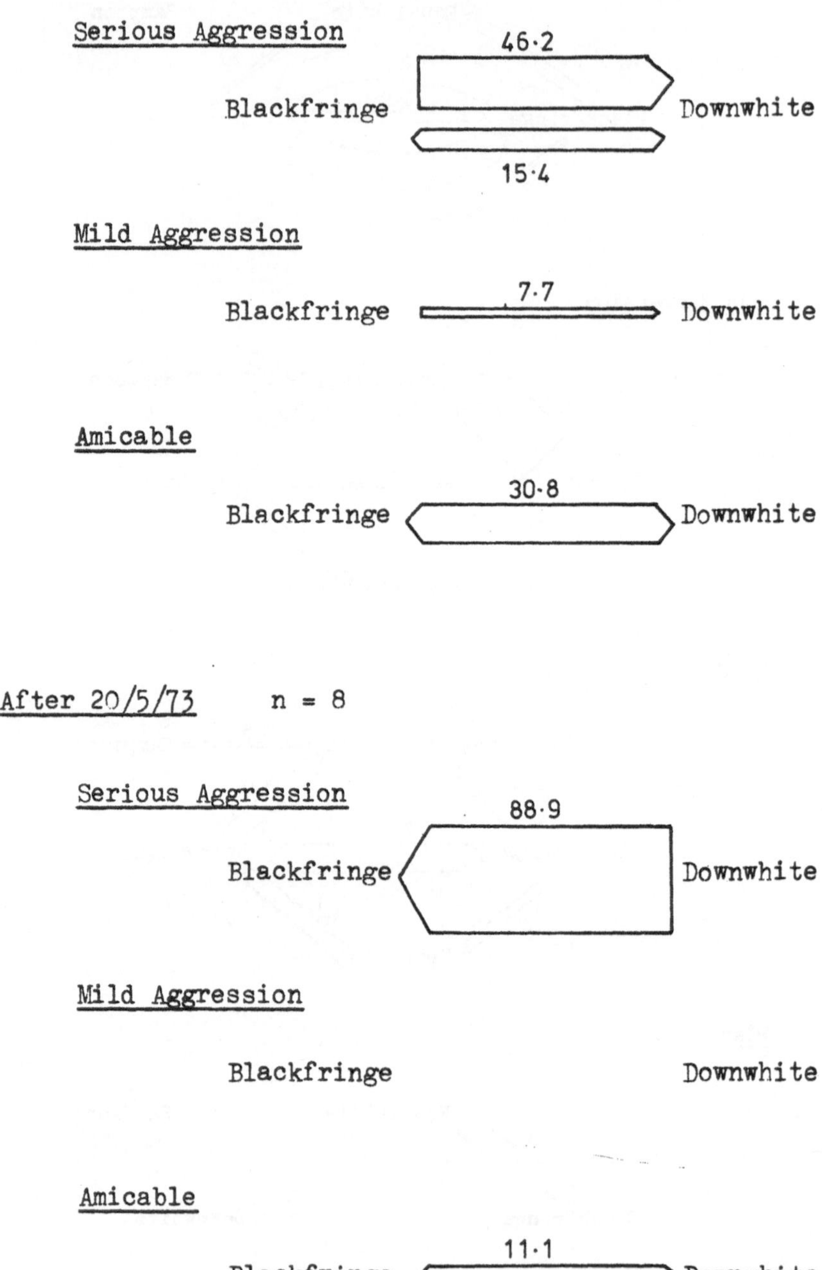

Fig. 4. Summary of relationship between Blackfringe ♀ and Downwhite ♀ during 13 interactions before and 8 interactions after 20/5/73.

say nothing about the extent to which given individuals may be seeking or avoiding others. For instance, on 10 observation nights before 20th May both Blackfringe ♀ and Downwhite ♀ were present at the feeding site on the same 9 nights. On 7/9 nights the two vixens were in the garden simultaneously and I was able to record direct encounters between them. That is, on 77.8% of the nights when both these foxes visited the feeding site they were observed together. However, for 30 observation nights following 20th May when foxes were seen, there were 13 occasions when during the course of one observation period both Downwhite ♀ and Blackfringe ♀ visited the area, but on only one of these nights did they have direct encounters, that is on 7.6% of the possible nights.

In fact, after Blackfringe's downfall she often approached the feeding site, detected Downwhite's presence and did not advance further. When Downwhite discovered Blackfringe's presence she attacked her e.g.:

On 17th August I saw Blackfringe ♀ sitting around the feeding site. At 0030 hours Blackfringe ♀ stood up and loped away. Downwhite ♀, sped towards her and while Blackfringe ♀ tried to make an escape, she was quickly overhauled and attacked by Downwhite ♀. A serious fight ensued with both vixens rearing on their hind legs and stabbing at each other with their forepaws. Blackfringe ♀ broke free and again tried to flee, but Downwhite ♀ caught her amidst much gekkering, and finally whimpering, whereupon Downwhite ♀ trotted back toward the feeding site; for the next 30 minutes, Blackfringe ♀ moved around the adjoining field but never approached the site.

In another group two vixens with litters visited a feeding site where one of them and her cubs were rather shy. The older, more confident vixen regularly carried food to her group mate and cubs who waited at a distance.

Non-breeding vixens and cubs

Field observations show that barren vixens within a family group typically behave amicably towards cubs, often playing with them. E.g. when Red ♀ (barren) appeared in a field where one of her group members' 3 month old cubs was foraging:

The cub stopped foraging and ran straight towards Red ♀, overshooting by 4–5 m and swerving round her, before throwing itself flat on the ground just in front of her. Red ♀ stared at the cub as it lay, ears flicking back and tail lashing. The cub leapt up and raced around her, repeatedly approaching and withdrawing. Red ♀ broke into a run and a play chase developed with both foxes running and chasing after each other for over 30 seconds. Afterwards Red ♀ began foraging for beetles. The cub followed close behind her, investigating each capture. During the next 15 minutes they played 3 times.

Normally I could not distinguish the behaviour of non-breeding vixens towards cubs from that of the real mother, although in one case the mother and another vixen of the same group behaved very differently:

Blackfringe ♀ (the mother) was foraging intently on the west side of a meadow when gekkering and squealing broke out from the adjacent wood where her cubs were. Instantly Blackfringe ♀ bolted straight towards the sound; she hit the thick vegetation of the ditch and did not alter course but rather crashed straight into the vegetation with such force that she rolled over as she emerged. Similarly she bombarded the vegatation at the edge of the wood and almost at once the noise stopped. Meanwhile, Equalwhite ♀ (barren) continued foraging although she looked towards the noise.

Observations such as these suggested that different barren vixens, even within one group, behaved differently towards cubs.

Radio tracking disclosed that barren vixens often visited the vicinity of earths harbouring cubs but the accuracy was rarely adequate to be certain that these vixens actually visited the earth itself. By watching natal earths all night it was possible to confirm that they do (see above); Murie (1936) made similar observations, and several gamekeepers have told me that they sometimes kill "dry" vixens around an earth in which they had previously killed a lactating vixen and, indeed, some have seen these vixens attempting to dig out gassed cubs and some of them have been carrying food.

The field observations outlined here suggest that within family groups only the minority of vixens rear cubs. Non-breeding vixens behave amicably towards the cubs of their group-mates and may even feed them. For those groups for which I have complete data, the younger animals (which were generally subordinate) were the non-reproductive ones. Why, in some groups, does only one vixen breed, while in others more do so? What is the adaptive significance of communal denning described above? What is the role of the non-reproductive vixens, why do they not breed and why do they remain in the group? To supplement field observations concerning these questions, I have watched captive family groups.

OBSERVATIONS IN CAPTIVITY

Enclosures

Two family groups were maintained in enclosures measuring just under 1,000 m^2 and which were surrounded by 2 m high chain link fencing topped by a further 1 m of 2″ (5.1 cm) wire mesh bent in at 45°. A metre of wire was sunk underground. Each enclosure supported natural vegetation and was supplied with artificial earths.

Methods

An estimated 4000 hours of observation were made within the enclosures, and from a camouflaged rostrum. By night I used either infra-red binoculars, (Old Delft, Holland) or infra-red video cameras. Most of the cubs reared within each enclosure grew up wild and observations of these animals were fraught with the same difficulties as watching wild foxes. To overcome this problem a remote-control infra-red video recording system was designed to observe actions from 200 m away while the foxes remained undisturbed and in complete darkness.

The camera unit

Image intensifiers were unsuitable because of the poor image quality and lack of relief, due to flat lighting from a cloudy sky and too large a contrast range in the moonlight between trees and undergrowth. Thus infra-red equipment was chosen. This equipment was designed to suit my fox watching requirements by J. Noakes and P. Townsend of the British Broadcasting Corporation.

Most cameras designed for 25 or 18 mm vidicon pick-up tubes would be capable of modification to accept the silicon-diode array vidicon tube necessary for working in infra-red. Silicon has the advantage as a photosensor, of having a wide sensitivity range from about 1.2 m in the near infra-red region through visible light and into ultraviolet. In this case a Link Electronics type 101 camera was used. Silicon-diode tubes usually have a life of over 2,000 hours and are more resistant than most TV camera tubes to damage by excessive exposure. Unfortunately direct control of tube sensitivity is not possible and so it was necessary to use a lens with adjustable aperture. The camera and tube were mounted on a servo system allowing operation of the lens iris, zoom, focus and camera pan and tilt from a control and viewing caravan over 200 m away. The zoom lens gave focal lengths between 15 and 150 mm, i.e. angles of view from 6 to 60 degrees.

Lighting

Infra-red lighting was provided by specially designed lamps. Four 1,200 watt floodlamps were positioned around the enclosure, each with a 45° beam (15" filter) which adequately illuminated about one-sixth of the area. These floodlamps were complemented by a "dimmable" 150 watt spot lamp mounted on and moving with the camera. All the lamps used cheap tungsten filament bulbs with a filter to remove the visible light (removing light of ≤ 800 nanometres). For low cost the filter was a composite of readily available theatrical lighting materials (two sheets of primary green no. 39 and one of primary red no. 9 cinemoid by Strand Lighting). Because of the spectral energy distribution of tungsten lamps the temperature of the cinemoid is minimised when radiation is filtered first by the red and then by the 2 green layers. By fan cooling the floodlamps, a filter life in excess of 200 hours was achieved. Air was drawn into the lamp between the layers of cinemoid and hence cooled them before passing around the lamps and out via a tangential fan.

RESULTS

The reproductive history of Group I was studied for five years, and that of Group II for three years. Summing the number of adult vixens which might have bred during the course of these observations gives a maximum productivity of 22 litters. In fact, only 7 litters were born (32% productivity of litters). As the two family groups grew, recruited young animals and developed a stratified age structure of the type observed amongst wild groups, only the dominant animals gave birth to cubs although dominance ranks changed during the study. All male offspring and some female offspring were removed in late Autumn to maintain the group's structure. Some non-reproductive vixens contributed to the welfare of the cubs of the reproductive vixen within their group, guarding and sleeping with them and provisioning them with food. On one occasion two vixens of comparable status conceived within one family group. Soon afterwards the vixen's relationship became more polarised; this influenced the more subordinate one's behaviour such that none of her cubs survived. After their death she nursed the cubs of the dominant vixen.

During five years I have monitored the social relationship underlying these results, and considering:

(i) what process underlies the selection of reproductive vixens within a group,
(ii) what role does the dog-fox play towards reproductive and non-reproductive vixens,
(iii) how do non-reproductive vixens react to their dominant's cubs?

First I will present an introductory açcount of Group I before going on to a detailed analysis of the social dynamics within Group II.

GROUP I (founded by Vixen I and Dog I, both born March 1974)

Dog I was introduced to Vixen I in December 1974, when both were 10 months old (and when radio-tracking indicated itinerant activity amongst wild dog-foxes). They mated (about 25th January, 1975) and reared 5 cubs, of which 4 vixens became integrated into the adult group. Of the 5 vixens next winter only Vixen I mated (23rd January 1976) and gave birth to 4 cubs (3 ♀♀, 1 ♂). The vixen cubs grew amongst the group while I hand reared the male for Group II. In late September 1976 one of the first year's vixens and two of the second were removed leaving the original pair (then nearly 3 years old), 3 of their 1975 female offspring (then nearly 2 years old) and 1 of their 1976 offspring, approaching 1 year old.

On 25th January 1977 Vixen I mated again and produced 4 cubs on 9th March (2 ♂♂, 2 ♀♀). In Easter 1977 some subordinate vixens were released leaving the original pair (nearly 4 years old), 2 of their 1975 offspring (nearly 3 years old) and one 1977 vixen (nearly 1 year old). Again, only Vixen I mated and in March 1978 gave birth to 3 cubs (2 ♂♂, 1 ♀).

Plate 1. During the few days around oestrous the male fox persistently follows the vixen, repeatedly following her, especially around the side of the head, as pictured here.

Relationship between vixens

Observations showed that Vixen I remained dominant to her daughters even when they were mature. For instance, in 1978 her original cubs were three years old, but still subordinate during play and encounters over food. Vixens within a family group are organised hierarchically in the wild (Macdonald, 1977a) and this was also true in captivity. Each year the three or four days prior to Vixen I's oestrus saw a marked increase in her aggression towards her subordinates, but this disappeared directly after oestrus finished.

Courtship (Plate 1)

Each year Dog I only courted Vixen I, the dominant and eldest member of his group. The pattern was always the same and involved continuous following of Vixen I and several copulations. As an example the observations made between 1–29 January 1976 are reported below. These observations were made largely by night through the remote-controlled infra-red equipment.

Figure 5 shows the average minimum distance between the dog fox and Vixen I each night during 30 × 10 minute periods. It can be seen that by

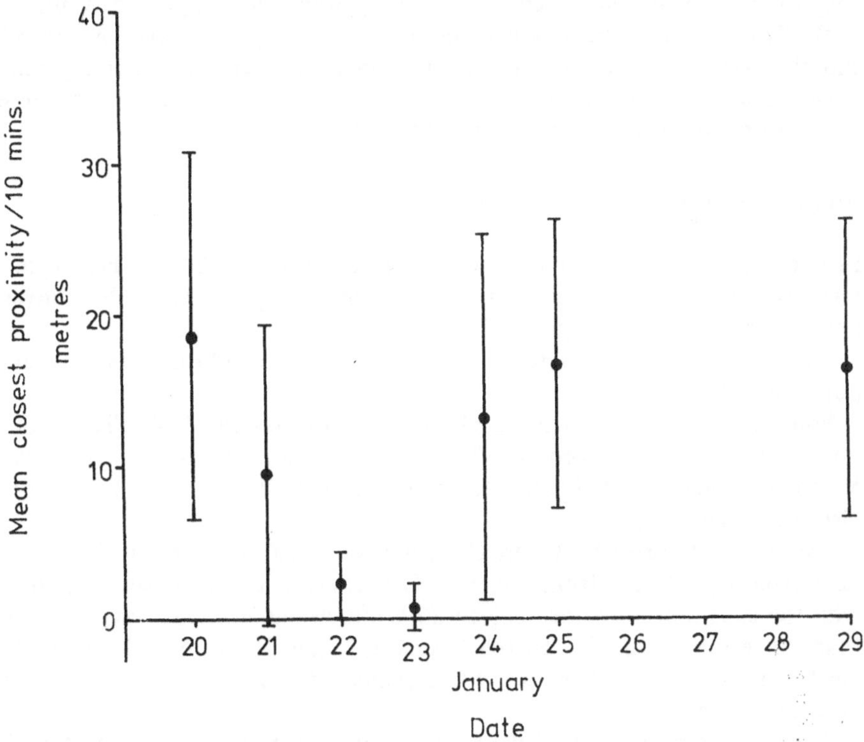

Fig. 5. Least distance (±S.D.) of dog-fox from vixen during 10 minute intervals for 3 hours each night.

23rd January he was always very close to her, while by the following night he had resumed his customary "indifference". It seems that mating occurred only during one night.

22/1/76: Dog I was atypically restive, continually moving about the enclosure. His brush was extended horizontally or just above the horizontal behind him in contrast to its typical drooping crescent (see also Tembrock, 1957). Vixen I slept at the north end of the enclosure and the dog-fox made regular trips to her (each ½–3 minutes). He approached to within 2–6 m of her but generally no closer, whereupon he would pause, stretch forward to sniff at her and then abruptly wheel round and trot off. When Vixen I moved he followed her. He frequently seemed about to mount the vixen, but retreated when she reacted aggressively.

23/1/76: Dog I followed Vixen I at never more than 3 m distance; she frequently rebuked him. Each time she paused he stood watching her and if she remained in the same place he carefully scraped a small depression in the ground, and curled up beside her. Vixen I ate, but the dog-fox did not, rather he sat the other side of the food watching her. By 0330 hours the dog-fox had attempted unsuccessfully to mount four times. Each attempt involved edging toward the vixen, craning forward and then gingerly placing a paw on her back – each advance was met with aggressive gaping. Again he carried his brush aloft. By dawn the foxes had mated.

24/1/76: Dog I slept uninterrupted until 2300 hours. Thereafter he paid no attention to Vixen I.

Hiroko (pers. comm.) watched a pair of foxes that mated on two consecutive days: 4 times on the first day and twice on the second. Three consecutive copulations lasted 45, 30 and 29 minutes.

While courting the dominant vixen the male interacted only rarely with the younger vixens and reacted aggressively when they approached him.

By 22nd January not only was the dog fox taking an interest in Vixen I but the younger vixens also frequently ran up to her, sniffing her genital region and even attempting to mount her. Vixen I responded to these pseudocopulations aggressively, (see Plate 2).

Response to cubs

In each of 5 years, Vixen I moved her sleeping earth the day before the birth of the cubs. Observations in the field also suggest that sudden moves before parturition are frequent. For the first week after birth the vixen spent at least 95% of her time actually lying with the cubs and on many days did not emerge at all.

Four cubs were born in the enclosure on 14th March 1976. Vixen I did not react at all aggressively to Vixen II's inspection of her newborn cubs. Vixen II seemed "excited" and ran round the enclosure playfully, repeatedly visiting the cubs.

For the next fortnight Vixen II visited the cubs every day without any detectable resentment from Vixen I. Often Vixen I and II were together with the cubs, but only once before 25th March did Vixen II stay for more than 30 seconds. On 25th March Vixens I, II and III were all in the main earth with the cubs and the latter two remained in the earth when Vixen I emerged to eat.

A similar pattern followed the birth of Vixen I's third litter on 9th March 1977. The cubs aroused considerable interest in the barren vixens, in particular Vixen II (still the most dominant subordinate vixen), but it was

(2a)

(2b)

Plate 2. Foxes were observed and filmed by infra-red light. Here a series of pictures taken from video-tape illustrate the breeding female's response to an attempted mounting by one of her daughters of the previous year. (2a) The younger vixen mounts her mother who gapes aggressively and (2b) twists around before rearing up to "box" with her daughter (2c). Both vixens then have their ears layed back and gape widely before the dominant mother pushes back her daughter (2d).

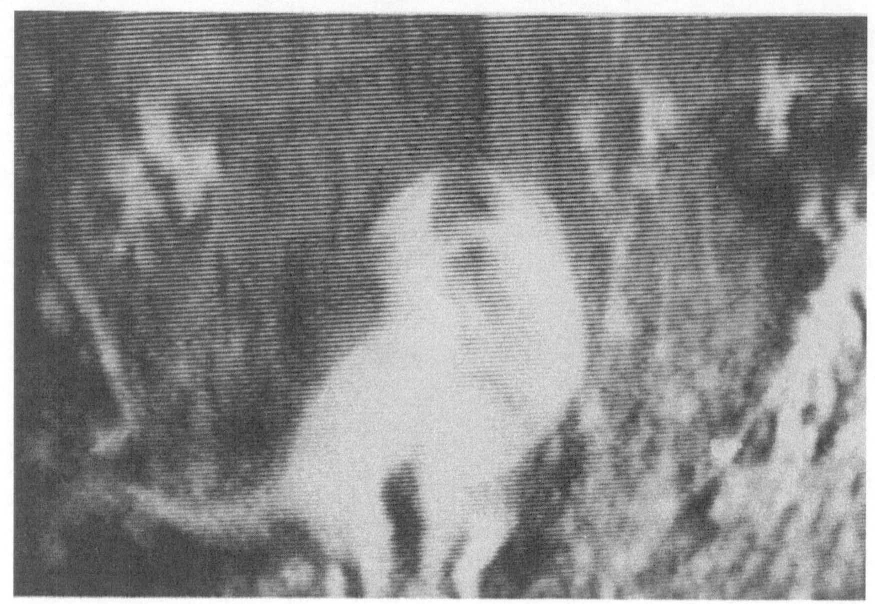

(2c)

(2d)

not until they were over 2 weeks old (25th March) that Vizen II slept with them for the first time, and did so every day from 31st March (age 22 days). In 1977 and 1978 Vixen II performed all maternal duties except suckling. Often, she guarded all the cubs alone, while Vixen I slept in an adjacent earth. The two vixens sometimes curled up together with the cubs, both grooming cubs and licking their anuses to remove faeces and urine. The subordinate vixen's "maternal" behaviour was diligent, e.g. on 29th April while Vixen I looked on Vixen II carried the same cub to an earth four times in quick succession and each time it ran out again to play.

Vixen I began taking food to the cubs when they were between 3 and 4 weeks old. After the birth of cubs she made many caches including storage of otherwise unfavoured foods (Macdonald, 1977b). Food was taken to the earth whereupon the vixen made the "warble" call, at which the cubs would run out and suckle or take food. Once, a cub got stuck in the undergrowth and must have separated for over an hour during which Vixen I had fed the others. When Vixen I found this cub, she managed to pull it free, returned it to the earth and immediately excavated a cache and took the contents to the errant cub.

Vixen I often nosed her cubs into small holes which an adult fox could not reach. From the end of April the cubs were increasingly mobile and often slept outside the earth, with other adults.

Allomaternal behaviour and cub survival

In early April Vixen I injured her front paw while fighting with a wild vixen through the wire. While Vixen I was incapacitated, her cubs were cared for in another earth by subordinates. It is hard to believe that they could have survived otherwise, but all did so.

ENCLOSURE GROUP II

Hierarchy amongst vixen cubs

Methods

Five vixen cubs (sisters) were hand reared from 4 days old in 1975. The form of interactions between them was recorded using a dictaphone. There has been considerable debate about the usefulness of dominance orders (e.g. Rowell, 1966; Jolly, 1966) although they have been defended by Richards (1977). So I have recorded two different measures in order to check their congruity.

(i) *Playful interactions*

(1) Amicable, where a fox approaches another, and they sniff or lick each other with slight ear and tail movements.

(2) Chase or pounce, the initiating fox chases its sib or stalks and pounces upon it.

(3) Attack, play that involves serious barging, rolling over and play fighting, together with playful encounters that developed into an aggressive bullying.

(4) "Lost" includes all interactions in which the fox that initiated them did not follow through the chase or attack but where roles were reversed. For all interactions other than "lost" it can be assumed that the fox initiating the encounter was the "victor".

(ii) *Access to prey*

Access to a food item (dead chicken or rabbit) involves a limited resource (see Reynolds and Luscombe, 1969). The foxes were presented with the prey and their order of access to it noted.

Results

(i) *Play*

During the 10 bouts of play, of more than 30 minutes duration each, recorded during May 1975 (cubs aged 8 weeks) 225 interactions were detailed; their direction and form are scored on Table 2 and Fig. 6. Of these Big Ears ♀ initiated 116 (51.6%) while Whitepaws ♀ initiated only 21 (8.2%). The most timid cub, Wide Eyes ♀, was incessantly attacked by her sibs and often had to be segregated. Ranking the foxes in order of play initiated (of any type) gives:

Big Ears – Pseudo Sickly – Sickly – Whitepaws – Wide Eyes

The same rank order is reached if a measure of the smallest percentage of play fights initiated that are lost is used. The measure of percentage of attacks received, which gives the inverse of this order, might be confounded by two opposing reactions – a fox with a low score might be so indomitable that it was seldom attacked but alternatively it might be so retiring that it was seldom available for attack.

In all these measures Big Ears ranks top and Wide Eyes bottom. Whitepaws was always second to bottom. Whitepaws, unlike Wide Eyes, did not give the impression of being subordinate but simply of not being playful.

(ii) *Access to food*

When I gave the cubs their food a furore ensued, in which the prey often changed hands several times before any individual had a chance to eat. Thereafter the fox in possession would be attacked and occasionally supplanted. Sometimes the displaced fox immediately retaliated before the newcomer had any chance to eat. Generally a fox that had successfully driven another from the food ate until it was sated, whereupon it would give way easily when another approached. The takeover might be followed by another furore before a second strong possessor took over. Scoring first

Table 2. Summary of 225 interactions during play between 4 vixen cubs aged about 2 months.

	Recipient																					
	Big Ears				Whitepaws				Pseudo Sickly				Sickly				Total play initiations		No. play initiations of each type			
	Amicable	Chase	Fight	Lost	Amicable	Chase	Fight	Lost	Amicable	Chase	Fight	Lost	Amicable	Chase	Fight	Lost	nos	%	Amicable	Chase	Fight	Lost
Big Ears					0	6	17	0	7	31	23	1	2	13	16	0	116	51.6	9	50	56	1
Whitepaws	0	1	0	2					0	0	2	3	2	2	4	5	21	8.2	2	3	6	10
Pseudo Sickly	6	13	3	5	0	0	14	0					0	4	12	0	57	25.3	6	17	29	5
Sickly	0	3	0	2	1	9	7	0	1	4	3	4					31	13.8	2	13	10	6
Total interactions received	6	17	3	9	1	15	38	0	8	35	28	8	32	19	32	5	225	Total	19	83	101	22
	35				54				75				60									

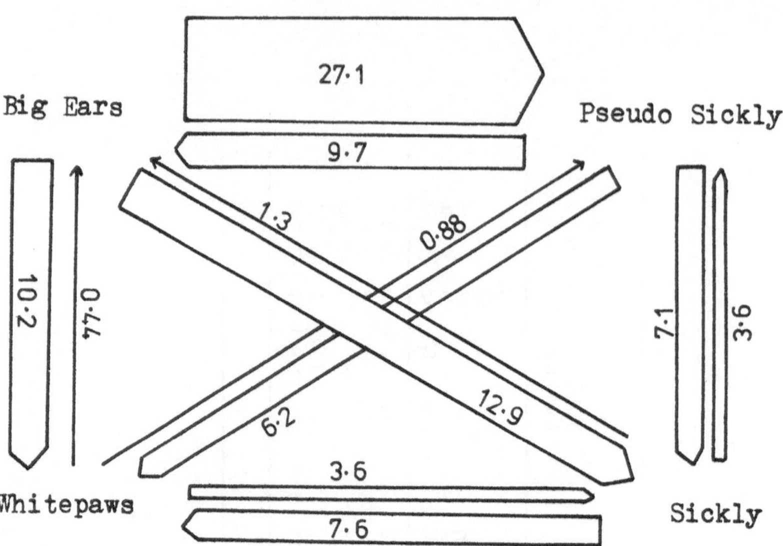

Fig. 6. Summary of direction of initiated play behaviours (including initiation rebuffed). As % of all initiations during 225 interactions.

possession of food during nine trials between 30th May and 27th July 1975 the order of foxes was:

> Whitepaws (55.6%) – Pseudo Sickly (33.3%) – Big Ears (11.1%) – Sickly and Wide Eyes (0%)

However, this lumps together foxes that had first access for only a few seconds during the initial melee, and those that actually managed to eat. If the foxes are ranked on the basis of which ate first the order changes to:

> Whitepaws (55.5%) – Big Ears (44.5%) – Pseudo Sickly, Sickly and Wide Eyes (0%)

During observations Big Ears was powerful, but tolerant; she often let other foxes take the food and then displaced them with little ado, while in contrast Whitepaws always fought furiously and repeatedly attacked the possessor. During 9 trials each fox had the potential of being ahead of four others per trial, a maximum score of 36. The percentage of the maximum score achieved gives the following rank:

> Big Ears (80.6%) – Whitepaws (77.8%) – Sickly (50%) – Psuedo Sickly (38.9%) – Wide Eyes (0%) (see Table 3).

Similar data were collected during 5 observation periods during August and September (when the cubs were 17 weeks old), and give a slightly different rank (Table 3):

> White Paws (80%) – Big Ears (65%) – Pseudo Sickly (50%) – Sickly (20%) – Wide Eyes (5%)

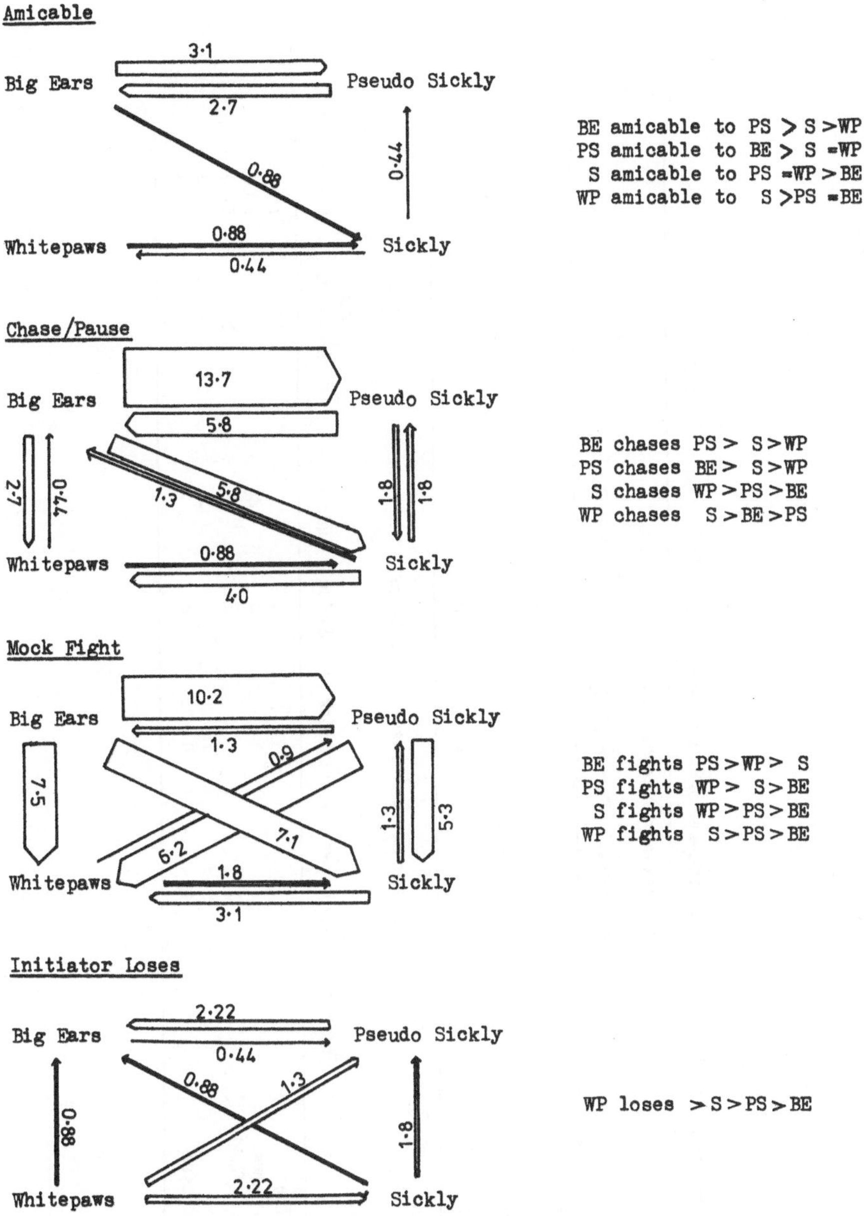

Amicable

BE amicable to PS > S > WP
PS amicable to BE > S = WP
S amicable to PS = WP > BE
WP amicable to S > PS = BE

Chase/Pause

BE chases PS > S > WP
PS chases BE > S > WP
S chases WP > PS > BE
WP chases S > BE > PS

Mock Fight

BE fights PS > WP > S
PS fights WP > S > BE
S fights WP > PS > BE
WP fights S > PS > BE

Initiator Loses

WP loses > S > PS > BE

Fig. 7. Percentage of 225 interactions during play between 4 vixens that took a given form and direction.

Table 3. Number of times when each cub gained meaningful possession of food before each of the other cubs.

	Late May–Mid June (age: 10–12 weeks) 9 trials						
Cubs with meaningful possession	Big Ears	Whitepaws	Pseudo Sickly	Sickly	Wide Eyes	Total max = 36	% max
Big Ears		4	8	8	9	29	80.6
Whitepaws	5		7	7	9	28	77.8
Pseudo Sickly	1	2		2	9	14	38.9
Sickly	1	2	6		9	18	50.0
Wide Eyes	0	0	0	0		0	0.0

	August–Mid September (age: 19–23 weeks) 5 trials						
Cubs with meaningful possession	Big Ears	Whitepaws	Pseudo Sickly	Sickly	Wide Eyes	Total max = 20	% max
Big Ears		1	3	4	5	13	65.0
Whitepaws	4		4	4	4	16	80.0
Pseudo Sickly	2	1		3	4	10	50.0
sickly	0	0	1		3	4	20.0
Wide Eyes	0	1	0	0		1	5.0

The three sisters whose social relationships seemed most polarised, and most different in character, became the nexus of enclosure Group II. Up until the winter of 1976 the same hierarchy prevailed: Whitepaws, Big Ears and Wide Eyes. These vixens were nearly 2 years old when I first introduced a year old male, Smudge, to their enclosure. This was to avoid the possibility that failure of any of them to breed could be attributed to immaturity.

ENCLOSURE GROUP II

Reproductive behaviour

The relationship between members of enclosure Group II have been detailed during over 600 hours observation between January 1977 and June 1978 and observed casually for an estimated further 400 hours. The observations have been distilled into summary from by categorising the initiator of each interaction as "amicable", "mildly aggressive" or "seriously aggressive". Similarly, the respondent might react amicably (type 0), aggressively (type I) or with mild (type II) or intense (type III) submission (see Plate 3a and b).

(3a)

Plate 3. Interactions between foxes were largely amicable. The male fox was normally dominant to the vixens who might submit to him slightly (3a) or intensively (3b). In Plate 3a the male fox (left) has upright ears during a greeting with a slightly submissive vixen (right) whose ears are layed back and who adopts a lowered posture. A third and more submissive vixen can just be seen between them in the background, with her ears more strongly layed back. In 3b a vixen has rolled onto her back with her tail lashing at the approach of the male fox (standing).

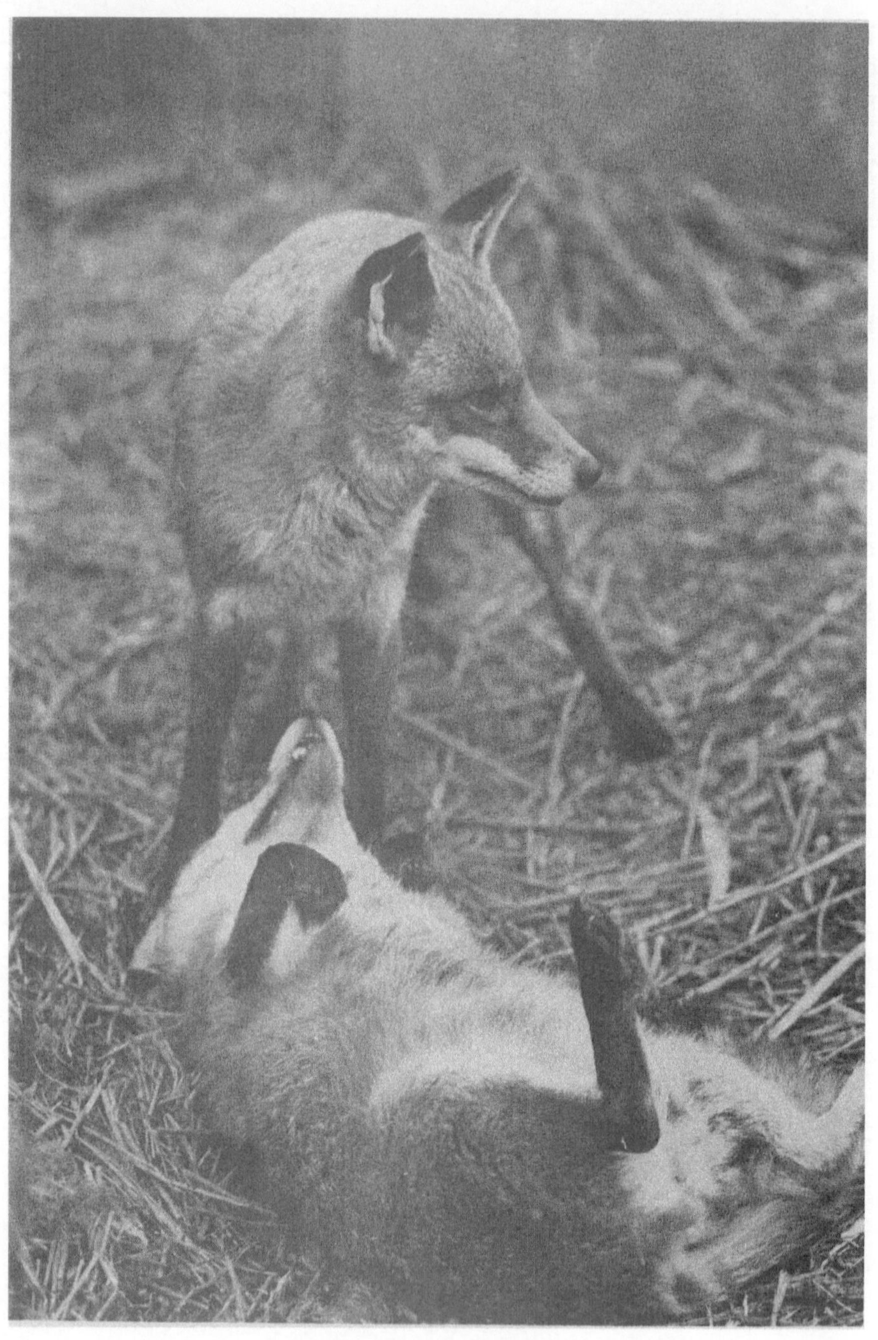

(3b)

RESULTS

Breeding season 1977

The observations have been split into five periods: courtship (6th January–25th January), gestation (26th January–18th March), suckling (19th March–25th April, weaning (22nd April–8th May) and increasing independence (9th May–29th May).

The summarised relationships for each period are presented on Tables 4–8, and Figs. 8–11. These data give broad insights into the social dynamics of this group of foxes, but it is necessary to stress that such summaries inevitably present a grotesquely impoverished impression of the subtleties of the behaviour involved. Even the delineation between "dominance" or "submission" distorts the complexity of the personal relationships involved. Sociograms as simple as those presented in Figs. 8–11 have some more general shortcomings: one animal may only rarely be the subject of serious attack by another "dominant" animal because, fearing the dominant, it ensures that their paths never cross, or because the relationship does not require reinforcement, or legion alternative explanations. Here, these summaries are complemented by descriptions of the animal's behaviour. On

Table 4. Summary of interactions between inmates of enclosure II, during courtship.

6 January–25 January

Initiator : Recipient (down) (across); Recipient's response

Amicable

	SM	WP	BE	WE		Smudge 0	I	II	III	IV	Whitepaws 0	I	II	III	IV	Big Ears 0	I	II	III	IV	Wide Eyes 0	I	II	III	IV
Smudge		20	15	9	Smudge						5	7	6	—	—	—	—	6	9	—	—	—	5	4	—
Whitepaws	2		4	5	Whitepaws	2	—	—	—	—						—	—	1	3	—	1	—	4	—	—
Big Ears	3	2		2	Big Ears	3	—	—	—	—	2	5	—	—	—						—	1	1	—	—
Wide Eyes	—	2	—		Wide Eyes	—	—	—	—	—	—	—	2	—	—	—	—	—	—	—					

Mild aggression

	SM	WP	BE	WE		Smudge 0	I	II	III	IV	Whitepaws 0	I	II	III	IV	Big Ears 0	I	II	III	IV	Wide Eyes 0	I	II	III	IV
Smudge		11	3	2	Smudge						—	4	6	1	—	—	—	—	3	—	—	—	—	2	—
White paws	—		6	9	Whitepaws	—	—	—	—	—						—	1	5	—	—	—	1	5	3	—
Big Ears	—	2		5	Big Ears	—	—	—	—	—	—	1	1	—	—						—	—	3	2	—
Wide Eyes	1	1	—		Wide Eyes	—	1	1	—	—	—	—	4	—	—	—	—	—	—	—					

Serious aggression

						0	I	II	III	IV	0	I	II	III	IV	0	I	II	III	IV	0	I	88	III	IV
Smudge		3	—	1	Smudge						—	—	—	1	3	—	—	—	—	—	—	—	—	1	—
Whitepaws	5		—	12	Whitepaws	—	—	5	—	—						—	—	—	—	—	—	—	—	12	1
Big Ears	—	—		2	Big Ears	—	—	—	—	—	—	—	—	—	—						—	1	—	2	—
Wide Eyes	—	—	—		Wide Eyes	—	—	—	—	—	—	—	—	—	—	—	—	—	—	—					

Summary of play and grooming initiations

Play	SP	WP	BE	WE	Grooming	Smudge	Whitepaws	Big Ears	Wide Eyes
Smudge		—	1	—	Smudge		14	1	2
Whitepaws	—		—	—	Whitepaws	1		—	1
Big Ears	—	—		—	Big Ears	—	—		—
Wide Eyes	—	—	—		Wide Eyes	1	—	1	

Table 5. Summary of interactions between inmates of enclosure II, during pregnancy.

26 January–18 March

Initiator : Recipient (down) (across) — Recipient's response

Amicable

	SM	WP	BE	WE		Smudge					Whitepaws					Big Ears					Wide Eyes				
						0	I	II	III	IV	0	I	II	III	IV	0	I	II	III	IV	0	I	II	III	IV
Smudge		32	37	29	Smudge						5	—	4	26	—	3	2	12	20	—	1	1	2	25	—
Whitepaws	—		29	16	Whitepaws	—	—	—	—	—						17	1	9	2	—	6	—	8	2	—
Big Ears	2	14		15	Big Ears	1	0	—	—	—	10	—	4	—	—						7	—	6	2	—
Wide Eyes	2	12	10		Wide Eyes	2	0	—	—	—	7	3	1	—	—	6	2	2	—	—					

Mild aggression

	SM	WP	BE	WE		Smudge					Whitepaws					Big Ears					Wide Eyes				
						0	I	II	III	IV	0	I	II	III	IV	0	I	II	III	IV	0	I	II	III	IV
Smudge		8	8	6	Smudge						—	—	—	8	—	—	1	1	6	—	0	1	—	5	—
Whitepaws	1		5	14	Whitepaws	0	1	—	—	—						0	2	2	1	—	0	1	4	9	—
Big Ears	—	—		5	Big Ears	—	—	—	—	—	—	—	—	—	—						—	—	2	3	—
Wide Eyes	—	1	2		Wide Eyes	—	—	—	—	—	0	1	—	—	—	0	1	1	—	—					

Serious aggression

	SM	WP	BE	WE		Smudge					Whitepaws					Big Ears					Wide Eyes				
						0	I	II	III	IV	0	I	II	III	IV	0	I	II	III	IV	0	I	II	III	IV
Smudge		1	3	4	Smudge						—	—	—	1	—	—	—	—	3	—	—	—	—	4	—
Whitepaws	—		1	12	Whitepaws	—	—	—	—	—						—	—	—	1	—	—	—	—	13	1
Big Ears	—	—		1	Big Ears	—	—	—	—	—	—	—	—	—	—						—	—	—	1	—
Wide Eyes	—	—	—		Wide Eyes	—	—	—	—	—	—	—	—	—	—	—	—	—	—	—					

Summary of play and grooming initiations

Play	SM	WP	BE	WE	Grooming	Smudge	Whitepaws	Big Ears	Wide Eyes
Smudge		10	12	16	Smudge		1	3	—
Whitepaws	—		2	—	Whitepaws	—		15	11
Big Ears	2	3		—	Big Ears	—	6		7
Wide Eyes	1	—	—		Wide Eyes	1	7	4	

Table 6. Summary of interactions between inmates of enclosure II, during suckling.

19 March–21 April

Initiator : Recipient (down) (across) — Recipient's response

Amicable

	SM	WP	BE	WE		Smudge					Whitepaws					Big Ears					Wide Eyes				
						0	I	II	III	IV	0	I	II	III	IV	0	I	II	III	IV	0	I	II	III	IV
Smudge		37	44	24	Smudge						1	2	11	23	—	8	—	18	18	—	4	0	4	16	—
Whitepaws	—		10	3	Whitepaws	—	—	—	—	—						7	—	3	—	—	—	—	—	3	—
Big Ears	11	16		29	Big Ears	10	1	—	—	—	9	1	4	2	—						19	—	7	3	—
Wide Eyes	—	—	16		Wide Eyes	—	—	—	—	—	—	—	—	—	—	14	—	2	—	—					

Mild aggression

	SM	WP	BE	WE		Smudge					Whitepaws					Big Ears					Wide Eyes				
						0	I	II	III	IV	0	I	II	III	IV	0	I	II	III	IV					
Smudge		—	2	7	Smudge						—	—	—	—	—	—	—	1	1	—	—	—	3	4	—
Whitepaws	—		3	8	Whitepaws	—	—	—	—	—						—	—	1	2	—	—	—	3	6	—
Big Ears	1	—		7	Big Ears	1	—	—	—	—	—	—	—	—	—						—	—	5	2	—
Wide Eyes	—	—	—		Wide Eyes	—	—	—	—	—	—	—	—	—	—	—	—	—	—	—					

Table 6 (continued)

Serious Aggression

	SM	WP	BE	WE		Smudge 0	I	II	III	IV	Whitepaws 0	I	II	III	IV	Big Ears 0	I	II	III	IV	Wide Eyes 0	I	II	III	IV
Smudge		—	1	—	Smudge						—	—	—	—	—	—	—	—	1	—	—	—	—	—	—
Whitepaws	—		—	17	Whitepaws	—	—	—	—	—						—	—	—	—	—	—	—	—	17	3
Big Ears	—	—		1	Big Ears	—	—	—	—	—	—	—	—	—	—						—	—	—	1	
Wide Eyes	—	—	—		Wide Eyes	—	—	—	—	—	—	—	—	—	—	—	—	—	—	—					

Summary of play and grooming initiations

Play	SM	WP	BE	WE	Grooming	Smudge	Whitepaws	Big Ears	Wide Eyes
Smudge		2	24	4	Smudge		3	1	—
Whitepaws	—		—	—	Whitepaws	—		4	—
Big Ears	11	—		4	Big Ears	1	—		11
Wide Eyes	—	—	—		Wide Eyes	—	—	9	

Table 7. Summary of interactions between inmates of enclosure II, during weaning.

22 April–8 May

Initiator : Recipient
(down) (across)

Recipient's response

Amicable

	SM	WP	BE	WE		Smudge 0	I	II	III	IV	Whitepaws 0	I	II	III	IV	Big Ears 0	I	II	III	IV	Wide Eyes 0	I	II	III	IV
Smudge		7	7	10	Smudge						—	—	3	4	—	—	1	3	3	—	2	—	4	4	—
Whitepaws	—		6	9	Whitepaws	—	—	—	—	—						5	—	1	—	—	2	—	5	2	—
Big Ears	—	10		7	Big Ears	—	—	—	—	—	7	—	2	1	—						3	—	1	3	—
Wide Eyes	15	—	3		Wide Eyes	5	—	—	—	—	—	—	—	—	—	4	—	—	1	—					

Mild aggression

	SM	WP	BE	WE		Smudge 0	I	II	III	IV	Whitepaws 0	I	II	III	IV	Big Ears 0	I	II	III	IV	Wide Eyes 0	I	II	III	IV
Smudge		—	1	2	Smudge						—	—	—	—	—	—	—	—	—	—	—	—	1	1	—
Whitepaws	—		—	12	Whitepaws	—	—	—	—	—						—	—	1	—	—	—	—	3	8	—
Big Ears	12	—		1	Big Ears	3	2	8	—	—	—	—	—	—	—						—	—	1	—	—
Wide Eyes	—	—	—		Wide Eyes	—	—	—	—	—	—	—	—	—	—	—	—	—	—	—					

Serious aggression

	SM	WP	BE	WE		Smudge 0	I	II	III	IV	Whitepaws 0	I	II	III	IV	Big Ears 0	I	II	III	IV	Wide Eyes 0	I	II	III	IV
Smudge		—	—	—	Smudge						—	—	—	—	—	—	—	—	—	—	—	—	—	—	—
Whitepaws	—		—	5	Whitepaws	—	—	—	—	—						—	—	—	—	—	—	—	—	5	—
Big Ears	3	—		1	Big Ears	—	—	2	1	—	—	—	—	—	—						—	—	—	1	—
Wide Eyes	—	—	—		Wide Eyes	—	—	—	—	—	—	—	—	—	—	—	—	—	—	—					

Summary of play and grooming initiations

Play	SM	WP	BE	WE	Grooming	Smudge	Whitepaws	Big Ears	Wide Eyes
Smudge		—	—	1	Smudge		—	—	2
Whitepaws	—		—	—	Whitepaws	—		1	—
Big Ears	—	—		1	Big Ears	—	2		3
Wide Eyes	—	—	—		Wide Eyes	5	—	—	

Table 8. Summary of interactions between inmates of enclosure II, with cub independence.

10 May–29 May

Initiator: Recipient (down) (across) — Recipient's response

Amicable

	SM	WP	BE	WE		Smudge 0	I	II	III	IV	Whitepaws 0	I	II	III	IV	Big Ears 0	I	II	III	IV	Wide Eyes 0	I	II	III	IV
Smudge		8	22	6	Smudge						—	—	4	4	—	3	—	9	10	—	—	—	1	5	—
Whitepaws	1		13	5	Whitepaws	—	—	—	—	—						10	1	2	1	—	1	—	4	—	—
Big Ears	2	19		24	Big Ears	2	—	—	—	—	13	—	5	1	—						11	—	12	1	—
Wide Eyes	—	3	4		Wide Eyes	—	—	—	—	—	1	—	1	1	—	2	1	1	—	—					

Mild aggression

	SM	WP	BE	WE		Smudge 0	I	II	III	IV	Whitepaws 0	I	II	III	IV	Big Ears 0	I	II	III	IV	Wide Eyes 0	I	II	III	IV
Smudge		1	5	2	Smudge						—	—	—	1	—	—	—	—	5	—	—	—	—	2	—
Whitepaws	—		1	6	Whitepaws	—	—	—	—	—						—	—	1	—	—	—	1	3	1	—
Big Ears	6	—		3	Big Ears	2	—	4	—	—	—	—	—	—	—						—	—	3	1	—
Wide Eyes	7	—	1		Wide Eyes	1	1	5	—	—	—	—	—	—	—	—	—	1	—	—					

Serious aggression

	SM	WP	BE	WE		Smudge 0	I	II	III	IV	Whitepaws 0	I	II	III	IV	Big Ears 0	I	II	III	IV	Wide Eyes 0	I	II	III	IV
Smudge		—	—	—	Smudge						—	—	—	—	—	—	—	—	—	—	—	—	—	—	—
Whitepaws	—		—	2	Whitepaws	—	—	—	—	—						—	—	—	—	—	—	—	—	2	—
Big Ears	5	—		3	Big Ears	—	2	2	1	—	—	—	—	—	—						—	—	1	2	1
Wide Eyes	14	1	—		Wide Eyes	—	2	9	3	—	—	—	1	—	—	—	—	—	—	—					

Summary of play and grooming initiations

	Play SM	WP	BE	WE		Grooming Smudge	Whitepaws	Big Ears	Wide Eyes
Smudge		2	2	—	Smudge		—	—	—
Whitepaws	—		—	—	Whitepaws	—		4	—
Big Ears	—	—		3	Big Ears	—	13		3
Wide Eyes	—	—	—		Wide Eyes	—	—	—	

Figs. 8–11 there were apparently few amicable approaches to the dog-fox by any of the vixens; this was not because he was shunned by them. In fact, everywhere the dog-fox walked, the nearest vixen would rush to greet him submissively. He was commonly amicable or playful with them, and only became aggressive when a vixen persistently rolled in front of him. Rather arbitrarily, I have recorded this type of interaction, as an amicable approach by the dog towards the vixen, and it is amongst the commonest form of interaction observed. The vixens were so excited by the dog's approach that even when he ignored them, calm "amicable" approaches by them towards him were rare.

My impression of the relationships between these foxes may help the reader: Smudge ♂ was undoubtedly dominant to all the vixens for most of the study period. Similarly, Wide Eyes ♀ was subordinate to both Whitepaws ♀ and Big Ears ♀, as she had been since she was a month old. The relative position of Whitepaws ♀ and Big Ears ♀ was more equivocal during everyday interactions, but Whitepaws ♀ continued to be dominant over food.

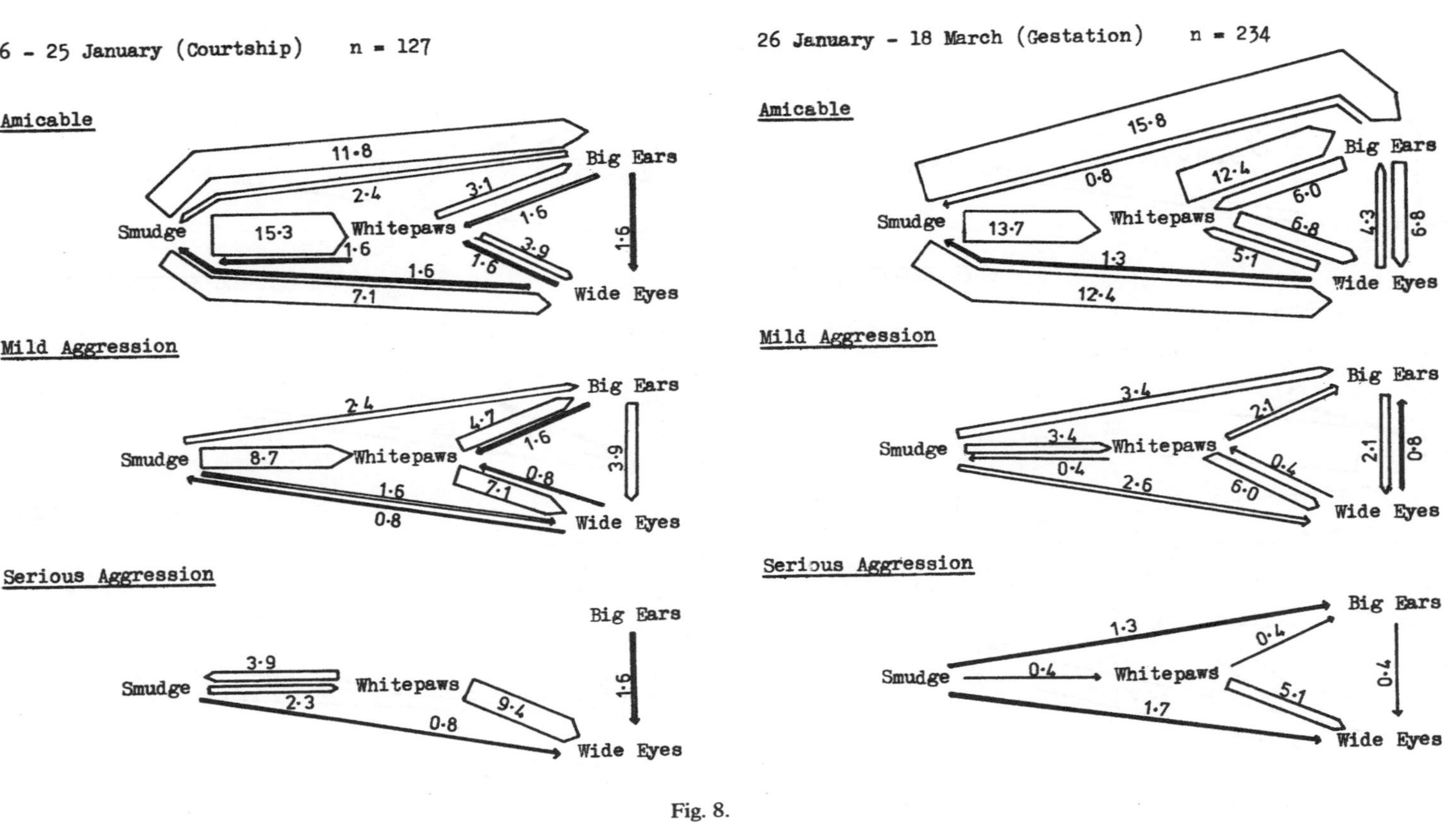

6 - 25 January (Courtship)   n = 127
Amicable
Mild Aggression
Serious Aggression
26 January - 18 March (Gestation)   n = 234
Amicable
Mild Aggression
Serious Aggression
Big Ears
Wide Eyes
Whitepaws
Smudge

Fig. 8.

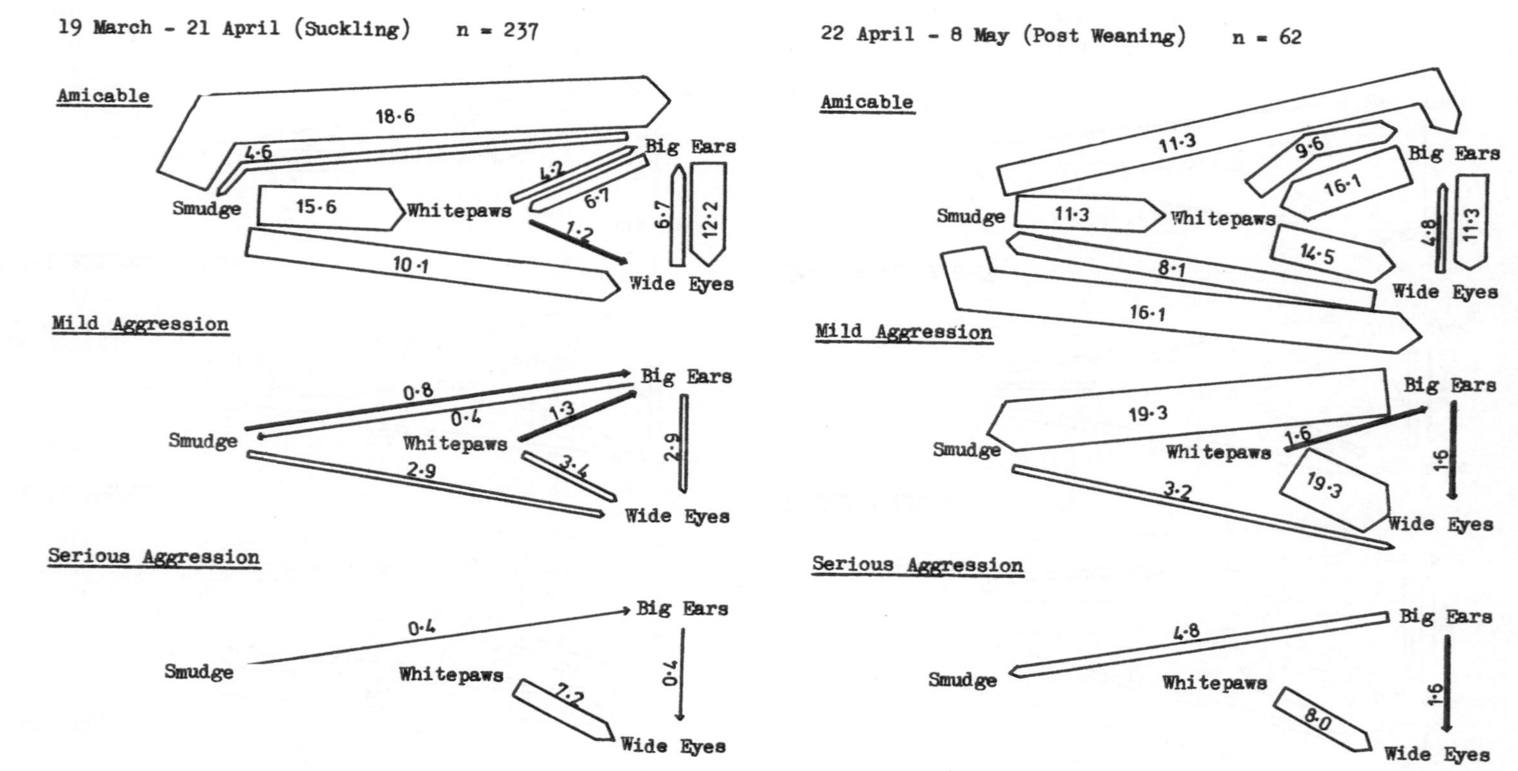
19 March - 21 April (Suckling) n = 237
Amicable
18·6
4·6
Big Ears
4·2
6·7
Smudge
15·6
Whitepaws
1·2
6·7
12·2
10·1
Wide Eyes
Mild Aggression
0·8
Big Ears
0·4
1·3
Smudge
Whitepaws
2·9
2·9
3·4
Wide Eyes
Serious Aggression
Big Ears
0·4
Smudge
Whitepaws
0·4
7·2
Wide Eyes
22 April - 8 May (Post Weaning) n = 62
Amicable
11·3
9·6
Big Ears
16·1
Smudge
11·3
Whitepaws
4·8
11·3
14·5
8·1
Wide Eyes
16·1
Mild Aggression
Big Ears
19·3
Smudge
Whitepaws
1·6
1·6
19·3
3·2
Wide Eyes
Serious Aggression
Big Ears
4·8
Smudge
Whitepaws
1·6
8·0
Wide Eyes

Fig. 9.

10 - 29 May (increasing independence) n = 164

Amicable

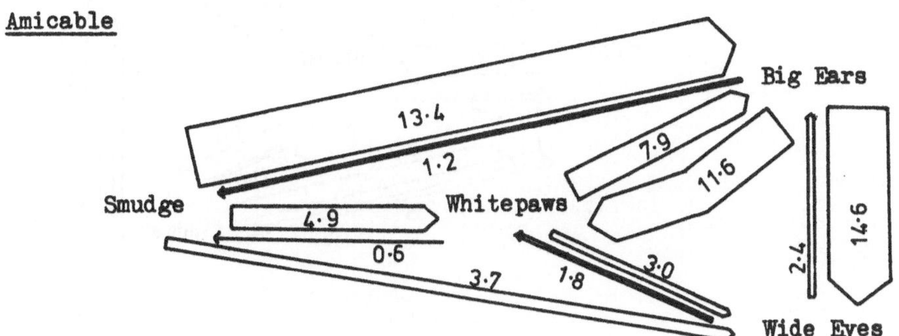

Mild Aggression

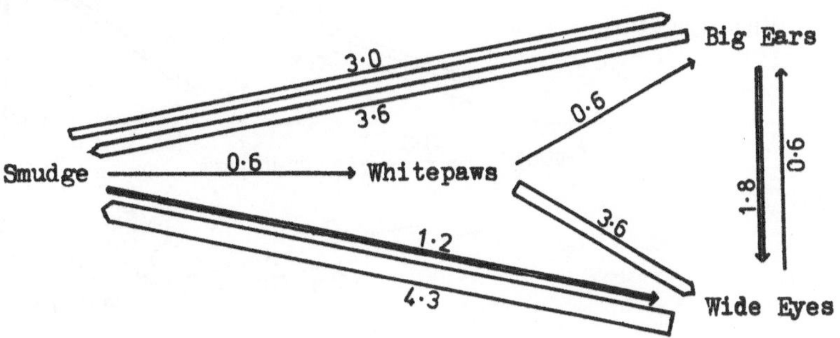

Serious Aggression

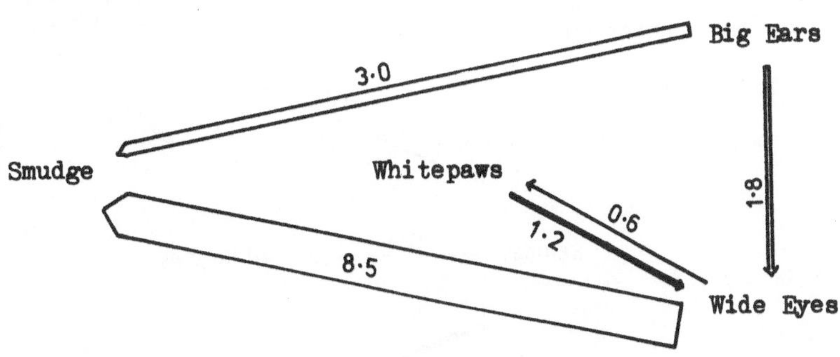

Fig. 10.

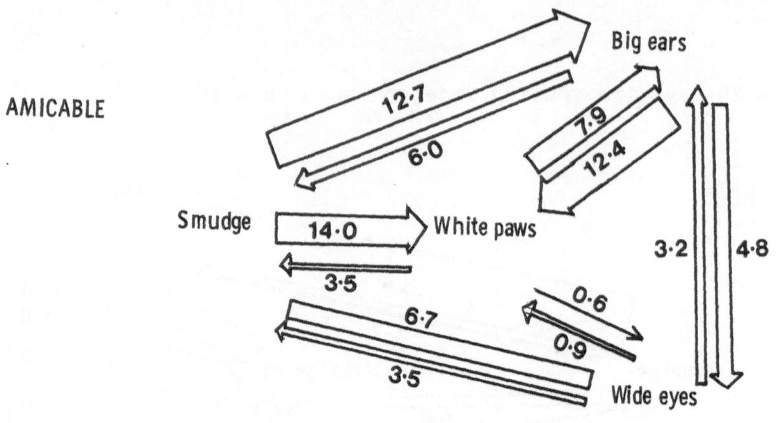

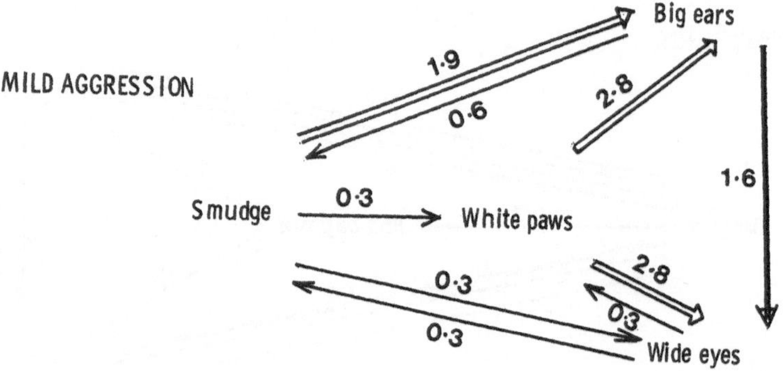

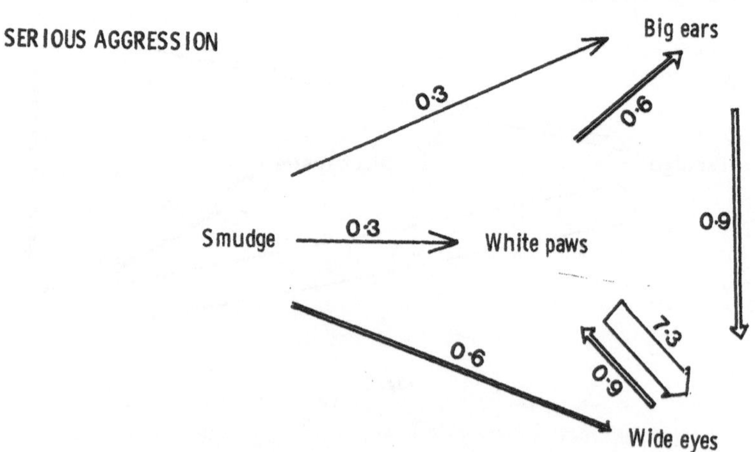

Fig. 11. 1st June–31st August, 1978, $n = 315$.

Courtship – 6th–25th January

The main feature of this period was Whitepaws' assertiveness towards Smudge ♂. Both Big Ears ♀ and Wide Eyes ♀ submitted lavishly to Smudge ♂ and he was normally amicable towards them, although savagely asserting his dominance over food when they tried (rarely) to pilfer from him. Whitepaws ♀ was attractive to Smudge ♂ and he frequently trotted up the enclosure to visit her and was rebuffed with at least mild aggression. This contrasts dramatically with Big Ears' and Wide Eyes' response:

E.g. on 15th January, Smudge ♂ trotted across to where Big Ears ♀ and Wide Eyes ♀ slept together and amicably sniffed at them both. Both vixens immediately rolled on their backs, lashing tails, ears flat back, gaping submissively and panting and squealing. Smudge ♂ viewed them without changing his posture as they squirmed towards him, until Wide Eyes ♀ was nibbling at his teeth and sniffing his maxillary skin and Big Ears ♀ sniffing his scrotum. Whitepaws ♀ was attracted to the noise and rushed up. Smudge ♂ looked towards them and all three vixens dropped in submission, whereupon he adopted a play posture towards Whitepaws ♀ who gaped at him aggressively and lunged forward, causing him to retreat.

After 2–3 days of atypically aggressive responses from Whitepaws ♀, Smudge ♂ became cautious of her: he would stare fixedly at Whitepaws ♀, drop his forelegs and chest to the ground with his haunches in the air as if soliciting play from her. Whitepaws ♀ responded by aggressive "snirking" and occasionally by attacking. On other occasions, Whitepaws ♀ groomed his lips and perioral region. By 23rd January Smudge ♂ began following Whitepaws ♀ (as the male of enclosure Group I followed Vixen I), grooming her whenever she permitted. Whitepaws ♀ had been token urinating within the enclosure during the previous weeks and on the 23rd on four occasions she appeared to walk deliberately in front of Smudge ♂ before doing so (see Macdonald, 1979b). Every time Whitepaws ♀ retreated into an artificial earth Smudge ♂ would jump on top of it and wait, sometimes for hours, while she slept below him.

Whitepaws ♀ reacted aggressively to Wide Eyes' presence during this period and on 23rd January attacked her three times during 30 minutes without provocation. Strangely, during one unprovoked attack on Wide Eyes ♀ on 24th January, Whitepaws ♀ ran towards the cowering Wide Eyes ♀ but came to an abrupt halt just before her and then dropped to the ground in apparent submission; Wide Eyes ♀ ran off. On the 25th Smudge ♂ mated with Whitepaws ♀. During that day both Big Ears ♀ and Wide Eyes ♀ repeatedly rushed to Smudge ♂ and rolled submissively in front of him, but he stepped over them and continued to prowl along behind Whitepaws ♀.

Gestation – 26th January–19th March

During Whitepaws' gestation relations between the foxes were predominantly amicable (Table 5 and Fig. 9). Whitepaws ♀ was no longer aggressive towards Smudge ♂. He took no further interest, other than a playful one, in any of the vixens and no further matings occurred.

The adult foxes played amongst themselves, although as always (see Fig. 8) Big Ears ♀ was the most playful. Table 5 shows that mutual grooming was not uncommon.

Wide Eyes ♀, on 28th February, approached Whitepaws ♀ and submissively crept forwards to groom her. Whitepaws ♀ stood, eyes closed and ears slightly held back while Wide Eyes ♀ groomed her head and neck. A few moments later Big Ears ♀ began to groom her; Wide Eyes ♀ also closed her eyes and quivered as Big Ears ♀ groomed. Thereafter all three vixens stood with their noses together, sniffing amicably.

On 15th February Big Ears ♀ sat for nearly 10 minutes pushing her head under Whitepaws' nose in an attempt to solicit grooming, which Whitepaws ♀ sporadically responded to. Big Ears ♀ kept her eyes shut throughout all grooming, and pushed toward Whitepaws ♀ whenever she paused.

It is noteworthy (for later comparison) that Whitepaws ♀ and Wide Eyes ♀ were often amicable towards each other during this period. For instance, on 3rd February they spent over 5 minutes in perioral sniffing and nibbling at each other's teeth. Similarly, all three vixens frequently slept together during the daytime, and they all submitted to Smudge ♂.

By 20th February much of Smudge's playfulness had returned, having been eclipsed as he doggedly trailed Whitepaws ♀. He would move playfully towards a vixen and, when she rolled submissively on the ground, he would pounce on her and drag the prostrate body along the ground by the neck before jumping back as if "expecting" a playful response. This behaviour was particularly directed towards Wide Eyes ♀ (see Plate 5).

The only real discordance within the group lay in Whitepaws' sporadic attacks on Wide Eyes ♀, which were infrequent but severe (see Fig. 8).

The mood between these vixens was often very relaxed; e.g. on the afternoon of 27th February there was a lot of grooming between the vixens; Whitepaws ♀ lay in the sun beside Big Ears ♀ and meticulously groomed her

Plate 4. The subordinate vixen (Wide Eyes) rolled submissively on the ground at the male fox's (Smudge) attempts to solicit play. He responded by dragging her along the ground, apparently trying to make her stand up.

head and ears for over 10 minutes. Whitepaws ♀ then stretched so that her paws lay either side of Big Ears' neck and continued to groom until both vixens fell asleep with heads on each other's flanks. After Big Ears ♀ woke and moved off Whitepaws ♀ sought out Wide Eyes' sleeping place and woke her by pawing at her head and then proceeded to thoroughly groom her head and ears also. At one stage Whitepaws ♀ moved about 1 m away and token marked and then returned to Wide Eyes ♀ (who submitted slightly) and continued to groom her. Later the same afternoon the four foxes contested over a rabbit and after much aggression they ate in the order Smudge ♂, Whitepaws ♀, Big Ears ♀ and Wide Eyes ♀.

Suckling – 19th March–21st April

Two aspects of the group's behaviour changed after the birth of Whitepaws' cubs. First, Smudge ♂ began carrying food to Whitepaws in the earth or else caching it nearby, and secondly, Whitepaws ♀ launched a ferocious vendetta against Wide Eyes ♀ which lasted for 2–3 weeks from the birth of the cubs (see Fig. 9, Table 6); Whitepaws ♀ remained amicable to Big Ears ♀.

Smudge's food provisioning involved gathering up as much food as he could, normally before he ate anything himself, and carrying it to Whitepaws ♀. He would pause at the entrance to her earth and "warble" a soft staccato call (otherwise similar to the wowwow call). Then he would poke the food into the mouth of the den, pushing it in with his nose. He would continue to take food vastly in excess of the vixen's possible requirements and the earth soon accumulated a large mound of prey. This resulted in a food shortage for the other vixens and although Big Ears ♀ could venture near the earth and pilfer occasional morsels, Wide Eyes ♀ was so victimised by Whitepaws ♀ that she avoided the half of the enclosure where the earth was. Wide Eyes' desire for food put Smudge ♂ in some conflict, e.g.

On 24th March, Smudge ♂ was ferrying back and forth with food to Whitepaws ♀. Wide Eyes ♀ had been cowering under a bush watching the food being carried past her. As the last chicken was carried towards the earth Wide Eyes ♀ bolted out and threw herself to the ground beside Smudge ♂, but he trotted past her. Wide Eyes ♀ dashed along beside him, casting glances towards the earth and squealing at Smudge ♂. He paused again, because the chicken was so big he could barely carry it. As Smudge ♂ eventually moved forward, Wide Eyes ♀ writhed on the ground in front of him and tripped him up so that both foxes and the hen tumbled to a heap on the ground. By then about half way to the earth, Smudge ♂ righted himself and purposefully looked first at the earth and then back at the squirming Wide Eyes ♀ and continued to look at each in turn for 15 seconds, whereupon he turned and trotted off, leaving Wide Eyes ♀ with food. Neither Big Ears nor Wide Eyes ♀ took any food to Whitepaws ♀ during this period.

During this period all the foxes took a great interest in the cubs. Both Smudge ♂ and Big Ears ♀ frequently went to the artificial earth and sniffed excitedly inside, both when Whitepaws ♀ was within and without.

Big Ears ♀ was still the most willing to play with Smudge ♂ (Table 6). Part of their play took a stylised form, centring around a particular grassy tussock in the enclosure. When Big Ears ♀ approached this tussock or pounced or rolled on it this seemed to be a signal for protracted play with Smudge ♂,

Weaning – 22nd April–8th May (see Plate 6)

Although the cubs had begun to squabble amongst themselves over scraps of meat on 8th April (3 weeks previously), only Whitepaws ♀ and Smudge ♂ had fed them or shown other parental behaviour before 22nd April. Both parents concentrated their caching activity in the vicinity of the earth. However, on 22nd April Big Ears ♀ moved into the earth with Whitepaws ♀ and thereafter shared all maternal duties except suckling. Big Ears ♀ retrieved cubs that wandered too far from the earth, groomed them, ate their faeces and, from 28th April onwards, also took them food. During the same period Whitepaws ♀ became rather less attentive and, indeed, Big Ears ♀ was often left in sole charge of the cubs and seemed more concerned than Whitepaws ♀ if they wandered. As Table 7 and Fig. 9 show, Big Ears ♀ became more aggressive in this role. In particular she began attacking Smudge ♂ when he played with the cubs, which he was always keen to do. Wide Eyes ♀ although hardly harassed then by Whitepaws ♀, remained cautious of the earth area, but took every opportunity to play with the cubs and from 7th May also began to feed them. Possibly because of her timidity Wide Eyes ♀ did not carry food directly to the earth, but "warbled" at cubs that followed her and made no objection when it tried to pilfer the food she was carrying – which in adult fox society would provoke serious rebuke.

On 5th May Smudge ♂, Whitepaws ♀ and Big Ears ♀ all carried food to the cubs. By this time the cubs were moving around the enclosure themselves and all three adult followed the cubs with food, as opposed to leaving the food at the earth. The same evening Wide Eyes ♀ played at length with the cubs who swung from her tail as she ran around. There were still occasional lapses in the relationship between Whitepaws and Wide Eyes ♀:

On 6th May Wide Eyes ♀ entered an artificial earth, apparently without noticing that Whitepaws ♀ had gone in before her – a fight broke out and Big Ears ♀, together with the cubs, were attracted by the commotion and ran in too. Almost at once they all came tumbling out as Whitepaws ♀ ejected Wide Eyes ♀ who ran to the far end of the enclosure, where Whitepaws ♀ caught her and continued to attack. Big Ears ♀ also joined in briefly, apparently joining forces with Whitepaws ♀. This fight was not of the same ferocity as Whitepaws' earlier attacks on Wide Eyes ♀.

By the 7th Big Ears ♀ warned Smudge ♂ away from the cubs on 8 occasions during 2 hours. Three times he played with the cubs for 5–10 minutes before Big Ears ♀ noticed. The cubs were running around him playing excitedly, biting his ears and tail before the commotion attracted Big Ears ♀. Smudge ♂ did not submit to her on these occasions, but withdrew.

Increasing independence – 9th–29th May

Throughout the previous period Wide Eyes ♀ became more confident and, as described earlier, began to play with the cubs without provoking an attack. This culminated on 10th May when she too began taking food to the cubs and by 11th May she was spending much more time with the cubs, and on 12th May she too began to caution Smudge ♂ when he played with them.

She slept unmolested with the cubs for the first time on 15th May. Whitepaws ♀ was by that time taking less interest in the cubs than either Big Ears ♀ or Wide Eyes ♀.

By 15th May Wide Eyes ♀ attacked Smudge ♂ each time he tried to play with the cubs in exactly the same way as Big Ears ♀ had done before. In fact 8.5% of the group interactions during this period involved serious attacks against Smudge ♂ by Wide Eyes ♀ (Fig. 10, Table 8), all of which followed Smudge ♂ attempting to play with the cubs. Figure 11 shows that Big Ears' attacks on Smudge ♂ dwindled, and latterly disappeared. Smudge ♂, however, continued to solicit play from the cubs (albeit sometimes rather furtively). For instance, a typical example on 13th May:

Smudge ♂ had been playing with a cub but noticed Wide Eyes ♀ approach and ran off. The cub seemed unaware of Smudge's attempts to be "unobtrusive" and play chased after him, pursued by Wide Eyes ♀. Smudge ♂ ran from Wide Eyes' aggressive attack with the cub still trying to bite his brush. Big Ears ♀ was attracted by the noise and also launched at attack on Smudge ♂. At this point the cub rolled on the ground in submission in front of the two vixens who paused while Smudge ♂ made his escape. Thereafter, both Wide Eyes ♀ and Big Ears ♀ simultaneously groomed the cub, but Whitepaws ♀ made no effort to become involved.

All the vixens frequently groomed the cubs and played with them. However, Big Ears ♀ and Wide Eyes ♀ were noticeably more attentive than Whitepaws ♀. Wide Eyes' behaviour when taking food to the cubs differed from that of the other two vixens: she would run to a cub, carrying food and "warble-calling" but then seemed unwilling to let go of the food. Thus she stood, summoning the cub to eat but refusing to let go of the food as it tugged at the food she held in her mouth. Eventually she let go, but sometimes after rushing off as if to cache the food before returning to give it to the cub. Sometimes, having summoned the cubs to eat, she would hold them at bay while she cached the food and then stand beside them amicably watching as they unearthed the cache and ate in front of her.

THE GROUP DURING 1977 AND MATING IN WINTER 1977–1978

During the summer of 1978 the social behaviour between the adult members of the group returned to a similar *status quo* to that which had existed prior to the birth of the cubs. The most significant feature other than the generally amicable relationships within the group was Whitepaws continuing aggression towards Wide Eyes who tended to avoid her whenever possible (Fig. 11, Table 9). For the first time Big Ears also became occasionally aggressive to Wide Eyes. All adults ceased to feed the cubs during the summer.

By Autumn and early winter changes in the basic structure of the group emerged. Whitepaws' dominance over Big Ears seemed to wane till the two vixens gave the impression of being either on a *par* or else that Big Ears was gaining precedence. This impression is supported by a comparison on the data on Tables 9 and 10 (and Figs. 11 and 12). Smudge's amicable interactions with these two vixens comprised 36.2% of the observed interactions between June–August and a very similar figure of 33.0% during September–December. However, during the first period 17.5% involved

Table 9. Summary of interactions between inmates of enclosure II.

June, July, August 1978

Initiator: Recipient (down) (across) — Recipient's response

Amicable

	SM	WP	BE	WE		Smudge 0	I	II	III	IV	Whitepaws 0	I	II	III	IV	Big Ears 0	I	II	III	IV	Wide Eyes 0	I	II	III	IV
Smudge		44	40	21	Smudge						2	—	26	16	—	5	—	16	18	—	1	—	7	13	—
Whitepaws	11		25	2	Whitepaws	10	1	—	—	—						18	1	8	—	—	—	2	—	—	—
Big Ears	19	39		15	Big Ears	14	4	1	—	—	22	1	12	2	—						6	—	8	3	—
Wide Eyes	11	3	10		Wide Eyes	11	2	—	—	—	3	—	—	—	—	9	2	—	—	—					

Mild aggression

	SM	WP	BE	WE		Smudge 0	I	II	III	IV	Whitepaws 0	I	II	III	IV	Big Ears 0	I	II	III	IV	Wide Eyes 0	I	II	III	IV
Smudge		1	6	1	Smudge						—	—	—	1	—	—	—	3	3	—	—	—	—	1	—
Whitepaws	—		9	9	Whitepaws	—	—	—	—	—						2	—	6	1	—	—	3	3	3	—
Big Ears	2	—		5	Big Ears	—	2	—	—	—	—	—	—	—	—						—	—	2	3	—
Wide Eyes	1	1	—		Wide Eyes	—	—	1	—	—	—	1	—	—	—	—	—	—	—	—					

Serious aggression

	SM	WP	BE	WE		Smudge 0	I	II	III	IV	Whitepaws 0	I	II	III	IV	Big Ears 0	I	II	III	IV	Wide Eyes 0	I	II	III	IV
Smudge		—	1	2	Smudge						—	—	—	—	—	—	—	—	1	—	—	—	—	2	—
Whitepaws	1		2	23	Whitepaws	—	—	—	1	—						—	—	—	2	—	—	1	2	19	1
Big Ears	—	—		3	Big Ears	—	—	—	—	—	—	—	—	—	—						—	—	—	3	—
Wide Eyes	3	—	—		Wide Eyes	—	—	3	—	—	—	—	—	—	—	—	—	—	—	—					

Summary of play and grooming initiations

	Play SM	WP	BE	WE		Grooming Smudge	Whitepaws	Big Ears	Wide Eyes
Smudge		20	19	15	Smudge		2	—	1
Whitepaws	—		4	—	Whitepaws	9		14	—
Big Ears	9	6		6	Big Ears	—	6		2
Wide Eyes	4	—	4		Wide Eyes	—	2	2	

Whitepaws and 18.7% involved Big Ears while during the second period only 12.0% involved Whitepaws and 21.0% Big Ears. This constitutes a 26.5% increase in the proportion of amicable interactions initiated by Smudge which were directed to Big Ears (during the same periods the percentages of his amicable initiations directed to Wide Eyes remained very constant at 20.0% and 20.5%).

The change in the dog fox's relationship with the vixens coincided with a changed relationship between them. The proportion of amicable interactions between Whitepaws and Big Ears remained rather constant between the summer and autumn at 18.3% and 20.3% but during the summer 3.34% of interactions involved Whitepaws directing aggression against Big Ears who directed none either in reciprocation or spontaneously against Whitepaws (see Table 9). However, by Autumn only 1.9% of interactions involved aggression by Whitepaws to Big Ears while 0.8% involved spontaneous aggression by Big Ears against Whitepaws. More importantly, Whitepaws was clearly avoiding Big Ears under circumstances which might lead to

Table 10. Summary of interactions between inmates of enclosure II.

September, October, November, December 1978

Initiator: Recipient (down) (across) — Recipient's response

Amicable

	SM	WP	BE	WE		Smudge					Whitepaws					Big Ears					Wide Eyes				
						0	I	II	III	IV	0	I	II	III	IV	0	I	II	III	IV	0	I	II	III	IV
Smudge		35	54	23	Smudge						2	5	20	9	—	13	1	26	14	—	8	1	13	1	—
Whitepaws	10		31	16	Whitepaws	7	—	3	—	—						19	1	8	1	—	3	1	11	1	—
Big Ears	25	38		27	Big Ears	21	3	1	—	—	17	—	17	4	—						9	—	14	3	—
Wide Eyes	5	7	8		Wide Eyes	5	—	—	—	—	1	3	2	1	—	7	—	1	—						

Mild aggression

	SM	WP	BE	WE		Smudge					Whitepaws					Big Ears					Wide Eyes				
						0	I	II	III	IV	0	I	II	III	IV	0	I	II	III	IV	0	I	II	III	IV
Smudge		1	3	5	Smudge						—	—	—	1	—	—	—	—	3	—	—	—	1	4	—
Whitepaws	1		7	25	Whitepaws	—	—	1	—	—						—	3	4	—	—	—	—	16	7	—
Big Ears	—	3		10	Big Ears	—	—	—	—	—	—	—	2	1	—						—	—	5	5	—
Wide Eyes	—	—	—		Wide Eyes	—	—	—	—	—	—	—	—	—	—	—	—	—	—	—					

Serious aggression

	SM	WP	BE	WE		Smudge					Whitepaws					Big Ears					Wide Eyes				
						0	I	II	III	IV	0	I	II	III	IV	0	I	II	III	IV	0	I	II	III	IV
Smudge		5	—	2	Smudge						—	—	—	5	—	—	—	—	—	—	—	—	—	2	—
Whitepaws	—		—	26	Whitepaws	—	—	—	—	—						—	—	—	—	—	—	—	—	26	1
Big Ears	—	—		8	Big Ears	—	—	—	—	—	—	—	—	—	—						—	—	—	8	—
Wide Eyes	—	2	—		Wide Eyes	—	—	—	—	—	—	—	—	1	1	—	—	—	—	—					

Summary of play and grooming initiations

Play	SM	WP	BE	WE	Grooming	Smudge	Whitepaws	Big Ears	Wide Eyes
Smudge		7	12	5	Smudge		5	16	3
Whitepaws	1		2	1	Whitepaws	2		5	—
Big Ears	4	4		—	Big Ears	4	3		5
Wide Eyes	—	—	—		Wide Eyes	1	4	6	

conflict and on 3/7 cases where Whitepaws was aggressive to Big Ears, she responded aggressively rather than submitting. Indeed, during the previous spring Big Ears had shown more deference to Whitepaws even during totally amicable meetings.

The apparent reversal of roles between these two vixens continued into January when foxes normally mate; on the 24th January Smudge mated with Big Ears. Of 56 interactions observed during January till that time there had been no aggression between Whitepaws and Big Ears, but Big Ears had initiated 21.4% amicable interactions towards Whitepaws while Whitepaws' amicable approaches to Big Ears comprised only 8.8% of interactions. In every case when Whitepaws approached her, Big Ears responded with no sign of submission while in just under half of the cases when Big Ears approached her, Whitepaws submitted. Coincident with Big Ears' rise she too began to harrass Wide Eyes more frequently (Table 10).

On February 3rd Smudge mated with White Paws, so that two of the three third year vixens within the enclosure were pregnant, while the dog

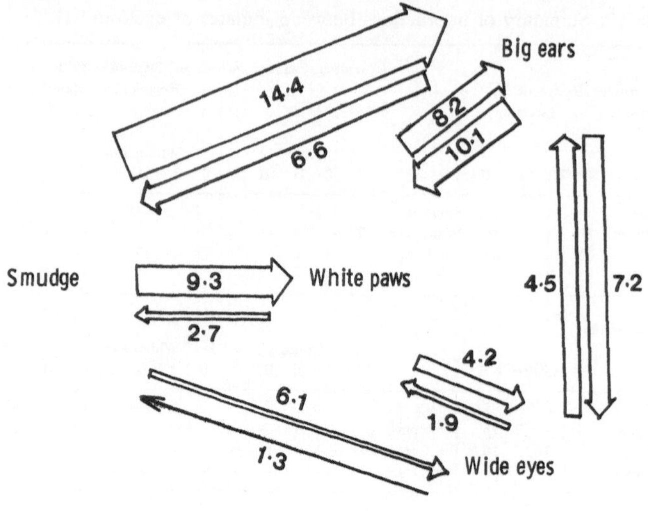

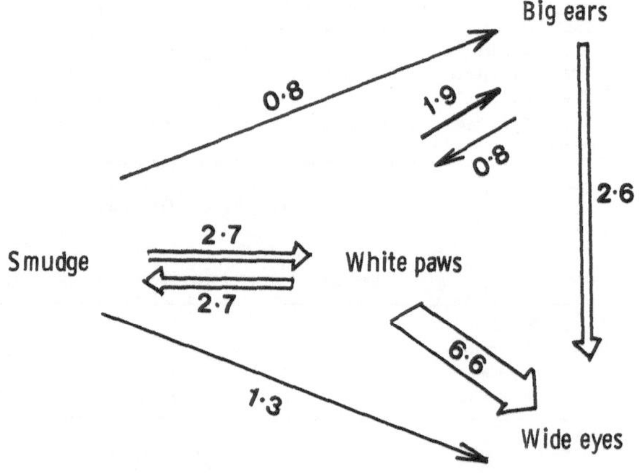

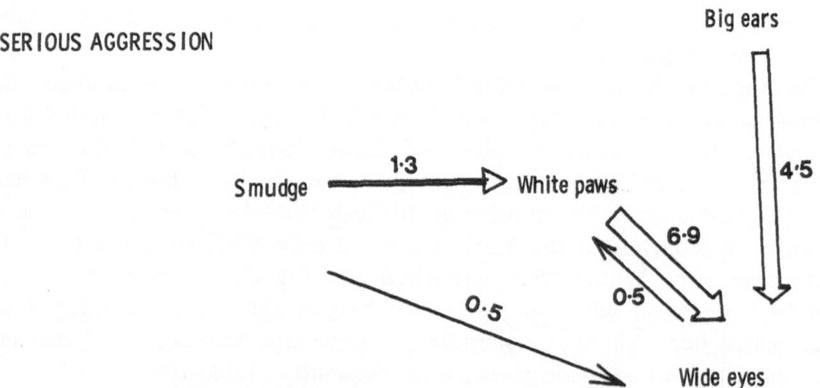

Fig. 12. 1st September–30th December 1978, $n = 376$.

fox had made no approach to either the remaining third year vixen or to this yearling daughter.

Big Ears and Whitepaws had cubs on the 28th March and 11th April respectively. Until that time relations amongst group members remained the same, with Whitepaws remaining slightly subordinate to Big Ears. Two changes are noteworthy: Whitepaws and Wide Eyes became unaccustomarily amicable, indulging in frequent mutual grooming and Wide Eyes began a series of savage attacks on the yearling vixen (beginning on the 13th March). Neither of the more dominant vixens nor the male paid much attention to this young vixen.

As soon as Big Ears' cubs were born Whitepaws became very submissive to her, at the same time (6th April) Big Ears began spontaneous serious attacks on Wide Eyes. This aggression had largely disappeared by 17th April and thus followed a similar pattern to Whitepaws' attacks on Wide Eyes the previous year.

Following the birth of her cubs, Whitepaws became extremely nervous, in contrast to her calm maternal behaviour the previous year. She frequently rushed from her earth carrying a cub and ran about the enclosure. Three of the four cubs died after two days of this treatment. Big Ears occasionally visited Whitepaws' artificial earth whereupon Whitepaws became frantically submissive and raced away carrying a cub. On the 19th the last cub died during transport and within an hour Whitepaws' behaviour was transformed. She became calm, moved into Big Ears' earth and began to suckle and care for Big Ears' cubs. Some details of her behaviour immediately following the death of her last cub are given below (19th April 1979):

> Whitepaws emerged from her earth carrying the corpse of her last cub. She ran to Big Ears' earth, but Big Ears was away. White Paws carried her cub into Big Ears' earth, then exited and dropped the cub outside. During the next 10 minutes she visited Big Ears' cubs ten times, during which time Big Ears had returned. Big Ears showed no aggression toward Whitepaws who submitted lavishly to her. Big Ears made three separate trips to Whitepaws earth and explored it thoroughly, whereupon she returned to her own earth where Whitepaws was lying with the cubs. Big Ears greeted Whitepaws and then wandered off, leaving her with the cubs.

By the next day both vixens were sleeping together in the same earth with one cub suckling on each. This pattern continued thereafter. Each vixen would frequently groom a cub suckling from her sister. On 25th April Whitepaws also began to attack Wide Eyes, as Big Ears had done as a "helper" the previous year.

DISCUSSION

In some habitats foxes live in social groups comprised of several vixens, whilst elsewhere pairs predominate (e.g. Macdonald, 1977, Storm et al. 1976). Complementary studies of wild and captive groups, described above, suggest that only the minority of vixens successfully rear young within larger social groups and that some of those which do not breed may act as helpers. The ecological factors which favour larger groups in some habitats will be discussed elsewhere (in prep.), but status-linked reproductive suppression within groups may explain some of the variation in vixen productivity found within and between habitats. Storm et al. (1976) recorded that 96% of vixens bred in the American prairies, in Wisconsin 59% of the yearlings and 89% of other vixens breed (Pils et al., 1978). Lloyd et al., (1976) summarised studies from Switzerland, Bavaria and Wales which indicated that 90% of vixens conceived, although Lloyd (1975, 1977) noted variations in fecundity in Wales varying between 71–86%. McIntosh (1963) found 97.4% of vixens conceived in the Canberra district of Australia. Niewold (pers. comm.) found that almost all wild vixens in the Veluwe region of Holland conceived. Ables (1975) implied that approximately 15% of yearling vixens failed to breed, as noted in fur farms. In Sweden, Englund (1970) found that yearling fecundity varied between 5–88%. Using different methodologies Sheldon (1949) and Layne and McKean (1956) recorded 4.7% and 2.1%–16.6% non-reproductive vixens in New York state.

Where barren vixens are common, most of them seem to be young. S. Harris (1979) found that 52% of yearlings were barren in one sample of foxes from London and in southern Swedish forests Englund (1970) recorded the proportion of barren yearlings to older foxes to be 81% : 27% and 78% : 36% during two years when he had large samples, while in agricultural land 40%, 60% and 64% of yearlings did not breed during three years when the corresponding figures for older vixens were 23%, 36% and 34%. These data are compatible with the hypothesis that some non-reproductive females are subordinate (young) group members. This hypothesis is strengthened by Englund's (1970) finding that annual fluctuations in food supply did not greatly affect the proportion of vixens breeding in agricultural habitats in Sweden. Englund (1970) concluded that some "social stress" influenced vixen productivity. My suggestion that this social stress is status-linked reproductive suppression within social groups is testable by further field studies in contrasting habitats. I would predict, for example, that in regions with annual fluctuations in prey abundance (e.g. rodent cycles) but where alternative stable prey were available, that more females would be recruited into groups in "good" years, but that they would not breed and hence that the overall vixen productivity would fall; one might also predict that the sex ratios of dispersing foxes would be female biased during crash years and male biased as the food supply recovered. Of course, some vixens may not breed because they are not residents, rather than because of low status within a group; this may apply where seasonal or annual fluctuations in food supply necessitate frequent readjustment of territory size and shapes, perhaps prohibiting the formation of stable related groups (and explaining why in the northern forests of Sweden Englund (1970) found vixen productivity to be more directly linked to rodent cycles).

The mechanism of reproductive suppression is unclear. In captivity, males made no attempt to mate with subordinate vixens, nor was there any indication of this being a consequence of interference from dominant vixens. Englund (1970) suggested that the low productivity of vixens in the northern forests of Sweden in "bad" rodent years was largely attributable to vixens which did not come into oestrus. The oestrous condition of non-reproductive vixens in this study was unknown although there were no signs of bleeding; Rasa's (1973) findings that male mongoose did not mate with oestrous subordinate females belies the assumption that lack of male interest indicates anoestrus. Similarly, J. Packard (pers. comm.) has found that some non-breeding female wolves nevertheless cycle normally. Elsewhere (e.g. Holland, F. Niewold, pers. comm.) almost all vixens conceive, but many lose their cubs either before or shortly after birth (one case of cannibalism has been documented, Macdonald, 1977b). It is thus possible that two mechanisms operate, both with the same end result, so that only a proportion of vixens rear cubs. The cause of neonatal deaths in Niewold's study is unknown, as is the subsequent behaviour of the bereaved vixens towards surviving cubs within their groups. This problem is further complicated by methodological difficulties. It is difficult to distinguish near full-term abortion from live-birth on the basis of placental scars, and indeed to be certain of the distinction of scars from different litters (E. Lindstrom, pers. comm.). Furthermore, assessment of the proportion of barren females has been on diverse grounds, e.g. those not lactating in late spring (Sheldon, 1949) or those with corpora lutea but no scars (Englund, 1970, McIntosh, 1963). How comparable these data are in terms of their sociological implications is unclear.

In one case in captivity, and one in the wild, two vixens bred within one group and on both occasions they were of comparable status during the mating period. The influence which status can exert on reproductive physiology may depend on factors such as frequency of direct or indirect contact, territory size and hence ultimately, food supply. Variations in these factors may underlie possible differences in the mechanism of reproductive repression between habitats. By comparison, crowded house mice, *Mus musculus*, may under different circumstances react either by many going into anoestrus or by a high litter mortality, and this may be habitat controlled (Crowcroft, 1966, Rowe et al. 1964).

Kin selection considerations: The pros and cons

Bertram (1976) has emphasised that different selective pressures operate on each sex, age and status class within a society. What advantages and disadvantages affect the members of fox groups?

1. *The dominant vixen*

There are few direct disadvantages of tolerating the presence of subordinates; they would not detract from the dominant's food resources since not only can she beat them during direct contests, but if serious food shortage occurs, the subordinates leave the group (Macdonald, 1977a). However, the dominant vixen should suppress the subordinates' reproduction since their cubs would compete with her own, both for food and paternal (or alloparental) care and future tenancy of the territory.

There are also tangible advantages to the dominant of the presence of the other non-breeding vixens in the group:

(a) *During the breeding season*

1. Additional food provided for the cubs;

2. more probable presence of an adult in the vicinity of the den should danger strike;
3. possibility of splitting the litter, almost certainly of antipredator value, and yet still having an adult with each group;
4. in captivity, at least, there is evidence that mothers may spend more time away from the cubs when "aunts" are active, which may not only give them respite, but may also reduce the risk to which they are subject since breeding earths are easily found by predators;
5. substitutes to rear the cubs in the event of the mother's and/or father's death.

(b) *Advantage outside breeding context*

1. Defence of the territory – both in co-operative attacks on intruders and in that an intruder is more likely to be discovered (described in Macdonald, 1977a);
2. on her death the successor will be related to her.

Similar factors have been argued to apply to other societies. Moehlman (1979) shows that helpers among jackal, *C. mesomelas*, groups increase cub survival. Bernstein (1969) has proposed that an advantage for co-operative breeding by pigtail macaques, *Macaca nemestrina*, is that a substitute is available in the event of the mother's death. Amongst birds, co-operative breeding is widespread; Skutch (1961) has reviewed instances of non-parents feeding chicks of many passerines, and other birds, both nidicolous and nidifugous. Allomaternal behaviour is almost invariably confined to related individuals. In Rowley's (1965) study of Australian blue wrens, *Malurus cyaneus*, supernumerary related males rear more fledgelings per participating adult than do nests without helpers. In many of these systems the helpers are males, for instance, the jay, *Aphalocoma ultramarina*, (Skutch, 1935) and the nuthatch, *Sitta pygmaea*, (Norris, 1958) and often it seems that there is a superabundance of males in the population so that these helpers could have little chance of breeding themselves if they left home.

2. *Subordinate vixens*

What are the advantages to the subordinate vixens of staying in the family group and tolerating reproductive suppression? Two ideas are relevant, either the helpers may gain through direct increase in their inclusive fitness (kin selection) by increasing the chances of survival of each to which they are related. Equally, the helpers may be "hopeful reproductives" gaining by helping to maintain a group structure which confers various benefits so that in the future they themselves get the chance to breed. These hypotheses are not mutually exclusive.

In one wild group studied in detail the average coefficient of relatedness between vixen group members was estimated at 0.38. Whether this estimate of mean relatedness is high is unimportant, the principle remains that vixen group members are probably related somewhere between a full sib and a cousin and so, *assuming* that a vixen has no cubs of her own, then selection could favour contributing to the upbringing of cubs to which she is likely to be related by about 0.38. But why do they not have any cubs of their own? Theoretically, it is because some factor or factors reduce their chances of rearing a cub of their own to less than 2.63 times that of the reproductive's chances of doing so, if they help her. To clarify this, imagine a barren vixen faced with the "decision" of whether to embark on pregnancy and rearing her own cubs or whether to devote her energies to rearing her group-mate's cubs. She is related, on average, by 0.38 to her group-mate. The group-mate will be related to her own cub by 0.5. Thus the non-reproductive vixen will be related to the cub by (0.38 × 0.5) or 0.19. By not having her own cub she is sacrificing (0.5–0.19) or 0.31 relatedness or doing (0.5/0.19) = 2.63 times less well. This reduction in inclusive fitness would be integrated into the vixen's lifetime expectancy of breeding if she stays or emigrates. In the real example Big Ears ♀ is an aunt to Whitepaws' cubs and thus related to them on average by 0.25 or half the amount she would have been related to her own cubs on average. Thus in a "decision" to opt to help rather than to breed, Big Ears ♀ was "assuming" that she would have less than half the chance of getting her own cubs to adulthood than would Whitepaws ♀ aided by Big Ears' help.

What factors could reduce a vixen's chances of successfully rearing cubs by a factor of 2–3 (rather a small amount) such that she opts for acting as a nanny instead? The evidence presented above suggests that within a family group either in captivity or on Boar's Hill, the non-reproductive vixens have no choice in the matter. While the mechanism of suppression is still uncertain, it seems that the presence alone of a dominant vixen may prevent them from breeding. Similar information on other canids also suggests that the subordinate female has no choice but to forego breeding if she remains in the group. Mech (1975) and Zimen (1976) report how alpha-bitches amongst wolves attack their subordinates when these come into oestrus and so prevent them copulating, although Zimen's account shows the complexity behind this generalisation. Similarly, L. Corbet (pers. comm.) tells me that subordinate bitch dingoes, *Canis dingo*, are prevented from mating by dominants and moreover, when the suppression fails, that the dominant kills and eats the cubs of the subordinate who then channels her energies into suckling the dominant's cubs. H. Van Lawick (pers. comm. and 1975) reports similar infanticides perpetrated by a dominant female hunting dog, *Lycaeon pictus*, against the cubs of her subordinate. Infanticide has been reported amongst primates by Sugiyama (1965) and Hrdy (1974), but in these instances, as with Bertram's (1973, 1975a and b, 1976) lions, *Panthera leo*, it is the males' genetic involvement at stake and hence the males that kill.

In the case of the fox, behavioural interference does not seem to be the mechanism of suppression but there is no shortage of precedents for social odours influencing the reproductive physiology of rodents, for instance, the Lee–Bruce effect, prolonging anoestrus (Lee-Boot, 1955) and the Bruce effect where a strange male's urine terminates pregnancy and induces oestrus (Bruce and Parrot, 1960). Such effects may be widespread, e.g. Schinckel (1954) reports oestrus synchrony in sheep.

If the subordinate's only hope of breeding in her first year is to leave the group, this could explain why she tolerates suppression. There is ample circumstantial evidence to suggest that an itinerant fox's chances of surviving at all, far less finding a new territory, are low. First, the territories around Boar's Hill were contiguous and group membership could apparently only be gained by birth. Especially in habitats that favour a larger group size, the chances of a vixen finding a slot must be low (a perfectly adapted fox might choose to emigrate on the basis of local group sizes, habitat and population turnover). In addition, it appears that she is likely to be at higher risk while travelling over unfamiliar ground. Of 5 experimental foxes liberated with radio collars, 4 were killed while travelling outside the area they knew, two of these were killed during one-night excursions out of a range they had lived in for months previously. There is circumstantial evidence that leaving a familiar territory involves greater risks (Storm et al.'s (1976) data do not necessarily support this, but their definition of "resident" and "transient" actually relates to the foxes' movements and not to their social status at the moment of death). Hirons (1976) showed that 44% of first year tawny owls, *Strix aluco*, die after leaving the parental range, mostly of starvation. Errington (1963) shows that emigrating muskrats, *Ondatra zibethicus*, are particularly subject to mink, *Mustela vison*, predation, while Christian (1970) summarises the data for rodents and states that the vast majority of transients die. Metzgar (1967) has demonstrated in the laboratory that *Peromyscus leucopus* is more vulnerable to the screech owl, *Otus asio*, when on unfamiliar terrain. Hawkins et al. (1971) thought that greater mobility contributed to heavy mortality of white tailed deer, *Odocoileus virginianus*. However, one cannot distinguish whether the higher mortality of transients results from their itinerant lifestyle or because transients are generally low quality animals in the first place.

A second factor encouraging subordinates to help would be if this experience improved their subsequent reproductive success. This possibility has been emphasised by workers considering the same issue for birds (e.g. Rowley, 1965, Skutch, 1961) and monkeys (e.g. Jolly, 1972, Horwick and Manski, 1971). There is no field or captivity evidence to suggest that vixens which have been helpers are subsequently better mothers, although the idea has intuitive appeal.

3. *The dog-fox*

While the male might increase his fitness by being polygynous, this is in conflict with the interests of the dominant vixen. The necessities for territorial defence may limit any promiscuous search for other females and so, even with 4–5 vixens in his territory he is effectively, reproductively, monogamous and so can contribute most to realising his reproductive potential by helping with the cubs.

It is interesting that in many instances of mixed-litters both are litters in one earth. Several selective forces might be thought to influence such clumping of litters:

(1) widely spaced litters may be subject to less likelihood of predation, less contact with disease etc.;
(2) both vixens in one earth may share maternal duties;
(3) if one vixen is killed the other could raise both litters;
(4) both vixens together may increase their chances of male aid.

We do not know how the dog-fox allocates his contributions between 2 breeding vixens if they den separately. Since vixens, at least those fed by their dogs, remain in the natal earth for more than a week, provision of food by the male may be particularly important. Cock bobolinks, *Dolichonyx oryzivorus*, normally do not feed their secondary hens, who manage by virtue of laying smaller clutches. However, where these secondary clutches are large the male does help, thereby altering his strategy when the secondary hen is overworked (Martin, 1974). If dog-foxes only fed one litter, then it might benefit vixens to den with or near each other, in the hope of getting some of the male's food.

4. *The young dog-fox*

Why do not at least some young males remain, like young vixens? This question can be tackled from the viewpoint of both the young dog and his father.

Advantages of staying:

To the young dog: (1) avoids higher mortality outside home territory;
(2) possibility of inheriting territory.

To the old dog: (1) has son to inherit territory;
(2) has another male to co-operate with group defence.

Disadvantages of staying:

To the young dog: (1) has to tolerate subordination;
(2) has to wait to inherit territory;
(3) may inbreed with sisters if inherits territory.

To the old dog: (1) son may compete with him for access to vixens (although females may resist advances from a brother);
(2) son not spreading his genes elsewhere.

The balance of these pros and cons will again depend on the chances of the young dog-fox establishing himself as a territory holder should he leave home. Since there is only one male per group these chances may be considerably higher (in habitats favouring groups) for a dog than a vixen. More important may be the risk of inbreeding which would favour the dispersal of one sex (males, as argued by Trivers, 1972). There are indications of brother–sister incest taboos in wolves (Zimen, 1976) and possibly in mongooses (Rasa, 1973). Itani (1972) presents evidence that avoidance of inbreeding accounts for males leaving groups of Japanese macaques, and

Table 11. The table lists a selection of species for which there is some evidence of social factors influencing reproductive success within a group. √ indicates that a certain phenomena has been recorded, x that it has not (which does not mean that it does not occur). Of course, carnivore social systems have been shown to be so flexible that because a species shows certain behaviour in one habitat does not mean that it will do so elsewhere, and *vice versa*.

Canidae	Communal denning of breeding ♀♀	Communal suckling	Non-breeding group members	Helping other than suckling ♂	♀	Infanticide by ♀	♂	Source
Red fox, *Vulpes vulpes*	√	√	√	x	√	√	√	see this paper.
Arctic fox, *Alopex lagopus*	x	x	√	x	x	x	x	P. Hersteinsson (pers. comm.)
Indian fox, *Vulpes velox*	√	√	x	x	x	x	x	A. Johnsingh (pers. comm.)
Bat eared fox, *Otocyon megalotis*	√	√	√	x	x	x	x	P. Moehlman (pers. comm.) Lamprecht (1979)
Coyote, *Canis latrans*	√	√	√	x	x	neighbouring pack		Camenzind (1978)
Dingo, *Canis dingo*	x	√	√	x	√	√	x	Corbett (1978)
Golden jackal, *Canis aureus*	x	x	√	√	√	x	x	P. Moehlman (1978)
Black backed jackal, *Canis mesomelas*	x	x	√	√	√	x	x	P. Moehlman (1979)
Wolf, *Canis lupus*	√	√	√	√	√	x	x	Zimen (1976); J. van Hoof (pers. comm.)
Dhole, *Cuon alpinas*	x	√	√	x	x	x	x	A. Johnsingh (pers. comm.)
Hunting dog, *Lycaon pictus*	x	x	√	√	√	√	x	Frame et al. (1979) Malcolm (1979)

Felidae								
Lion, *Panthera leo*	√	√	x	√	√	x	√	Schaller (1972), Bertram (1976)
Domestic cat, *Felis catus*	√	√	x	x	√	x	√	Macdonald and Apps (1978); Macdonald (pers. obs.)
Mustelidea								
European badger, *Meles meles*	√	√	√	x	√	√	x	Neal (1977); H. Kruuk (pers. comm.)
Viveridae								
Dwarf mongoose, *Heligale undulata*	√	√	√	√	√	x	x	Rasa (1973); Rood (1978)
Procyonidae								
Raccoon, *Procyon hotor*	x	x	√	x	x	x	x	Fritzell (1978)
Coati, *Nasua narica*								
Hyaemidae								
Spotted hyaena, *Crocuta crocuta*	√	x	√	x	x	x	√	Kruuk (1972)

male olive baboons, *Papio anubis*, transfer between groups, apparently without compulsion. Packer (quoted in Clutton-Brock and Harvey, 1976) makes the critical observation that female olive baboons present more frequently to immigrant males than to those that were born in their own troop. Shaffer (pers. comm.) has demonstrated "inbreeding" inhibitions amongst children of peer groups from kibbutzim and certain societies in Taiwan. Hill (1974) has shown that inbreeding in rodents does reduce their fitness.

The phenomenon of pseudopregnancy, common amongst domestic dogs, also occurs in foxes (e.g. a previously fecund vixen isolated from her group prior to mating subsequently came into milk). Zoo keepers tell me that the phenomenon is widespread amongst canids. I suggest that this is an adaptive mechanism, preparing non-breeding females for a role as helpers.

CONCLUSIONS

Many of the phenomena recorded for foxes in this paper have recently been reported amongst other carnivore societies (Table 11). In most cases circumstantial evidence suggests that helpers are relatives. Rood (1978) records an exception where an immigrant female dwarf mongoose was a more assiduous helper than related group members. The phenomenon of "aunties", first documented by Fraser-Darling in 1938 seems widespread amongst Canids. In foxes "helping" is confined to females, while P. Moehlmann (pers. comm. and 1979) notes that the sex ratio amongst black-backed jackal helpers mirrors that in the population at large. This contrast is noteworthy since most other aspects of the social biology of foxes and jackals are apparently similar. Why is the balance of pros and cons affecting young male foxes and jackals apparently different?

The list of species exhibiting reproductive suppression and helpers seems to grow with each additional study of carnivorous species. A. J. T. Johnsing (pers. comm.) recently observed a pack of dholes, *Cuon alpinus*, during two consecutive years. During the first year the pack comprised five males and three females and in the second year 7–8 males and 2–3 females. In both cases only one female bred. Johnsingh (in press) also records observation of the Indian fox, *Vulpes velox*, where two females jointly suckled a litter of four cubs. The parentage of these cubs was not known, and the difficulty of distinguishig cubs of different mothers in the field bedevils many studies, e.g. J. Rood (pers. comm.) notes that on the rare occasions when a subordinate dwarf mongoose conceives that both she and the dominant will eventually suckle all the cubs, but that by the time the cubs are first seen there are about the same number as normally comprise one litter, not two. Several documented cases do exist of dominant females killing the offspring of subordinates who then become helpers (e.g. dingoes and hunting dogs) (Corbett, pers. comm. and van Lawick, 1975) and maternal inability leading to cub mortality was described above for a subordinate vixen red fox.

Malcolm (1979) describes how the "help" given by non-breeding yearling hunting dogs to pups is dependent on plentiful food, and indeed that in times of food shortage the older hunting dogs invest more in their yearling offspring than in their younger pups, perhaps because the yearlings have a greater chance of survival for a given investment. Of 26 hunting dog dens watched by Frame et al. (1979), at 20 the breeding female was a dominant animal. Of the 6 subordinate females who bred, only one litter survived.

Table 11 illustrates a contrast between canids and felids: social repression of reproduction seems to be widespread amongst canids, but absent in the two social felids. Female lions and domestic cats apparently produce kittens irrespective of their age or status and under circumstances where their survival is improbable (Schaller, 1972, Bertram, 1970 and Macdonald and Apps, 1979), and in both cases females will co-operate in rearing kittens.

Accepting all the obvious shortcomings of enclosure studies, the combined results of observations in the field and in captivity, although bedevilled by sample size, suggest that social pressures limit vixen productivity within social groups and that some non-breeding vixens act as helpers (Macdonald 1979a). Furthermore, maternal ability seems to be transient and to depend upon current status. However, this study highlights many more questions than it answers: more intensive field studies in contrasting habitats are required, and the effects of human hunting pressure and disturbance disentangled. The possibilities of female immigration and of both male promiscuity and incest taboos must be further studied to test some of the kinship assumptions mentioned above. At a more mechanistic level, the physiology of suppression must be investigated and the deplorably woolly concept of dominance within fox groups made rigorous.

ACKNOWLEDGEMENTS

I am grateful to Drs. B. C. R. Bertram, N. B. Davies, H. Kruuk, P. Moehlman and H. N. Southern for helpful discussions and comments on an earlier manuscript. My wife has helped invaluably with both fieldwork and in looking after the captive foxes. I am grateful to the many farmers and householders who have tolerated my nocturnal ramblings on their property and to R. Spiro on whose land the enclosures were sited.

BIBLIOGRAPHY

Ables, E. D. 1975. Ecology of the red fox in America. In : The Wild Canids, 216–236. Ed. Fox, M. W., van Norstrand Reinhold Co., New York and London.

Bernstein, S. S. 1969. Stability of the status hierarchy in a pigtail monkey group (*Macaca nemestina*). *Anim. Behav.* 17: 452–458.

Bertram, B. C. R. 1973. Lion population regulation. *E. Afr. Wildl. J.* 11: 215–225.

Bertram, B. C. R. 1975a. Social factors influencing reproduction in wild lions. *J. Zool.* (*Lond.*) 177: 463–482.

Bertram, B. C. R. 1975b. The social system of lions. *Sci. Am.* 232; 54–65.

Bertram, B. C. R. 1976. Kin selection in lions and in evolution. In: Growing points in Ethology. Eds. Bateson, P. P. G. and Hinde, R. A. Cambridge Univ. Press. pp. 281–301.

Bruce, H. M. and Parrott, D. M. V. 1960. Role of olfactory sense in pregnancy block by strange males. *Science*, 131: 152b.

Christian, J. J. 1970. Social subordination, population density and mammalian evolution. *Science*, 168: 84–90.

Clutton-Brock, T. H. and Harvey, P. H. 1976. Evolutionary rules and primate societies. In: Growing points in Ethology, Eds. Bateson, P. P. G. and Hinde, R. A. Cambridge Univ. Press. pp. 195–237.

Crowcroft, W. P. 1966. Mice all over. Faulis. London.

Englund, J. 1970. Some aspects of reproduction and mortality rates in Swedish foxes (*Vulpes vulpes*) 1961–63 and 1966–69. *Viltrevy* 8: 1–82.

Errington, P. L. 1963. The phenomena of predation. *Amer. Sci.*, 51; 180–192.

Fox, M. W. 1970. A comparative study of the development of facial expressions in canids, wolf, coyote and foxes. *Behaviour* 36: 49–73.

Frame, L. H., Malcolm, J. R., Frame, G. W., and H. van Lausck (1979). Social organisation of the African Wild Dog (*Lycaon pictus*) on the Serengeti National Park. Tanzania, Z. Tierpsychol. (in press).

Harris, S. (1979) Age-related fertility and productivity in red foxes (*Vulpes vulpes*) in suburban London *J. Zool. Lond.* 187, 195–199.

Hawkins, R. E. 1971. Dispersal of deer from Crab Orchard relational Wildlife Refuge. *J. Wildl. Mgmt.* 35: 216–220.

Hill, J. L. 1974. *Peromyscus*: effect of early pairing on reproduction. *Science* 186: 1942–1044.

Hirons, G. J. M. 1976. A population study of the tawny owl, *Strix aluco*, L. and its main prey species in woodland. D.Phil. thesis, Oxford.

Horwick, R. H. and Manski, D. 1975. Material care and infant transfer in two species of colubus monkeys. *Primates* 16: 49–73.

Hrdy, S. B. 1974. Male–male competition and infanticide amongst langurs (*Presbytis entellus*) of Alsi Rajasthan. *Folia Primat.* 16: 49–73.

Itani, J. 1972. A preliminary essay on the relationship between social organisation and incest avoidance in non-human primates. In : Primate Socialisation. Ed. Poirier, F. E. pp. 268–171. Random, New York.

Jennrich, R. I. and Turner, F. B. 1969. Measurement of non-circular home range. *J. Theoret. Biol.* 22: 227–237.

Jolly, A. 1966. Lemur behaviour. Chicago Univ. Press, Chicago.

Jolly, A. 1972. The evolution of primate behaviour. Macmillan, Lond. pp. 397.

Lawick, H. van. 1975. Solo. Collins. London.

Layne, J. N. and McKeon W. H. 1956. Some aspects of red fox and gray fox reproduction in New York. *N.Y. Fish Game J.* 3: 44–74.

Lee. S. U. and Boot, L. M. 1955. Spontaneous pseudopregnancy in mice. *Acta Physiol. Pharm. Neer.* 4: 442–443.

Lloyd, H. G. 1975. The red fox in Britain. In: the Wild Canids, 207–215. Ed. Fox, M. W., van Nostrand Reinhold Co., New York and London.

Lloyd, H. G. et al. 1976. Annual turnover of fox populations in Europe. *Zbl. Vet. Med.* 3. 23: 580–589.

Lloyd, H. G. 1977. Wildlife Rabies, prospects for Britain. In: Rabies, the facts. Ed. Kaplan, C. Oxford Univ. Press. 91–103.

Macdonald, D. W. 1977a. The behavioural ecology of the red fox, *Vulpes vulpes*: a study of social organisation and resource exploitation. D.Phil. thesis, Oxford.

Macdonald, D. W. 1977b. On food preference in the red fox. *Mammal Rev.* 7: 7–23.

Macdonald, D. W. 1978. Radio-tracking: some applications and limitations. In: Recognition marking of Animals in Research. Ed. Stonehouse, B. 192–204, Macmillans, London.

Macdonald, D. W. 1980. Rabies and wildlife: a biologist's perspective. Oxford Univ. Press.

Macdonald, D. W. 1979c. The flexible social system of the gold jackal, *Canis aureus. Behav. Ecol. Sociobiol.* 5, 17–38.

Macdonald, D. W., Ball, F. and Hough, N. G. 1980. The evaluation of home range size and configuration from radio-tracking data. In: A Handbook on Biotelemetry and Radio-tracking. Eds. Amlaner, C. J. and Macdonald, D. W. Pergamon Press. Oxford.

Macdonald, D. W. and Apps, P. J. 1978. The social behaviour of a group of semi-dependent farm cats. *Felis catus*: a progress report. Carnivore Genetics Newsletter 3: 256–269.

Malcolm J. (1979). The African Wild Dog, Lycaon pictus Ph.D. thesis, Harvard University.

Macdonald D. W. 1979a. "Helpers" in fox society. *Nature, Lond.* 282, 69–71.

Macdonald D. W. 1979b. Observations and field experiments on the urine marking behaviour of the red fox. *Z. Tierpsychol.* (in press).

Macdonald D. W. and Amlaner, C. J. 1980. Practical Radio Tracking In: A Handbook on Biotelemetry and Radio Tracking. Eds. Amlaner C. J. and Macdonald, D. W., Pergamon Press, Oxford.

McIntosh, D. L. 1963. Reproduction and growth of the red fox in the Canberra district *CISRO Wildl. Res.* 8: 132–141.
Mech, L. D. 1970. The Wolf. *Nat. Hist. Press.* pp. 385:
Metzgar, L. H. 1967. An experimental comparison of screech owl predation on resident and transient white footed mice (*Peronyseus leucopus*). *J. Mammal.* 48: 387–391.
Moehlman, P. 1979. Jackal kelpers and pup survival. *Nature* 277. 382–383.
Murie, A. 1936. Following fox trails. *Univ. Mich. Misc. Publ.* 32: 7–45.
Norris, R. A. 1958. Comparative biosystematics and life history of the nuthatches, *Sitta pyrgmaea*, and *Sitta pusilla. Univ. Cal. Publ. Zool.* 56: 119–300.
Pils, C. M. and Martin, M. A. 1974. Dog attacks on a communal fox den in Wisconsin. *J. Wildl. Mgmt.* 38: 359–360.
Pils, C. M. and Martin, M. A. 1978. Population dynamics, predator-prey relationships and management of the red fox in Wisconsin. Dept. Nat. Resources, Madison, Wisconsin, Tech. Bull. 105, pp. 56.
Rasa, O. A. E. 1973. Intra-familial sexual repression in the dwarf mongoose, *Helogale parvula. Naturwissenschaften* 60: 303–304.
Reynolds, V. and Luscombe G. 1969. Chimpanzee rank order and the function of displays. In: Proc. 2nd congr. Primat. Atlanta 1968, 1, 81. Ed. Carpenter, C. R. Harper, New York.
Richards, S. M. 1974. The concept of dominance and methods of assessment. Anim. Behav. 22: 914–930.
Rowell, T. E. 1966. Hierarchy in the organisation of a captive baboon group. *Anim. Behav.* 14: 430–443.
Rowe, F. P., Taylor, E. J. and Chudley, A. H. J. 1964. The effect of crowding on the reproduction of the house mouse (*Mus musculus*). Living in corn ricks. *J. Anim. Ecol.* 33: 477–483.
Rowley, I. 1965. White winged choughs. *Australian Nat. Hist.* 15, 81–85.
Sargeant, A. B. and Eberhardt, L. E. 1975. Death feigning by ducks in response to predation by red foxes (*Vulpes fulva*).
Schaller, G. B. 1972. The Serengeti Lion: a study of predator–prey relations. Univ. Chicago Press. Chicago, pp. 480.
Schinckel, P. G. 1974. The effect of the presence of the ram on the ovarian activity of the ewe. *Aust. J. Agric. Res.* 5: 465–469.
Sheldon, W. S. 1949. Reproductive behaviour of foxes in New York State. *J. Mammal.* 30: 236–246.
Skutch, A. F. 1935. Helpers at the nest. *Auk*, 52: 257–273.
Skutch, A. F. 1961. Helpers among birds. *Condor* 63; 198–226.
Storm, G. L., Andrews, R. D. Phillips, R. L., Bishop, R. A., Siniff, D. B. and Tester, J. R. 1976. Morphology, reproduction, dispersal, and mortality of midwestern red fox populations. *Wildl. Monog.* 49: 1–82.
Sugiyama, Y. 1965. Behavioural development and social structure in two troops of Hanuman langurs (*Presbytis entellus*). *Primates* 6: 213–247.
Trivers, R. L. 1972. Parental investment and sexual selection. In: Sexual selection and the descent of Man. Ed: Campbell, B. G. Heineman, London, pp. 378.
Zimen, E. 1976. On the regulation of pack size in wolves. *Z. Tierpsychol.* 40: 300–341.

11 THE RED FOX IN A SMALL GAME COMMUNITY OF THE SOUTH TAIGA REGION IN SWEDEN

IN BETWEEN NUTRITIONAL AND SOCIAL REGULATION? (THE THEORETICAL BACKGROUND OF A PROJECT IN PROGRESS)

Erik Lindström*

INTRODUCTION

The context and demographic hypothesis of a study concerning fox population dynamics and predation is described. The study is part of a cooperative small game community project, where the dynamics of the community as a whole and of certain species, is considered in relation to food availability and predation.

A major feature of the research area in question is its location on the steep geomorphological, climatical and biogeographical gradient referred to as *limes norrlandicus*. Due to this, and with other indications given, the fox population is thought to be directly dependent upon vole abundance during vole crash years, but socially regulated other years.

During the 20th century many scientists and laymen have discussed the role of red fox (*Vulpes vulpes* L.) predation in natural ecosystems. A vast number of articles have been published on fox diet, qualitatively and semiquantitatively, all over the world (see Jensen and Sequeira, 1978, and Witt, 1976, for bibliographies). Yet only a few (e.g. Ryszkowski et al., 1973) have assessed the quantitative impact of fox predation on the population dynamics of its prey. To make such an assessment, an inevitable prerequisite apart from the quantified diet, is good quantitative data on fox population variables and parameters. However this data too is often lacking, and indexes, such as bag records, juvenile frequencies etc. must be relied upon. Some valuable remarks on one of the causes of this situation, the lack of hypothesis testing by crucial experiments, have been made by Lockie (1977).

Obviously, a major step in fox research today, is the attempt to quantify the dynamics of defined populations and the role played by the predation of these populations. This means simultaneous, long term studies of the dynamics of both the predator and prey to be elucidated. Probably, it has to include natural or induced experiments. The study outlined here is such an attempt.

The framework of this investigation is the small game community study of Grimsö. Within this cooperative community project, we try to assess the

* Grimsö wildlife research station, S-77031 Riddarhyttan, Sweden.

impact of predation and food resources upon the dynamics of the small game community as a whole and of certain species: Goshawk (*Accipiter gentilis* L.), Red fox, Blackcock (*Lyrurus tetrix* L.) and Mountain hare (*Lepus timidus* L.). Other species involved are: Marten (*Martes martes* L.), Stoat and Weasels (*Mustela* spp.), Capercaillie (*Tetrao urogallus* L.), Hazel hen (*Tetrastes bonasia* L.), European jay (*Garrulus glandarius* L.), Squirrel (*Sciurus vulgaris* L.), Voles (*Microtus agrestis* L. and *Clethrionomys glareolus* Schreb.) and browse species, Birch (*Betula verrucosa* Ehrh. and *B. pubescens* Ehrh.), Willows (*Salix* spp.), Rowan (*Sorbus aucuparia* L.) and Bilberry (*Vaccinium myrtillus* L.).

Our vole populations fluctuate with cycles of three to four years and we consider this as a natural experiment. Our basic hypothesis is the cyclic bufferspecies – one, which backdates through especially Keith (1974) to Lack (1954) and Hagen (1952). That is, the vole cycles are reflected in the functional and numerical response of the predators (e.g. red fox) and, through the dict of the predators, in the dynamics of alternative prey species (e.g. mountain hare).

In this paper I will consider the hypothesis of and background of the population demographic part of the fox project.

STUDY AREA

Grimsö wildlife research area is situated in central Sweden, some 50 km to the north of the latitude of Stockholm (Fig. 1.) Biogeographically it is located within the range of the south taiga (phytogeographically Lindquist (1966) and at the border between the highboreal and the south Scandinavian region (zoogeographically Ekman (1922).

Using another terminology, it is situated at the steep geomorphological, climatical and biogeographical gradient, which is referred to as *limes norrlandicus.* (Sjörs, 1967). The situation is more unique than realized when the station was established. This border separates the mountainous, coniferous regions to the north from the flattened out country of mixed coniferous/deciduous forests and arable land to the south. It is often illustrated by the northern limit of the oak (*Quercus robur* L.). In this part of the country the *limes* almost coincide with the former highest marine level with its implications for deposits and soil fertility.

Sjörs (1967) describes the most striking climatical feature of the land to the north of *limes* as the hard and prolonged winters with deep reliable snowcover. The different requirements of the regions to the north and to the south of *limes* is well illustrated by the number of plant and animal species, which here have their northern or southern limit of range or which here drastically change in frequency. See Sjörs (1967) and Pehrson (1958).

According to Fransson (1965), the steepest floristic gradient is found some 20 km to the south of the topographically determined border. Grimsö is situated in the interface between these two. From our own experience of the border character of the research area, we can observe that the winters of 1974 to 1976 were almost snowfree, while we experienced more than one

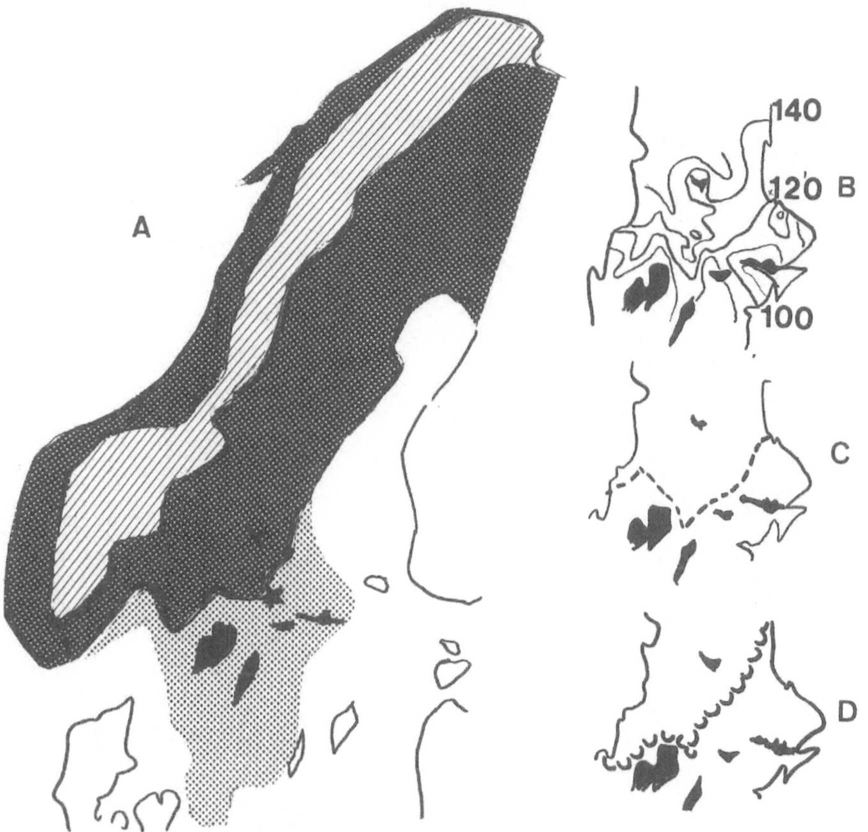

Fig. 1. A. Zoogeographic regions of Sweden and Norway redrawn from Ekman-22. ☰ Arctic-alpine region, ■ High boreal region, ⁞⁞⁞ South Scandinavian region, ★ Grimsö. B. Mean duration of snowcover (days). Redrawn from Sjörs-65. C. The northern limit of the distribution of (*Quercus robur* L.). Redrawn from Hultén-71. D. The geomorphological steep gradient. Redrawn from Fransson-65.

meter of snowcover during the winter of 1976/77. Consequently, we find both the heath hare (*Lepus timidus canescens* Nilsson) which is bluish-grey in winter, and the mountain hare (*Lepus timidus timidus* L.) which has a pure white winter pelage, within the area. The same is true of the weasel (*Mustela nivalis* L.), which has been observed both in the white and the brown winterphase.

Turning to the red fox, Englund (1970) found a pronounced dependence of the reproduction of this species upon the vole fluctuations in northern Sweden, but a stable situation in the southern parts of the country. This is considered as an indication of the possibility that the foxes are limited by food availability all over Sweden, but socially regulated at high densities in the southern parts of the country. The scarcity of food in northern Sweden should then never allow the population to build up to such a high level.

However, Englund's Värmland material, which is from areas close to Grimsö, showed a confusing picture and deceived penetration.

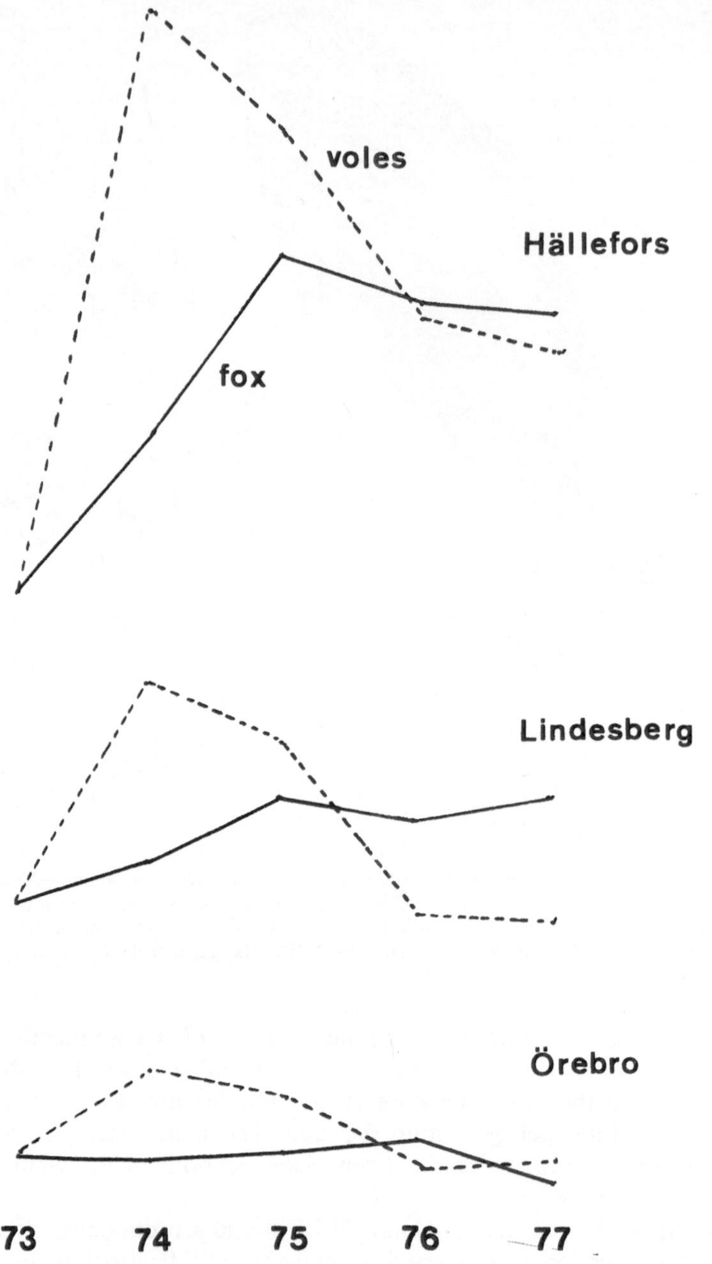

Fig. 2. Results concerning voles and red fox from a yearly questionnaire programme. Material divided according to area of origin (see text). Grimsö inventorial programme.

All answers = increase of density

Mean answer = no change

All answers = decrease of density

SOME RESULTS

Throughout the years since Grimsö was established as a research station in 1973/74, a programme concerning most game and wildlife species has been carried out. This comprises, among other methods, a questionnaire in which some 300 experienced hunters are asked their opinions of wildlife population trends on a yearly basis. The limitations of such a questionnaire are obvious, but for a qualitative comparison between areas, it should be sufficient. In Fig. 2 the results concerning voles and red fox are shown. There is a marked covariation of the two curves. The material is divided according to area of origin.

The Örebro data is from the arable land some 50–100 km to the south of Grimsö and well below *limes.* The Lindesberg data comprises the Grimsö research area. Finally the Hällefors material is from the boreal forests 50–100 km NW of Grimsö, well to the north of *limes.* Along this gradient we can note a change in the amplitude of the fluctuations. That is, both voles and fox seem to be fairly stable in the Örebro data while they show large annual variations in the Hällefors area. The Lindesberg material is intermediate.

Other parts of the inventorial programme concern vole populations and active fox dens of the actual research area. From the vole inventories we know that they were at their lowest level during the spring of 1975 and at peak levels in the autumn of 1973 and 1977 (Hörnfeldt pers. comm.).

The active number out of 78 dens controlled each year is shown in Fig. 3. We observe almost the same number each year irrespective of vole densities except the crash year of 1975. This was the state of knowledge in the autumn 1977. The fox project had then been running since the spring of 1975, that is during a phase of increasing vole densities. Among other things we had been collecting carcasses during the winters. A heavy dominance of males in the subadult cohort was striking and confusing. In conformity with Nellis and Keith (1976) the collected carcasses were considered a sample of

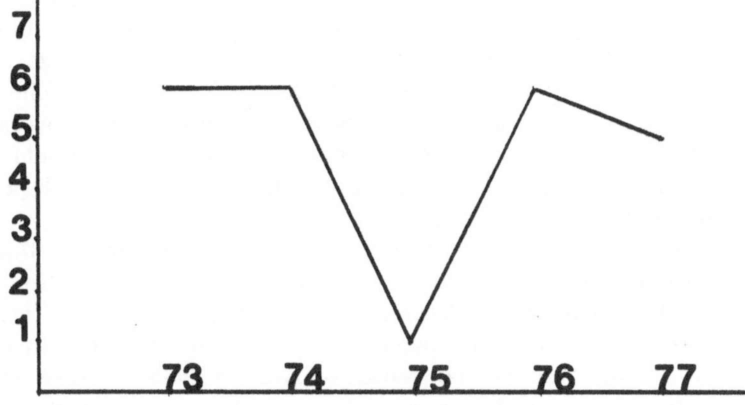

Fig. 3. Active dens out of 78 controlled each year.

the dying. (Their treatment of the data accords very well with our own experience. For example the foxes we trap most frequently are the inexperienced juveniles and hungry individuals such as lactating females. Most foxes in the collected material are shot at a bait or trapped. From marking of foxes we also know that hunting constitutes a large part of mortality). This should thus imply a dominance of females in the living fraction, a situation not much more clearly understood.

A HYPOTHESIS OF THE DYNAMICS OF THE BORDERLINE RED FOX POPULATION

In the autumn of 1977 David Macdonald visited Grimsö and presented his findings on fox social behaviour (Macdonald this volume). The most relevant fact in this context is that a dense stable fox population is made up of: (a) territorial family-groups of one male and several females; and (b) a transient cohort predominately male, suffering high mortality. By applying his ideas to the situation at Grimsö a hypothesis could be devised, which could explain both the threshold value of voles for the den count and the excess of females. The hypothesis is:

Starting in a year with low vole densities, the heavy mortality and low reproduction of the foxes cause only a few litters to be born. Let us suppose all territories are not occupied, and where they are, the number of adult foxes is one male and one female (the alpha-female). When vole populations increase, all territories become occupied and a litter is born in each. In those territories, which already were inhabited, young females are allowed to stay while young males have to emmigrate. Still only the alpha-female raises a litter. As long as vole densities increase, this continues, causing a population structure comparable to that of the stable, dense English populations described above. When the voles crash once again, the high mortality of all foxes causes the initial situation to reappear.

The implications of this hypothesis are more accurately described in Fig. 4.

The frequency of females having litters should obviously be highest in the year following the crash year. This can be controlled by analysing uteri for placental scars. The sex ratio among the subadults should stabilize at the same as in litters (56% ♂♂) during the crash year. This is also easily checked.

Another implication of the hypothesis is that among the recoveries of foxes marked as juveniles during the increase phase, there should be a marked dominance of males in the fraction considered as having dispersed, but not so among those that stayed "at home".

Thus the mean feature of this hypothesis is the assumption that we consider a fox population in between social and nutritional regulation. This implies social regulation (as described by Macdonald) for the foxes of southern Sweden and a total dependence upon voles in the northern parts of the country. Only here at the border – *limes norrlandicus* – one should obtain this kind of data on what is happening to the population. Further to the north there might always be new territories to occupy and males and females

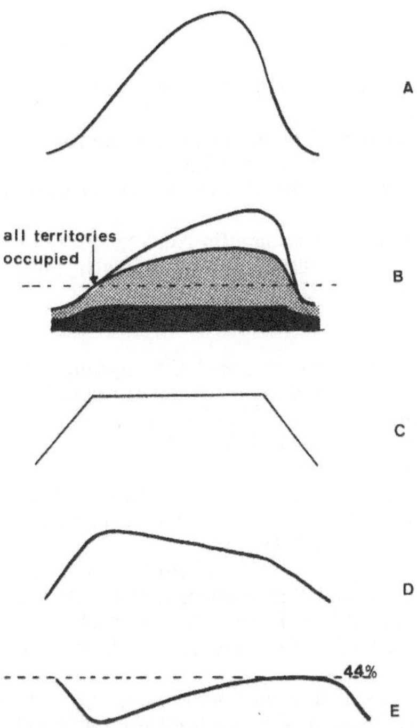

Fig. 4. A hypothesis of fox population dynamics throughout a vole cycle. See text. A. Voles numbers/area, B. Red fox numbers/area, ▓ resident males, ⁞⁞⁞⁞ resident females, □ transient cohort. C. Litters/area. D. Frequency of females having litters. E. Frequency of females among subadults in sample of dying.

disperse at the same rate. To the south the stable populations might cause the sex ratio of the subadult cohort to remain the same as in litters all years.

The fox project is now carried out according to this hypothesis, and hitherto *nothing has been found which contradicts it.*

The field work will be finished in 1979 and the results published as soon as possible, together with the predation part of the project.

Whether the attempt to quantify the impact of fox predation on prey population dynamics, is successful, remains to be seen. This also makes comparison with results from the other projects within the small game community study, to be revealed.

ACKNOWLEDGEMENTS

I would like to thank Jan Englund for planning and starting the den inventories in 1973 and Lloyd Keith and Jan Englund for initially planning the fox project. Their plan has been revised several times, but is still relevant in its fundamental parts. I also thank David Macdonald for providing the sociobiological background to the hypothesis and together with Torbjörn v. Schantz for invaluable discussions on the subject. My coworkers at Grimsö for innumerable reasons. The project has been supported by grants from stiftelsen Olle Engkvist, Helge Axelsson Johnson stiftelse and The Swedish Environment Protection Board.

REFERENCES

Ekman, S. 1922. Djurvärldens utbredningshistoria på Skandinaviska halvön (in Swedish). 614 p. Stockholm.

Englund, J. 1970. Some aspects of reproduction and mortality rate in Swedish foxes (*Vulpes vulpes*) 1961–63 and 1966–69, *Swedish Wildlife* 8: 1.

Fransson, S. 1965. The borderland. In: The Plant cover of Sweden. *Acta Phytogeogr. Suec.* 50: 167–175.

Hagen, Y. 1952. Rovfuglene og viltpleien (in Norwegian), Oslo: Gyldendal Norsk Forlag.

Hultén, E. 1971. Atlas of the distribution of vascular plants in north western Europe. Generalstabens litografiska anstalts förlag/Stockholm.

Jensen, B and Sequeira, D. M. 1978. The Diet of the Red Fox (*Vulpes vulpes* L.) in Denmark. *Dan. rev. game. biol.* 10: 8.

Keith, L. B. 1974. Some features of population dynamics in mammals. XI international congress of game biologists. Statens naturvårdsverks publikationer 1974: 13E.

Lack, D. 1954. The natural regulation of animal numbers. London: Oxford University Press.

Lindquist, B. 1966. Vegetation belts and floral elements. in: Atlas of Sweden. Generalstabens litografiska anstalts förlag. pp. 43–44. Stockholm.

Lockie, J. D. 1977. Studying carnivores. Mammal Rev. 7: 1, 3–5.

Nellis, C. H. and Keith, L. B. 1976. Population dynamics of coyotes in central Alberta, 1964–68. *J. Wildl. Manage.* 40: 3 389–399.

Pehrson, T. 1958. Sveriges djurvärld (in Swedish) 552 pp. Svenska Bokförlaget. Stockholm.

Ryszkowski, L., Goszynski, J. and Truszkowski, J. 1973. Trophic relationships of the common vole (*Microtus arvalis*) in cultivated fields. *Acta ther.* 18: 7.

Sjörs, H. 1965. Features of land and climate in: The Plant Cover of Sweden. *Acta phytogeogr. Suec.* 50: 1–12.

Sjörs, H. 1967. Nordisk växtgeografi (in Swedish) 240 pp. Svenska Bokförlaget. Stockholm.

Witt, H. 1976. Untersuchungen zur Nahrungswahl von Füchen (*Vulpes vulpes* L.) in Schleswig-Holstein. Zool. Anz. 197: 516, 377–400.

12 ASPECTS OF THE SOCIAL STRUCTURE OF RED FOX POPULATIONS: A SUMMARY

Freek J. J. Niewold*

INTRODUCTION

There were at least two reasons in the Netherlands to start a research program on the ecology and behaviour of the red fox, in 1968. In the first place there was the intention of controlling the fox as the main vector of rabies and of creating a fox-free zone along the border with the Federal Republic of Germany. During the subsequent discussions it became evident that our knowledge about the ecology and social behaviour of the red fox was too scanty to give a good insight in the management of fox populations in relation to the control of rabies. Secondly, we were very anxious about the still unprotected state of the fox and other small predators in contrast to the Birds of Prey.

Research on the social behaviour of the fox was emphasized because the rabies virus is spread by mutual contacts. Data was also collected about the population structure by means of body autopsies, food, mortality, natality and recent rabies outbreaks. In the course of the research period we became more and more interested in the social structure of this solitary predator with its enormous distribution pattern. The red fox is one of the few species which has kept its ground in Europe very well, notwithstanding heavy pursuits by men and a changing environment.

METHODS

In this report I will shortly describe some characteristics of the social structure as so far analysed from our data. These are mainly obtained from the radio tracking of about 150 foxes, old and young, during a 5 year period. Radiolocations of the foxes were made with a mobile receiving unit and mapped approximately each half hour, by means of crossbearings.† After intensive tracking the activity areas could be defined within a week using only nocturnal fixes. During tracking at night we attempted to observe the foxes, wherever possible, in spotlights or with infrared binoculars. In connection with the radio tracking, observations were made of one fox family on artificial feeding sites and near burrows during a period of nearly two years, by using a video camera with remote control, a monitor and infrared

* Research Institute for Nature Management, Arnhem, The Netherlands.

† See S. Broekhuizen, C. A. van 't Hoff, M. B. Jansen and F. J. J. Niewold: Application of Radio Tracking in Wildlife Research in the Netherlands; A Handbook on Biotelemetry and Radio Tracking, editors: C. J. Amlaner and D. W. Macdonald, Oxford 1979, Pergamon Press. pp. 65–85.

and red spotlights. Most observations were carried out in the National Park "The Veluwezoom" on the Veluwe near Arnhem, where no fox hunting occurred. Other data were collected from different places of the Veluwe and from a heath and forest area with a low fox density near Dwingeloo in the North.

RESULTS

Spacing

One of the important results of social behaviour is the distribution of individuals on the available surface, in other words on the available food resources and covers. This will largely depend on mutual competition. In the National Park we looked intensively for fox litters each spring, with the purpose of marking. In some adjacent areas litter locations were known by gamekeepers who were very keen to find and kill the cubs. Irrespective of the fact that more litters were found in wooded areas than in heather moor and culture fields, the found litter locations showed a more or less even distribution pattern.

When we look at the activity areas of individual foxes in one particular study area during one reproductive season we find a similar distribution pattern (Fig. 1). The activity areas were constructed by the minimum area

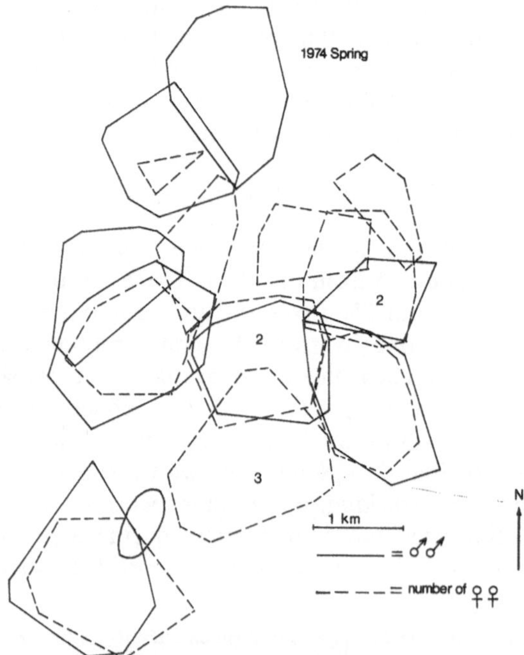

Fig. 1. The distribution of the activity areas of radio tracked adult foxes during spring 1974 in the National Park "The Veluwezoom".

method, all excursions excluded, and only for the foxes with a rather fixed activity area. It should be remembered that not all foxes were tracked during the same period and the same length of time. Nevertheless for most foxes we had enough fixes to draw up a more or less realistic activity area. Notwithstanding the shortcomings in the construction of individual activity areas, it seems obvious that all the available area is divided up in a very regular pattern. One male shared the same activity area with more females, up to a maximum of four. This distribution pattern seems well in accordance with that of the found litter locations. In fact these results strongly support the thought of the existence of a social territorial system and we shall now consider on what such a system is based and how it works.

Territorial behaviour

Territorial behaviour is mainly described in bird studies. I suppose, however, that the general basic components will hold for mammals and other animals as well. Territorial settlement has at least three important characteristics:

a. Isolation and restriction of some or all types of behaviour within more or less clearly defined areas;
b. Intolerance, with respect to similarly motivated conspecifics, resulting in the defence of their territories;
c. Advertising their presence by sound, scent or display within that area.

It is not simple, however, to prove the presence of these characteristics for the hardly observable nocturnal mammals, with an often complex social display. In many cases we can only observe or measure the results of probably displayed behaviour, without knowledge of the behaviour itself. We can now check our data to demonstrate territorial behaviour for the red fox.

Isolation

Firstly, we can exactly analyse all boundaries between neighbouring adult foxes tracked during the same period. The activity areas of adjacent pairs were considered monthly, neglecting excursions and some isolated fixes. The results of comparing 393 monthly combinations of 88 different pairs, proved in 90% of the combinations that the boundaries were composed of a well defined zone with occasionally fixes of both neighbours or even without any fixes. Especially when many fixes were determined and in the case of small territories, this boundary zone, greater than 50 m and very often more than 100 m, could be very well recognized (Fig. 2). It became evident from the analyses of continually tracked individual foxes that they visited these zones more frequently than was shown in the plotted data, but these visits were very short and quick.

Only a small number of the monthly neighbour combinations showed some overlap or a zone smaller than 50 m. This was mainly caused by: a

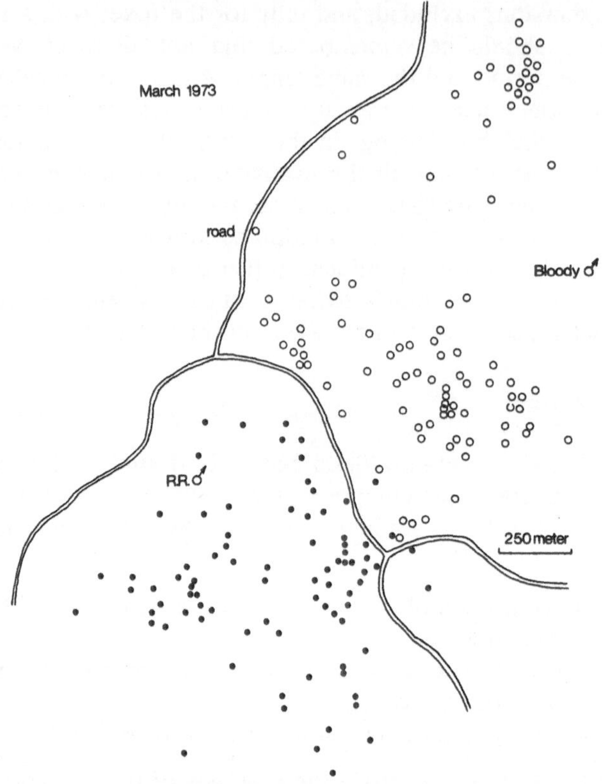

Fig. 2. Example of a boundary between two neighbouring adult dog foxes, as revealed by plotting all half hour radio fixes.

small number of fixes; boundary shiftings within the particular month; a place with easily obtainable food just near the boundary, such as refuse containers of restaurants and our trapping cages; and closely related neighbours especially females. Neighbour combinations which could be tracked during a long period showed an overlap or sharply defined boundaries only for merely one or two months in succession. These data prove that indeed foxes isolate themselves in certain activity areas, while most behaviour was restricted to those areas.

Intolerance

Many times foxes could be observed in spots or with the infrared binoculars, while they were walking along roads or paths, and crossing open places like meadows and fields. A number of observations were also made nearby special feeding sites and refuse containers. In most cases the observations involved were of solitary foxes and young with or without their parents. Sometimes adult foxes could also be seen together. Although many aggressive interactions were observed, the participating foxes either belonged to the same territory or at least one of them was not wearing a transmitter collar so that its social status was unknown. Once we observed that a male, just outside or on the border of its territory, was chased back into its own territory, most likely by another male.

During the whole study-period we could, with certainty, only twice notice an interaction between neighbouring foxes along the boundary zone. We based our conclusions on the many barking and aggressive sounds we heard and on the observed nervous movements of the radio tracked animals concerned. At least more than two foxes were involved in both interactions. Prior to the moment of the interaction, the boundaries were not well established or showed some overlap, while after that they became very well defined. These data demonstrate that probably physical aggressive interactions do not belong to the regular behaviour pattern of the territory-owners. Nevertheless they are important in determining and stabilizing the boundaries between the neighbours. Boundaries are probably formed where the two opposing components of retreat and attack are in balance. Topographical peculiarities could be important by determining the boundaries. This is indicated by the relatively high proportion (42% out of 57 monthly male-combinations, with easily recognizable topographical features along the boundaries such as roads and forest edges (Fig. 2). This does not exclude other possibilities of boundary determinations, among them boundary marking with faeces or urine and the use of odours for reinforcement.

Intolerance, with respect to conspecifics and similarly motivated foxes, was also found by the many shiftings of territory boundaries. This was always accompanied by shiftings in the opposite direction by or even a complete disappearance of the neighbour. They occurred within a few days and were directed towards members of the same sex. This explained partly the greater percentage of overlapping boundaries for the male-female neighbour combinations against the intra-sexual combinations (15%–6%).

Among the tracked adult animals we also found foxes that moved around in a large area, concentrating themselves sometimes in certain small localities. Probably they were not able to establish a territory and they were often radiolocated nearby boundary-zones in between two or more adjacent territories, or in temporarily suitable areas. They were composed of one or two year old males and females and some older males.

Intruders such as wanderers or territory-owners, making regular excursions outside their territory, crossed other territories often very fast and in a straight line. Such behaviour can greatly reduce the contact frequencies with the territory-owner.

Advertisement

The red fox has an extensive range of possibilities for advertising display such as the production of different sounds, scents and movement patterns. All these behaviour patterns are used for communication and orientation, and problems arise when we try to understand the different functional aspects. Some of the behavioural features probably deal with more functions

Plate 1. The mobile tracking unit.

Plate 2. Adult fox with transmitter.

and it is very difficult to distinguish them in the field. Time, place, circumstances and the reactions of conspecifics are the most crucial parameters for determining these functions.

Up till now it was neither possible for us clearly to determine the possible advertising functions of the use of sound and scent-marking behaviour with

Plate 3. Fox cub with small transmitter.

respect to reactions of intruders and neighbours, nor did we exclude these functions.

However, as to movement patterns, there was a great amount of evidence that foxes were continuously walking through their territory, visiting a great part of it each night. Males showed this behaviour more clearly than females and owners of the larger territories had a greater walking speed than owners of the smaller territories. This sort of patrolling in combination with special glands between the toes of the paws, offers good facilities to demonstrate the presence within the territory. As shown by birds, males are normally more involved in self-advertising and this also seems the case with foxes.

Composition and size of the territories

Especially at our feeding sites we could observe the different relationships between the inhabitants of one common territory. It was obvious that such a group was composed of one unrelated adult male and one or more closely related females with their offspring of that year. In our research-area near Dwingeloo we found territories of about 1000 ha, with always one female per territory. In the different areas of the Veluwe-region with smaller territories (Fig. 3), there were regularly two adult females, always a mother with her daughter. During January 1976 our group under observation was composed of four females, namely one adult female with two of her daughters of 1974 and a young female from one of these daughters. Here we were confronted with a real family-territory.

Only once did we find two adult males who shared the same territory. In this, probably exceptional, case we were confronted with a father and his

Fig. 3. The mean monthly territory size of some radio tracked adult red foxes and the average number of fox litters per km² found in various regions in The Netherlands. In parentheses numbers of animals concerned.

		Monthly territory size		Litters/km²
		Males	Females	
Valkenberg	(Veluwe)	63 ha (3)	54 ha (6)	1.0
Nat. Park Veluwezoom	(Veluwe)	127 ha (4)	116 ha (6)	0.6–0.8
Deelen	(Veluwe)	304 ha (1)	238 ha (3)	0.3–0.5
Dwingeloo	(Drenthe)	1000 ha (2)	880 ha (3)	0.1

subordinate very docile, one–two year old son. It was evident from our observations that the adult male was strongly dominant over the other foxes within his territory, the latter often showing sub-missive greeting-ceremonies and begging behaviour towards him. We could also demonstrate a linear dominance rank-order between the adult females, depending on age and the caring of cubs. Older females and females with cubs dominated other females. Only the one female that was obviously coupled with the male, was able to raise her cubs successfully. If two litters could be brought together, the young of both could be raised.

Although the females occupied a common territory, they clearly showed individual preferences for certain localities. Probably forced by a newly settled male, this could result sooner or later in a division of the territory.

One important point in assessing the biological significance of territoriality seems to be the determination of the territory size. I will not go into detail because we have not yet carefully analysed the data so far, but density expressed by the number of litters per km² seems to be inversely proportioned to territory size, as indicated by comparing some data of various areas (Fig. 3). We can state that all the available ground is divided up into territories in all these areas. Except for the two areas with the smallest territories, hunting pressure was very high.

Social relations and territorial behaviour

In foxes the territorial structure is not at all a fixed system but has a very dynamic nature with several adjustments to more specific conditions. During the reproductive period from December till June, there was a kind of pair formation between the male and one of the females. They were regularly tracked together at night and during the daytime they inhabited the same burrows. Observations with the video camera near a breeding earth confirmed the belief that the male carried food to the female just before and after parturition. During the same period most males and females, who shared a common territory, did maintain the same boundaries. Usually the female was found within the area of the male. Later on, during the summer months, they went their own way more and more with respect to boundary shiftings, which frequently resulted in the occupancy of completely separated territories with other partners during the next breeding season. As a consequence we found that no pair formation lasted any longer than one breeding season in succession. Also boundary shiftings and new settlements were frequently found during the period of June till December. All these data suggest that the seasonal variations of the various features of the individual fox territories, are related to the co-operation of male and female for the investment in the offspring.

Another adjustment to special conditions was the remarkable tolerance shown by the territory-owner towards intruders at certain moments. As mentioned before, intruders traversed very fast and in a straight line through other territories, avoiding contacts with the owners. Territorial males especially made excursions outside their territories during the mating time, probably looking for females in heat. A female could be observed closely followed by two males. The hind male was continually attacked and chased off by the closest follower of the female. I suppose that the territorial male at that moment, is not able to chase intruders completely out of his territory, as the female would then have left. Especially when more females are present, the chance for the territorial male to mate with the non-favourite females, seems to be diminished. Probably this has to deal with the shown investment of the male in only one litter. Therefore one possible function of the territory system in foxes seems to be the reduction of disturbances of reproductive performances, like copulations by rivals.

Many times we found territorial foxes during the daytime lying in suitable covers, such as extensive burrows and dense plantations, outside their territories and within others. This was especially the case when foxes were disturbed out of their normal shelters or when cover was lacking within their own territories. There were even a male and a female with a common territory in a rather open sandy heather dune area, that remained during the day in a plantation, 500 m to the north, while they were regularly visiting culture fields, one or two km to the south of their territory at night. These animals were tolerated by the territory-owners at least during the inactive daytime period.

A remarkable situation was created by closely related females (mother and daughter) who were occupying adjacent territories. Although the males were clearly separated, the females could freely visit each other after the division of their former common territory. Sometimes the litters were found even outside the male-territory. As already mentioned there seems to be a great tolerance in behaviour towards the opposite sex.

It may be true that territorial behaviour in the solitary hunting red fox has ultimatedly evolved to reduce intraspecific competition. However, the modifications of the clearly demonstrated territorial structure seem to be highly adjusted to the various ecological and behavioural conditions. This is probably one reason for the great success of this elusive species.

13 YEARLY VARIATIONS OF RECOVERY AND DISPERSAL RATES OF FOX CUBS TAGGED IN SWEDISH CONIFEROUS FORESTS

Jan Englund*

INTRODUCTION

In a study of the population dynamics of the red fox *Vulpes vulpes* (L.) in different parts of Sweden, productivity was measured from litter size and frequency of barren females. The mortality rate among cubs was also estimated (Englund, 1970). As part of the same project, an attempt was made to measure productivity, mortality and dispersal rates of foxes in a restricted area in the south of the northern coniferous zone during the years 1966-69. For lack of financial support, this part of the study could not be properly realized. It is felt, however, that the data obtained, while inconclusive on most points, may furnish the basis for a discussion of certain general problems related to projects of this kind.

STUDY AREA AND METHODS

Foxes were captured, marked and released in north Hällefors, about 200 km west of Stockholm (59°.50 N and 14°.30 E). This area mostly 200–400 m above sea level, forms part of the north coniferous zone with a forest mostly consisting of pine *Pinus silvestris* and spruce *Picea abies.* A large part of the area consists of peat land and there are plenty of lakes and small tarns. In 1944, arable land made up 7%, meadows 2.5% and woodland 73% (Anrick, 1953), and since that time some meadows and fields have been replanted with pine and spruce.

From 1966 through 1969, between about 20th May and 6th July, fox cubs were caught at dens, tagged and released. They were either dug out or caught by hand when they emerged from the dens by night, and in a few cases were lured out by a repeated faint nasal snort (Tembrock, 1957). By night cubs up to about 1.5 kg reacted very quickly to that sound, especially when hungry. Some litters were caught in small baited box traps inserted into den entrances. A couple of foxes were caught with a pole-net when we met them on our walks to new dens. One female cub was trapped in a large baited box trap (18th August 1967).

Each den was normally visited for only a short period of time. Any attempts to get cubs out of a den were discontinued after three days, and irrespective whether more cubs were inside or not such lengthy action took place only

* Swedish Museum of Natural History, Section for Vertebrate Zoology, S-104 05 Stockholm, Sweden.

when the cubs were rather large. In these cases, we fed the young by leaving plenty of meat inside and around the den. The high frequency of recoveries of the few foxes that were confined in the dens for 2–3 days indicates that this treatment did not decrease their chances of survival.

Having been sexed and weighed, the cubs were tagged with a numbered monel metal tag in each ear. The tags used in 1967 were size 4 by the National Band and Tag Co Newport, Kentucky. Subsequently, I used similar tags supplied by I. Ö. Mekaniska in Bankeryd, Sweden. For field identification one or both tags on each fox were fitted in addition with a small piece of coloured plastic fabric. All young were released at their dens or close to the place where they were caught. Notes were made on food remains around dens occupied by foxes and scats were collected for later analysis.

NUMBER OF LITTERS

A total of 92 juvenile red foxes were ear-tagged during the years of 1966 to 1969 inclusive (Table 1). One of these died within an hour of capture (1967) and another is known to have lost both his tags within two weeks (1968). One third of the cubs have been recovered, of which 28 were shot or

Table 1. Food situation, number of litters found, number of cubs tagged, percentages of recoveries and frequency of one-year old vixens during the different years. The number of scats or stomachs analysed, the number of dens, and vixens examined, and the number of litters where the number of cubs are known are given in brackets.

	1966	1967	1968	1969
Mean number of rodents per stomach (Jan–April)[1]	0.7–1.0 (26)	0.9–1.0 (25)	0.1–0.3 (114)	0.4 (121)
Mean number of rodents per scat (May–June)	1.0 (247)	0.7 (122)	0.2 (76)	0.8 (248)
Number of small prey with meat found at active dens	28	0	0	0
Number of large prey with meat found at dens	5	0	0	0
Number of small prey without meat found at dens	7	4	2	7
Number of large prey without meat found at dens	20	15	6	11
Mean number of rodents per scat (Sept–October)[1]	1.7 (11)	0.2 (19)	0.6 (13)	?
Number of litters found[2]	7 (257)	12 (362)	2 (280)	7 (295)
Mean litter size	5.3 (6)	4.6 (9)	4.5 (2)	4.3 (3)
Number of cubs ear-tagged and released	30	37	9	15
Percentage of recoveries[3]	50	16	44	33
Hunting bag during next hunting season	200	240	140	120

[1] Englund, 1970; [2] in some cases no cubs were caught; [3] 50% in 1968 if the male known to have lost both tags is excluded.

trapped and another 2 killed by traffic. Furthermore some cubs were observed or re-trapped during their first summer and released again.

Most foxes were killed during their first year of life, only 2 during the second and 1 during the third year. No more recoveries will be obtained since the last one known to be killed was reported nine years ago.

Unfortunately there was no possibility to observe all dens in all years within the study area; therefore it is impossible to give exact figures on the variation in the number of litters born during the years studied. No doubt, however, fewer litters were born in 1968 than in the other years (Table 1).

Loss of ear-tags

For calculating the risk of ear-tag loss and the frequency of animals that probably had lost both tags, Fairley (1969) used a simplified version of a fairly complicated formula for this purpose. He found that 12% and 8% of the tags were lost from adult and young foxes respectively. Assuming all tags have the same chance to be lost, this means that about 1.4% and 0.7% of the foxes had lost both tags.

According to Jensen (1973) about 4.5% of cubs tagged with aluminium tags will loose both tags during their first year of life while the corresponding figure for monel tags will be 0.3%. The frequency of double tag loss has been reported to increase with time from 0.8% during the first year after tagging for monel metal tags, to 1.6% during the second year and thereafter to 8.1% (Hubert et al., 1976).

In 28 of my recoveries the number of monel tags left is known and in 9 of these one tag was lost. Applying the formula given by Hubert et al. one finds that 3.7% had lost both tags (nearly all cubs were recovered within one year). This is much higher than found by others but is close to what Lord (1956) found for gray foxes. In that case, however, neither the type of tags used nor the age of the foxes tagged is known.

One possibility of the higher frequency of tag loss in my study area may be a worse food situation there resulting in a higher risk of tearing off the tags during fights for food. In 1966 for example when the food situation was extremely good only two foxes out of 14 had lost one of the tags, while in 1967 three foxes out of six and in 1968 three foxes out of four had lost one of them.

From the above one may ask if there is any reason to believe in such calculations? One assumption here is that all tags are at roughly the same risk, which can hardly be true. When cubs get too little food for example, either because of an overall scarcity of food or because the litter is extremely large, the chance of losing both tags must be higher. And if a tag is lost because of an allergic reaction, for example, the same fox will probably react on the other tag also, resulting in a higher than chance frequency of total (double) loss.

It seems possible, therefore, that some foxes run a greater risk of losing tags than do others, which restricts the value of calculations of the kind shown in Table 2. When comparing the recovery rate from different years or

Table 2. Percentage recoveries of cubs tagged at different sizes. Number of cubs tagged shown in brackets.

	1966	1967	1968	1969	All years
Weight in g.					
480–600	0 (1)	0 (3)	0 (2)	0 (2)	0 (8)
601–1000	38 (16)	6 (16)	57 (7)*	100 (1)	33 (40)
1001–1500	67 (9)	36 (11)	—	30 (10)	43 (30)
1501–1900	75 (4)	14 (7)†	—	50 (2)	38 (13)

* if one cub known to have lost both tags within 3 weeks is excluded, the recovery rate will be 67% instead.
† the weight of one of these cubs was 2.7 kg when tagged.

areas, however, such calculations must be valuable to see if part of the problem can be explained by differences in tag loss. But these comparisons will be meaningful only in case of large number of recoveries.

RECOVERY RATES IN DIFFERENT YEARS

The recovery rates for cubs tagged during 1966 to 1969 inclusive and recovered from the end of August exhibit substantial annual differences. The percentages were 50, 16, 44 (50 if the one known to have lost both his tags is excluded) and 33, respectively. These differences are statistically verified ($0.05 > P > 0.025$) and the significance is largely caused by the low recovery rate for the 1967.

We may ask whether the low recovery from the 1967 markings was caused by a higher frequency of double loss of tags or by a higher mortality among cubs in that year. Moreover should the latter be the case we may ask whether such mortality was achieved because more cubs were tagged when very small that year, i.e. at an age where the mortality may be higher, or if it really was a higher overall mortality among cubs in that year. Table 2 shows the percentages of recoveries of cubs tagged at different ages. The subdivision in weight-classes makes the groups hopelessly small. It is clear, however, that these cubs which were marked at a size of 0.6–1.0 kg were not recovered as often in 1967 as in other years. From the end of August onwards, only 6% of these were recovered as compared with 38% and 57% from the other years.

I therefore believe that the low recovery rate in 1967 mostly is caused by a higher overall mortality in that year, although part of the answer might be a higher proportion of total tag loss that year.

No other data to support my conclusion are at hand. However, it should be mentioned that among the 22 litters examined more closely during the four years (7, 9, 2 and 4), I have found 4 young dead or dying. One case was in 1966, in a litter of 11 (mean weight of the 10 live cubs 735 g). Two others were in 1967, viz., one in a litter of 6 (mean weight of the 5 live cubs 690 g), and one in a litter of 4 (mean weight of all cubs 480 g) who was still alive when caught but died within a couple of hours. The last case in 1969, belonged to a litter of 6 (mean weight of 5 live cubs 1190 g). In that case, skeleton remains showed that the cub died when very small.

DISPERSAL

As can be seen in Fig. 1 and Table 3, most foxes were killed close to the place where they were tagged. Out of the 30 recoveries from the latter part of August and onwards, 56% were killed within 4 km and 70% within 8 km from the place of tagging.

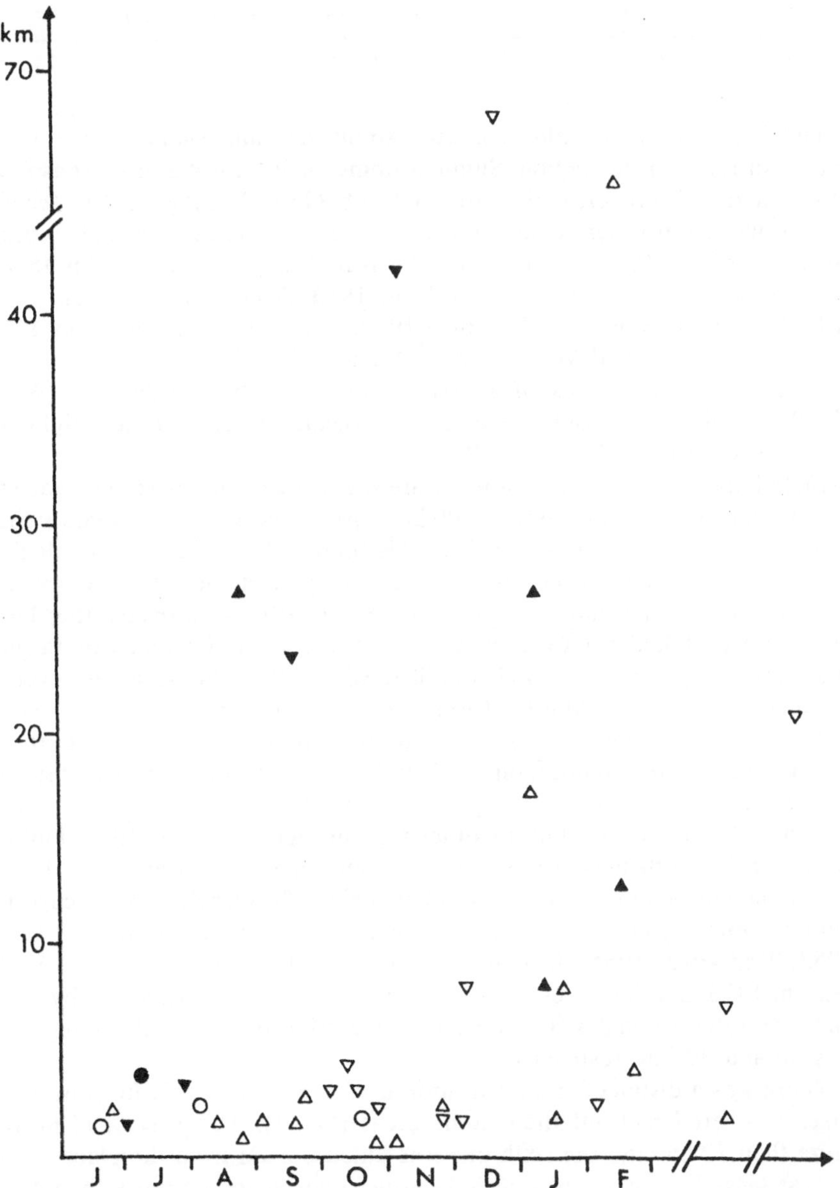

Fig. 1. Distances between first and last captures (or field observations) for foxes recovered during the 1st, 2nd or 3rd hunting season after tagging. Filled signs refer to foxes tagged in 1967.
▽▼ = ♀♀, △▲ = ♂♂, ○● = sex unknown, ▼▲● tagged in 1967.

Table 3. Number of foxes killed from late August and onwards, and the distances travelled (km).

Year of tagging	0–8		9–15		16–20		21–25		26–30		41–45		65–70		Total	
	♂	♀	♂	♀	♂	♀	♂	♀	♂	♀	♂	♀	♂	♀	♂	♀
1966	8	5	—	—	—	—	—	1	—	—	—	—	1	—	9	6
1967	1	—	1	—	—	—	—	1	2	—	—	1	—	—	4	2
1968	2	1	—	—	1	—	—	—	—	—	—	—	—	—	3	1
1969	1	3	—	—	—	—	—	—	—	—	—	—	—	1	1	4
Total	12	9	1	—	1	—	—	2	2	—	—	1	1	1	17	13

Unfortunately very little is known about size and shape of the home ranges of foxes in this region. Summer home-ranges for 4 adult vixens in an adjacent area, however, varied from 3.2 to 8.8 km^2 (Ryding, 1977). According to data from other regions however, the area varies from 0.2 to more than 10 km^2 and the maximum diameter from less than 1 up to 8 km (Seton 1929, Murie 1936, Hamilton 1939, Scott 1943, Arnold and Schofield 1956, Schofield 1960, Storm 1965, Ables 1969, Sargeant 1972, Niewold 1973, Storm et al. 1976, and Macdonald 1978 and 1980b).

Factors affecting the size of the territories may be habitat diversity (Ables 1969) as well as fox population density (Sargeant 1972), and the stability of the food resources (Macdonald 1980b).

Field indications suggest that foxes are rather scarce in my study area. The annual bag of foxes on about 1200 km^2 over a period of 20 years varied from 76 to 299 specimens, which equals 0.06 to 0.25 per km^2. Ecological diversity is low and food resources are very unstable. It is reasonable therefore to assume that foxes living here regularly move over rather large areas. Foxes killed up to a distance of 8 km from the place of tagging therefore have been regarded as killed within their home areas, even if distance itself will not tell if the foxes crossed the border of this area or not.

Accepting the figure of 8 km as a maximum range of action for a sedentary fox in this habitat, one finds that some individuals start to disperse as early as late August or early September (Fig. 1), which is about one month earlier than reported for other regions (Jensen, 1968, 1973, Phillips et al., 1972). Both these cases (one male and one female) are from 1967.

Also in contrast to reports from other regions (Tables 4, 5) my recoveries did not show any marked sexual difference in dispersal rate (Sheldon, 1950, 1953, Tchirkova, 1955, Arnold and Schofield, 1956, Phillips et al., 1972, Jensen, 1973, and Storm et al., 1976). The mean distance travelled by the 5 males and the 4 females that, on our above criterion, left their home areas was 30 and 39 km, respectively.

There was a distinct annual variation in dispersal. In 1967, five out of six foxes recovered had left their home areas (Fig. 1). These five had moved more than 12 km and the sixth one was killed 8.2 km from the place where he was tagged. Among cubs tagged during other years, no more than 4 out of 24 ($\frac{2}{15}$, $\frac{1}{4}$ and $\frac{1}{5}$) had moved more than 8 km, and most of the other were killed within 4 km. The figures are admittedly small; however, they certainly indicate that differences do occur between years.

Table 4. Percentage of recoveries of foxes tagged as juveniles and killed more than 8 km from the place of tagging (in most cases the recoveries are within one year). Number of recoveries within brackets. Recoveries of cubs received before the latter part of their first August are excluded (before October in Iowa and Illinois).

	Males	Females	Sex unknown	Author
Europe				
Ukrainian SSSR	89 (9)	14 (7)	—	Tchirkova, 1955
Germany (age not known)	—	—	37 (49)	Behrendt, 1955
Norway	—	0 (1)	100 (1)	Lund, 1967
Sweden (agricultural area)	12 (8)	0 (4)	0 (1)	Marcström, 1968
Sweden (north. conif. forest)	33 (3)	0 (1)	100 (1)	Marcström, 1968
Ireland	44 (9)	20 (5)	—	Fairley, 1969
Denmark	35 (65)	20 (49)	—	Jensen, 1973
Sweden (north. conif. forest)	29 (17)	31(13)	—	Present study
North America				
Iowa	—	—	100 (2)	Errington and Berry, 1937
New York	100 (6)	25 (4)	100 (1)	Sheldon, 1950, 1953
Minnesota	0 (1)	100 (1)	—	Longley, 1962
Iowa and Illinois	80 (312)	37 (223)	—	Storm et al., 1976
Wisconsin	88 (23)	58 (19)	—	Pils and Martin, 1978

Table 5. Percentage of recoveries beyond 8 km distance of foxes tagged as juveniles and recaptured more than one year thereafter. Number of recoveries within brackets.

	Males	Females	Author
Denmark	29–81 (31)	13–35 (31)	Jensen, 1973
Iowa and Illinois	96 (49)	58 (33)	Storm et al., 1976

CHANGES IN THE FOX POPULATION IN THE STUDY AREA

During the period of tagging the number of foxes killed in a hunting district of about 120,000 hectares, inside which this study was undertaken, increased from 197 in the hunting season of 1966–67 to 235 next year. Thereafter it dropped markedly to 133 and in 1969–70 down to 117. These changes may be explained in the following way.

As pointed out earlier both mortality and dispersal rate was higher among cubs born in 1967 than in other years, but why? And how could the hunting bag increase then next year? According to the mean number of rodents per scat collected in the spring of 1967 the abundance of rodents seemed to be rather good at that time. However, this index is only relative and is not so informative about the absolute amount of food. A better index of that is the amount of food found at the dens.

As seen in Table 1 there were remnants from 19 different prey around 12 dens in 1967; that is 1.6 prey per den. The corresponding figures for the other years were 8.6, 4.0 and 2.6 prey per active den. Furthermore, in the spring of 1966 there was still meat on most of the prey remnants (Table 1). The high mortality rate in 1967 therefore is supposed to have been caused by starvation. This has not necessarily been caused only by a general lack of food this spring, but can also have been caused by a higher proportion of parents with a lower capability to find food. Among vixens with placental scars that year collected in Värmland and adjacent areas, 5 out of 14 were yearlings (36%) compared with 9 out of 45 (20%) for the other three years altogether.

Sometimes during the summer of 1967 the number of rodents decreased drastically, and that year most of the recoveries had left their home areas. Furthermore one of them left as early as late August which is much earlier than other authors have found. Although the data is very small it indicates that fox cubs in these habitats disperse earlier and more often when prey availability is low.

Thus there are reasons to believe that the lack of food, perhaps in combination with the high proportion of inexperienced parents in the spring of 1967, caused the high mortality rate among the cubs, and that the food situation also brought the young foxes to disperse earlier in the autumn. In spite of that more foxes were killed in the winter of 1967–68 than other years. This problem may be explained in two ways. One is that the high productivity and the high survival of cubs in 1966 (perhaps in combination with few foxes shot in 1966–67 caused by the high abundance of rodents; see below) resulted in an increase of foxes in the spring of 1967. The higher number of vixens should then produce many more young (12 litters were found in 1967 compared to 7 in 1966) and in spite of the higher natural mortality among them during the summer period the population should have increased to the autumn of 1967. The extreme rarity of rodents during the winter of 1967–68 might also in a higher degree than usual have forced the foxes to the baited traps or to the baits where hunters waited in the moonlight nights. The end result of the above might have been a reduction of the population to the spring of 1968 although no direct figures are at hand. This in combination with an increased frequency of sterility caused by the scarcity of rodents (Englund, 1970) resulted in the very few litters born in 1968 (only 2 were found). The number of foxes killed next hunting season therefore decreased drastically. The decrease of foxes killed may also be explained by an increase in the rodent population which may make foxes more difficult to trap again.

In the spring of 1969 the number of foxes should be at a low. Regarding the low productivity in 1968 most foxes should be more than one year old. This in combination with the increase of the rodents must have resulted in an increase of the number of foxes to the autumn of 1969. In spite of that the hunting bag dropped even further down to around 117. This drop could be explained by a decreased hunting effectiveness caused by the fact that the rodents increased furthermore.

I have tried to explain data gathered from both shooting statistics and the tagging project. Although it is possible to explain data that first look contradictory, it is very unsatisfactory not to have more and better data. However, this is common in studies of population dynamics and demonstrates that projects on populations should be carried out on a much larger scale.

DISCUSSION

Whether dispersal is initiated by social behaviour such as overt aggression, changes in odour released during scent marking or a combination of several such factors is not known, due to lack of observations (Storm et al., 1976). However, Storm et al., state that their study produced three things that apparently contradict the hypothesis that young foxes are "forced" from natal ranges by their parents (pages 61 and 62). "It also seems unlikely that limited food initiates dispersal in foxes, at least in the midwest" they say. The question why foxes disperse therefore in most cases is unknown.

In most papers no or very few data exist about the food situation when the cubs were tagged. And daily weight increase of the cubs which indirectly should inform of the food situation is not at hand either. Data of population density which must be of importance here are non-existent too. In spite of the incompleteness of important data a comparison of the results of different authors is very interesting.

One then finds that the American foxes disperse more often than the European ones (Tables 4 and 5). Data from a rather stable population in Wales, England confirm this conclusion since 7 foxes out of 8 tagged when young were recovered within 13 km (Lloyd, 1968). And another material from Michigan, USA showed a high tendency of dispersal since the mean distance travelled by 9 males and 7 females tagged when young was 68 and 14 km respectively (Arnold et al., 1956). How do we explain the differences between the continents? Could it after all be that food was scarce every year in Iowa and Illinois or are the population densities there much higher than in Europe? This is not known, but that the areas in the midwest are not always overpopulated is pointed out by Sargeant (1972).

Another detail of interest is that 95% of the vixens in the midwest reproduce, and that the litter sizes are large too with a mean of about 7 (Storm et al., 1976). In Wisconsin 59% of the yearlings and 89% of older vixens reproduce, and the mean litter size is around 6.0 (Pils et al., 1978) and in Ontario nearly all reproduce with a mean of about 8 cubs per litter (Johnston, pers. comm.). This is quite different from the situation in Sweden where a large proportion of vixens is sterile. In the more unproductive taiga areas in the northern part of Sweden for example the frequency of productive vixens varies depending on the amount of rodents (12 to 67% among yearlings and 33 to 87% among older vixens). And in the agricultural provinces of Uppland and Södermanland there is always a rather low proportion of productive vixens and also with small variations between years (35 to 60% among yearlings and 64 to 83 among older ones) (Englund 1970). The reason for that may be the higher density of the fox population in these productive areas which in turn may be caused by the richer and also more stable food supply there.

The productivity in most European countries, on the other hand, is as high as in the midwest foxes in US. In The Netherlands the pregnancy rate is close to 100% and in Switzerland, Bavaria and Wales about 90% of the vixens reproduce (Lloyd et al., 1976). In a non-hunted population living in a very diverse habitat outside Oxford in England on the other hand most vixens do not reproduce, at least during some years (Macdonald, 1980a).

Summing up one can say that the situation seems to be the following. In some areas of Europe a high proportion of the vixens are unproductive, even when food abundance is high, but in other areas nearly all vixens reproduce. And the few data we have from tagged foxes indicate that the European foxes seldom disperse. American foxes on the other hand reproduce strongly and disperse to a very high degree. How are these differences to be explained?

I have no idea at which time of the year foxes are killed in different parts of the world. If for instance, a very high proportion are killed early in autumn in most European countries, but very few are killed before midwinter in the midwest in US but many are killed close before the mating season then the data can fit in a theory.

In Europe, most foxes have probably been tagged where hunters have shown where the dens are, that is in areas where many foxes are killed. If hunting in these areas starts early in the autumn many tagged foxes will be killed before time of dispersal. And when the others decide to disperse, then so many foxes have already been killed that good and empty areas are available close to their parents territory. This means that most foxes do not have to go far to find an area of their own. Being owners of good areas also means that they can reproduce. Only so called "innate dispersers" have to move far.

If in the USA very few foxes are killed during the autumn but the hunting or trapping pressure is very intense during a short midwinter period, then the midwest data will fit the theory too. Most juveniles will have time to disperse during the autumn and they too must go far, since the populations are dense. And a high reduction well in time before the next mating season will result in territories enough for nearly all vixens who therefore reproduce.

In a small area outside Oxford with a high and stable food situation and where there is no hunting, foxes live in groups with one adult male, 2–5 adult vixens producing one or at most two litters. The population density here is extremely high, most vixens are sterile and all young males disperse. In these circumstances, Macdonald argues (1980a), there is a high survival value for young females who, if allowed, stay at home even if they suffer perhaps lifelong sterility; and it may benefit in the long run even the parents. When the territory is filled with adults, the new young foxes have to disperse. When the amount of food falls well below what was available during the time the group was built up, some of the adult sterile vixens may be chased away by the productive vixens. There are at least some indications of this (Macdonald, 1977, 1980b).

What about the Swedish data? In the northern area many foxes disperse during the autumn–winter months when rodents are extremely few, and in the following spring most foxes do not reproduce. Dispersal during these years is caused by food scarcity. The low productivity in the following spring may be caused either directly by food scarcity or indirectly, either through increased physical stress or perhaps by the relatively high population densities around rubbish dumps favoured by the dispersing foxes.

A combination of high hunting mortality during such winters, low productivity next spring and rather high mortality rate among cubs born will result in few foxes next autumn. With growing rodent populations there will then be many good areas. Therefore no foxes have to move far, all females will get their own territories with plenty of food and therefore most vixens will reproduce. There is no survival advantage for young vixens to stay at home either and therefore no large social groups would be built up. But we know nothing about this latter question.

As long as rodents are abundant the number of foxes will increase. If there is time enough, that is how many years it will take before the rodent population will crash, the number of foxes that remained after the last crash and the effectiveness of the hunting, there will finally be no suitable areas left for new foxes. In that situation social groups of "Oxford-type" may be formed and thereafter all young have to disperse. It is not known however if this ever happens in the northern coniferous forests in Sweden.

What about the situation in the agricultural areas in Sweden such as the provinces north and south of Stockholm? The amount of rodents varies greatly in these areas too, but will never be so low as in the north. The amount of alternative food is also better. Nevertheless a high proportion of the vixens is sterile and the mortality rate among cubs seems to be very high too, even in years with plenty of rodents (Englund, 1970). Unfortunately very little is known about dispersal here. Some few recoveries from a newly started project indicate, however, that at least males disperse. Four males out of five tagged as juvenile and killed after October had dispersed at least 19 km from the tagging place, and one of them as far as 248 km, but none of four juvenile females were killed outside two km from where they were tagged. How do we explain the situation in this area?

During years with mass abundance of rodents even otherwise marginal areas should be good for the foxes, and therefore I doubt if the high frequency of sterility and cub mortality really is caused by food scarcity. Other explanations seem more probable.

If the fox populations in the agricultural areas are not depressed as they are in the northern area and if the hunting pressure is rather low, which it seems to be (according to shooting statistics), then data here too will be understandable. The high sterility frequency should then be socially induced caused by high population densities like that around Oxford. And the high mortality rate of cubs might be explained in the same way, since it is known from fox-farming that many vixens will not take care of their young but rather kill them (Johansson, 1938, Pearson et al., 1946). According to Pearson et al., this behaviour is not correlated with factors such as diet, weather or time of whelping but must be caused by something else. The curious behaviour of a low-ranking vixen resulting in total loss of her litter supports this idea (Macdonald, 1980a).

In spite of the low recruitment of young foxes to the autumn population, all areas may be occupied. Young males therefore have to disperse far. If rodents are very common young females will probably stay at home and thus build up social groups. If the rodent population crashes, probably all young of the year as well as old sterile vixens have to disperse, perhaps forced away by the productive pair. However, no data are at hand supporting these hypotheses.

Summing up one can say that both productivity, mortality, and rate of dispersal varies in different areas. The reason for these differences are not known but may be explained by differences in amount and stability of food resources, in population densities, in hunting pressures or the time of the year most foxes are killed. Better data for all relevant parameters are needed, however, if we really want to understand how the populations are regulated.

ACKNOWLEDGEMENTS

I wish to express my appreciation to Rune Pettersson who escorted me to all the dens examined, and to Lennart Pettersson who helped me to catch the cubs. I also want to thank David Macdonald for discussions on fox social behaviour. The study was supported by grants from the Natural Environment Protection Board and Helge Axelsson Johnsons stiftelse.

REFERENCES

Ables, E. D. 1969. Home-range studies of red foxes (*Vulpes vulpes*). *J. Mamm.*, 50(1): 108–120.

Anrick, C. J. 1953. Arable land area 1944, in Atlas över Sverige, 65–66. 6 pp.

Arnold, D. A. and Schofield, R. D. 1956. Home-range and dispersal of Michigan red foxes. Papers of the Michigan Academy of Science, Arts and Letters, 41: 91–97.

Behrendt, G. 1955. Beiträge zur Ökologie des Rotfuchses (*Vulpes vulpes* L.). *Zeitschrift für Jagdwissenschaft*, 1 (4): 161–183.

Englund, J. 1970. Some aspects of reproduction and mortality rates in Swedish foxes (*Vulpes vulpes*), 1961–63 and 1966–69. *Viltrevy* (Swedish Wildlife), 8 (1): 1–82.

Errington, P. L. and Berry, R. M. 1937. Tagging studies of red foxes. *J. Mamm.*, 18 (2): 203–205.

Fairley, J. S. 1969. Tagging studies of the red fox *Vulpes vulpes* in north-east Ireland. *J. Zool., Lond.* 159: 527–532.

Fairley, J. S. 1970. More results from tagging studies of foxes *Vulpes vulpes* (L.). *Irish Naturalist Journal*, 16 (12).

Hamilton, W. J., Jr. 1939. American mammals: their lives, habits, and economic relations. N.Y. XII + 434 pp.

Hubert, G. F., Storm, G. L., Phillips, R. L. and Andrews, R. D. 1976. Ear tag loss in red foxes. *J. Wildl. Mgmt.*, 40 (1): 164–167.

Jensen, B. 1968. Preliminary results from the marking of foxes (*Vulpes vulpes* L.) in Denmark. *Danish Review of Game Biology*, 5 (4): 3–8.

Jensen, B. 1973. Movements of the red fox (*Vulpes vulpes* L.) in Denmark investigated by marking and recovery. *Danish Review of Game Biology*, 8 (3): 3–20.

Johansson, J. 1938. Reproduction in the silver fox. *The annals of the agricultural college of Sweden* 5: 179–200.

Lloyd, H. G. 1968. The control of foxes (*Vulpes vulpes* L.). *Ann. appl. Biol.*, 61: 334–345.

Lloyd, H. G. and Englund, J. 1973. The reproductive cycle of the red fox in Europe. *J. Reprod. Fert., Suppl.*, 19: 119–130.
Lloyd, H. G., Jensen, B., Vanhaaften, J. L., Niewold, F. J. J., Wandeler, A., Bögel, K. and Arata, A. A. 1976. Annual turnover of fox populations in Europe. *Zbl. Vet. Med.* B. 23: 580–589.
Longley, W. H. 1962. Movements of red fox. *J. Mamm.*, 43 (1): 107.
Lord, R. D. 1956. The loss of ear tags in the gray fox and raccoon. *J. Mamm.*, 37 (4): 548.
Lund, Hj. M.-K. 1967. Om merking av rev. *Fauna, Norsk Zoologisk Förenings Tidskrift*, 20 (1): 7–17.
Macdonald, D. W. 1977. The behavioural ecology of the red fox, *Vulpes vulpes*; a study of social organisation and resource exploitation. D.Phil. thesis. Oxford University.
Macdonald, D. W. 1978. The sociable fox. *Wildlife Magazine*, June: 272–277.
Macdonald, D. W. 1980a. Factors affecting reproduction of the red fox, *Vulpes vulpes.*
Macdonald, D. W. 1980b. Rabies and Wildlife: A Biologists Perspective. Oxford. Univ. Press. Oxford.
Marcström, V. 1968. Tagging studies on red fox (*Vulpes v.*) in Sweden. *Viltrevy* (Swedish Wildlife) 5 (4): 103–117.
Murie, A. 1936. Following fox trails. *Univ. Mich. Mus. Zool. Misc. Publ.* 32: 1–45.
Niewold, F. J. J. 1974. Irregular movements of the red fox (*Vulpes vulpes*), determined by radio tracking. In XI International Congress of Game Biologists, Stockholm, September 3–7, 1973: 331–337.
Pearson, O. P. and Bassett, Ch. F. 1946. Certain aspects of reproduction in a herd of silver foxes. *Amer. Nat.*, 80: 45–67.
Phillips, R. L., Andrews, R. D., Storm, G. L., and Bishop, R. A. 1972. Dispersal and mortality of red foxes. *J. Wildl. Mgmt.* 36 (2): 237–248.
Pils, Ch. M. and Martin, M. A. 1978. Population dynamics, predator-prey relationships and management of the red fox in Wisconsin. Technical Bulletin No. 105. Dept. of Natural Resources, Madison, Wisconsin.
Ryding, J. 1977. Home-range storlek, revirhävdande och biotopval hos rödräv, *Vulpes vulpes* (L.) under grytperioden i sydtaigan. SNV. PM 1002.
Sargeant, A. B. 1972. Red fox spatial characteristics in relation to waterfowl predation. *J. Wildl. Mgmt.*, 36 (2): 225–236.
Schofield, R. D. 1960. A thousand miles of fox trails in Michigan's ruffed grouse range. *J. Wildl. Mgmt.*, 24 (4): 432–434.
Scott, Th. G. 1943. Some food coactions of the northern plains red fox. *Ecol. Monog.*, 13 (4): 427–479.
Seton, E. Th. 1929. Lives of game animals. Vol. 1. Doubleday Company, Garden City, New York. 640 pp.
Sheldon, W. G. 1950. Denning habits and home range of red foxes in New York state. *J. Wildl. Mgmt.*, 14 (1): 33–42.
Sheldon, W. G. 1953. Returns on banded red and gray foxes in New York state. *J. Mamm.*, 34 (1): 125–126.
Storm, G. L. 1965. Movements and activities of foxes as determined by radio-tracking. *J. Wildl. Mgmt.*, 29 (1): 1–13.
Storm, G. L., Andrews, R. D., Phillips, R. L., Bishop, R. A., Siniff, D. B., and Tester, J. R. 1976. Morphology, reproduction, dispersal, and mortality of midwest red fox populations. Wildlife Monographs No. 49. 82 pp.
Tchirkova, A. F. 1955. Tagging foxes. *Canadian Wildl. Serv.*, Translations Russian Game Reports 3 (1958): 208–214.
Tembrock, G. 1957. Zur Ethologie des Rotfuchses (*Vulpes vulpes* L.) unter besonderer Berücksichtigung der Fortpflanzung. *Zool. Garten*, 23: 289–560.

14 ZUR FEINDVERMEIDUNG FREILEBENDER ROTFÜCHSE

Felix Labhardt

SUMMARY

This paper discusses some behavioural patterns of foxes concerned with the avoidance of enemies. Before escaping from a presumed enemy, foxes try to find out about it. In such an ambiguous situation, between the desire to escape and the desire to explore, characteristic behaviour patterns are shown, such as lifting a foreleg (Figs. 1 and 2), alternating positions of the ears (Figs. 2 and 3) and bowing and lifting the head (Fig. 5). There are also characteristic expressions of fear, such as "squint-eyed" (Fig. 4). For acoustic location of quiet sounds, the head is held in a slanting position (Fig. 6).

Young foxes, having recognised a danger, always flee into the den (strong attachment to the birthplace). On the other hand in most observed cases adult foxes no longer behave in this way, which is perhaps a behaviour pattern adapted to the specific hunting methods of man.

Young foxes exposed to danger are warned by two different cries from their parents: Repeated warning cries made from a distance and a single barking-like call from nearby.

The phenomenon of higher trustfulness at night, which can be easily observed in the behaviour of young foxes, is discussed. As a hypothetical conclusion it is supposed, that the feeling of security given by the darkness of the night is equivalent to that experienced in the darkness of the den during the juvenile phase.

EINLEITUNG

Viele Tiere dienen anderen als Beute. Das wichtigste Anliegen eines jeden Individuums, sich selbst zu erhalten, ist mit der Nahrungsaufnahme allein noch nicht garantiert. Ebenso wichtig ist das Bestreben, nicht selbst seinen Fressfeinden zum Opfer zu fallen. Das meist schwächere Beutetier versucht durch meidende Verhaltensweisen sich dem Zugriff des Feindes zu entziehen. Aber bei weitem nicht jede fremde Tierart ist für die andere ein Feind. Eine energiezehrende Flucht lohnt sich also nur, wenn der mutmassliche Feind als solcher erkannt oder die von ihm ausgesendeten Störreize zumindest genau lokalisiert sind, um die Fluchtrichtung bestimmen zu können. Wie sehen solche feindvermeidenden Verhaltensweisen beim Fuchs aus? Die hier diskutierten Beobachtungen machte ich an freilebenden Füchsen im Raume Basel. Wie in den meisten Gebieten Europas sind auch

* Naturhist. Museum, CH-4051 Basel, Switzerland.

Abb. 1

Abb. 2

Abb. 3

Abb. 4

Abb. 5

Abb. 6

hier die natürlichen Feinde des Fuchses, wie Adler, Luchs und Wolf, längst durch den Menschen vernichtet worden. Der sich ausgesprochen technophil verhaltende Fuchs hat noch als einziger Feind den Menschen, wenn wir von tödlichen Krankheiten, wie die Tollwut absehen.

Meine Absicht, die Tiere möglichst unbemerkt zu beobachten, schlug häufig fehl. Die vorsichtigen Tiere entdeckten mich oft und ihr Fluchtverhalten protokollierte ich nebenbei, sozusagen unfreiwillig. Den hier geschilderten Verhaltensweisen liegt also keine systematische Beobachtungsweise zu Grunde, mit anderen Worten, ich habe mich nie absichtlich den Tieren als Feind zu erkennen gegeben und sie somit zu Schreck- und Fluchtverhaltensweisen provoziert.

DAS SICHERN

Umweltreize, die als Störung empfunden werden und von einem möglichen Feind stammen können, sind in der Regel optischer, akustischer oder olfaktorischer Art und werden vom Tier durch die entsprechenden Sinnesorgane Auge, Ohr und Nase registriert. Häufig gelingt es dem Tier aber nicht auf Anhieb, einen solchen Störreiz zu lokalisieren oder mangels Erfahrung die Bedeutung oder gar den Verursacher zu erkennen. Das Tier nimmt die orientierende Haltung des Sicherns ein, es befindet sich, zumindest aus der Sicht des Beobachters, im Zustand zwischen Angst und Neugier, es fühlt sich unsicher.

Der verunsicherte Zustand eines Fuchses kann vor allem an folgenden Merkmalen deutlich festgestellt werden:

(a) Vorderbeinheben in gespannt verharrender Körperhaltung (Abb. 1, 2).
(b) Alternierende Stellung der Ohrmuscheln (Abb. 2, 3).
(c) Angstausdruck: Vibrissen angelegt, Augen weit geöffnet, manchmal verdreht (Abb. 4).
(d) Kopfpendeln (Heben und Senken des Kopfes, Abb. 5).

Zu(a) Das Vorderbeinheben ist bei Säugetieren weit verbreitet. Die Bedeutung dafür ist unklar. Es wird sowohl bei überraschend auftretenden akustischen wie optischen Störreizen gezeigt.

Zu(b) Die wechselnde Stellung der Lauscher kann man als Bereitschaft zur Aufnahme von akustischen Reizen deuten. Mit der Ausrichtung der Ohren, das eine nach vorne, das andere seitlich, wird der grösstmögliche Empfangswinkel erreicht. Die wechselnde Stellung der Ohren wird auch beim blossen Erblicken eines fremden, unvertrauten Objektes angenommen (Abb. 3, das Tier betrachtet einen Handschuh). Wird der akustische Reiz aber klar lokalisiert, so sind die Ohren zur Reizquelle hin gerichtet (Abb. 1).

Zu(c) Der auf Abbildung 4 gezeigte Jungfuchs wurde in der Nacht fotografiert. Er befand sich gerade in einer Ruhephase, als er von einem Störreiz stark beunruhigt wurde. Das Tier wendete den Kopf aufgeregt nach allen Seiten, die Lauscher wurden immer wieder in wechselnde Stellung gebracht. Den Grund für seine Beunruhigung konnte ich nicht ausmachen. Schon Tembrock (1957) hat das Verdrehen der Augen als Ausdruck der Angst bei seinen Füchsen in Gefangenschaft festgestellt ("Blick ist ungerichtet").

Zu(d) Das merkwürdige Senken und Wiederheben des Kopfes in mehr oder weniger rascher Folge lässt sich beim Fuchs hauptsächlich dann beobachten, wenn das Tier ein fremdes Objekt oder eine unvertraute Bewegung wahrgenommen hat. Blaser (1975) glaubt, dass das Tier damit eine stereoskopische Sicht anstrebt. V. Braunschweig (pers. Mittl.) bestätigt diese Ansicht.

Die beschriebenen Verhaltendweisen der Unsicherheit lassen sich in paralleler Weise in denselben Situationen auch beim Reh feststellen (Labhardt, 1977).

Selbstverständlich wird nicht jeder Umweltreiz als Störung empfunden. Vernehmen Jungfüchse einen leise piependen Ton, den der Beobachter selbst erzeugt, so nehmen die Tiere eine eher neugierige Haltung ein. Sie drehen ihre Köpfe um den Achsialpunkt mal nach der einen, mal nach der anderen Seite, um den Lautpunkt akustisch genau zu lokalisieren (Abb. 6). Auch adulte Füchse auf Mäusejagd lokalisieren vor dem Sprung ihre Beute auf dieselbe Weise.

FLUCHTVERHALTEN IM FELD

Die Flucht zwecks schleunigster Distanzvergrösserung zum Feind geht bei vielen Wildtieren, soweit vorhanden, in ein Versteck, wie Wald, Dickicht,

Baue usf. Damit versucht das bedrängte Beutetier, sich in erster Linie mal der optischen Kontrolle des Feindes zu entziehen, was durchaus sinnvoll ist, denn das Raubtier ist schlussendlich beim Ergreifen der Beute allein auf seine Augen angewiesen. Aber auch gegenüber dem Menschen, der als einziges Wesen andere Lebewesen auf Distanz mittels der Gewehrkugel töten kann, wird dasselbe Fluchtverhalten angewendet.

Ein flüchtender Fuchs springt oft in Zick-Zack verlaufender Richtung davon, auch wenn er nicht unmittelbar verfolgt wird.

FLUCHTVERHALTEN AM BAU

Jungfuchs

Der Fuchs gebärt seine jungen in der Regel in einem Erdbau, den er selber gräbt oder, zumindest bei Basel, meist vom Dachs übernimmt. Der Bau is die erste vom Fuchswelpen wahrgenommene Umwelt. Die Dunkelheit der Höhle, die mütterliche Fürsorge, der wärmende Kontakt der Geschwister lassen den Jungfuchs seine Geburtsstätte als Ort der Sicherheit und Geborgenheit erleben. So ist es nur verständlich, dass Jungfüchse, solange sie noch an den Bau gebunden sind, diesen bei Gefahr stets als Fluchtort aufsuchen. Im Erwachsenenalter jedoch scheinen Füchse den Bau häufig nicht mehr als Fluchtort anzuerkennen. Dieses Paradoxon soll im Folgenden näher diskutiert werden.

Altfuchs

Abb. 7: Der Fuchs kommt aus dem Bau und setzt sich vor die Höhle. Er sichert, indem er um sich schaut. Plötzlich erblickt er den Menschen und rennt augenblicklich davon, flüchtet also nicht in den Bau zurück.

Abb. 7

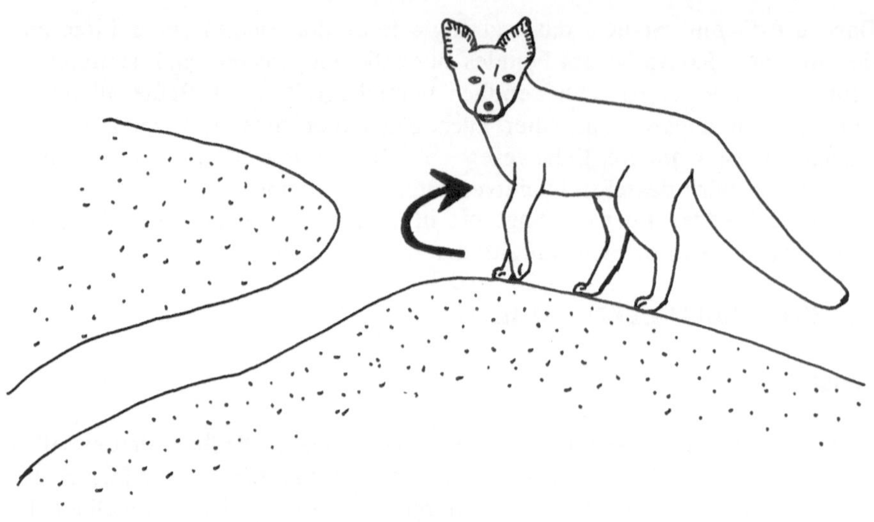

Abb. 8

Abb. 8: Der Fuchs kehrt am frühen Morgen zum Bau zurück, entdeckt den Beobachter und flüchtet in den Wald davon, also wieder nicht in den Bau.

1. *Beobachtungsbeispiel.* Ein Rüde kehrt zum Bau zurück. Die Jungen befinden sich schon im Bau. Obwohl das Tier sich bereits unmittelbar an der Röhre befindet, flieht es nicht in den Bau, als es mich gewahr wurde.
2. *Beobachtungsbeispiel.* Ein Rüde scharrt an einer Röhre und befindet sich mit seinem Körper derart in der Höhle, dass nur noch die Schwanzspitze herausragt. Als ich in einer Entfernung von ca 30 Metern einen Schritt seitwärts mache und dabei ein Kieselstein unter meiner Schuhsohle knirscht, kommt der Fuchs sofort zum Bau heraus, sichert kurz in meine Richtung (er schaut schräg an mir vorbei, hat mich also bestimmt nicht gesehen) und rennt in den Wald davon. Obwohl das Tier sich in der Röhre befand und ein feindlicher Zugriff nicht mehr möglich war, zog es den Bau als Fluchtort nicht vor. Lamprecht (pers. Mittl.) erwähnt, dass Schakale bei Gefahr niemals in den Bau flüchten. Uhlmann (pers. Mittl.) glaubt, der Fuchs fliehe deshalb nicht in den Bau, um sein Heim dem Feind nicht zu verraten. Ich vermute, dass dieses Fluchtverhalten eine Anpassung an die Ansitzjagdmethode des Menschen darstellt. Der Fuchs weiss offenbar, dass wenn er in den Bau flieht, er einmal doch wieder durch den Hunger gezwungen sein wird, den Bau zu verlassen und dann vom Menschen überrascht werden kann. In zwei Fällen wurde eine Ausnahmesituation beobachtet. Es handelte sich dabei jeweils um eine Fähe, die Junge hatte. Ihre starke Bindung zu den Welpen veranlasste sie offenbar, trotz Anwesenheit des Menschen, den Bau aufzusuchen. Eine andere Fähe, die tagsüber nicht mehr bei ihren Jungen im Entwöhnungsalter war, sondern einen Nebenbau bewohnte, floh nun hier niemals in den Bau.

Gegenüber natürlichen Feinden, wie Wolf, Luchs und Adler wäre es für den Fuchs zweifellos von Vorteil, in den Bau zu flüchten, denn jene können wegen ihrer grösseren Körpermasse dem Fuchs nicht in den Bau nachfolgen. Es ist auch nicht anzunehmen, dass diese Feinde dem Fuchs am Bau nach Art der Hauskatze am Mauseloch auflauern. Der Mensch aber betreibt diese Jagdmethode schon seit langer Zeit. Anders als der Mensch jedoch können die natürlichen Feinde den flüchtetenden Fuchs verfolgen und gegebenenfalls einholen; die Flucht in den Bau wäre dann seine Rettung.

Springen aus dem Bau

Füchse, aber auch andere Höhlentiere, die ihren Ort der Geborgenheit verlassen und in die Aussenwelt übertreten, die sie mit ihren Feinden teilen, zeigen an der Schwelle dieser beiden Zonen, nämlich an der Mündung des Höhleneinganges, das Verhalten des Sicherns. Mit den Hauptsinnesorganen prüfen die Tiere, ob "die Luft rein ist". In dieser Situation befinden sich die Tiere im Zustand höchster Aufmerksamkeit und Schreckhaftigkeit. Was tut nun ein Fuchs, wenn er abends zur Jagd aufbricht und in diesem Moment, wo er seinen Kopf zur Höhle rausstreckt (Abb. 9), den Menschen bemerkt? Die weitaus meisten Füchse verhielten sich folgendermassen: Wieder Sich-Zurückziehen in den Bau und Abwarten der nächtlichen Dunkelheit. Dann Springen aus dem Bau mit hoher Geschwindigkeit, wobei das Tier unmittelbar vor der Bauröhre eine Kehrtwendung macht und entgegengesetzt zum Röhrenverlauf, sozusagen nach rückwärts, davonspringt. (Abb. 10). Die Tiere befinden sich in derartiger Spannung, dass sie oft während des Sprunges einen keckernden Angstschrei ausstossen. Viele Füchse sprangen auch dann, wenn ich mich nur 2 Meter neben der Höhle befand. Merkwürdigerweise kamen manche nicht auf die "Idee", eine andere Röhre

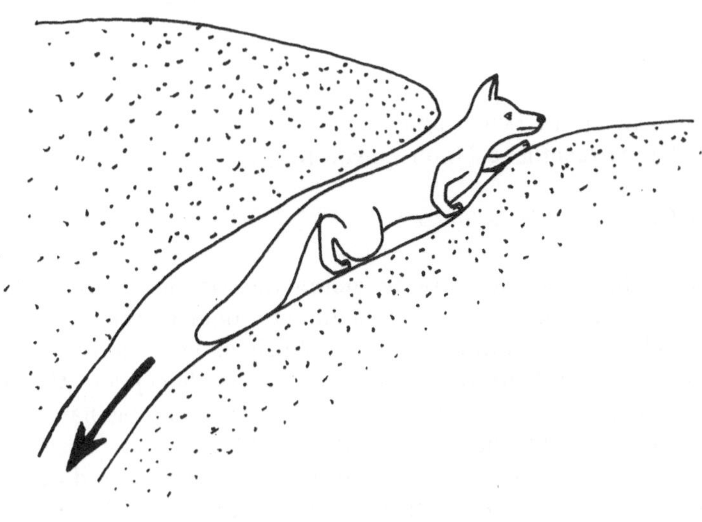

Abb. 9

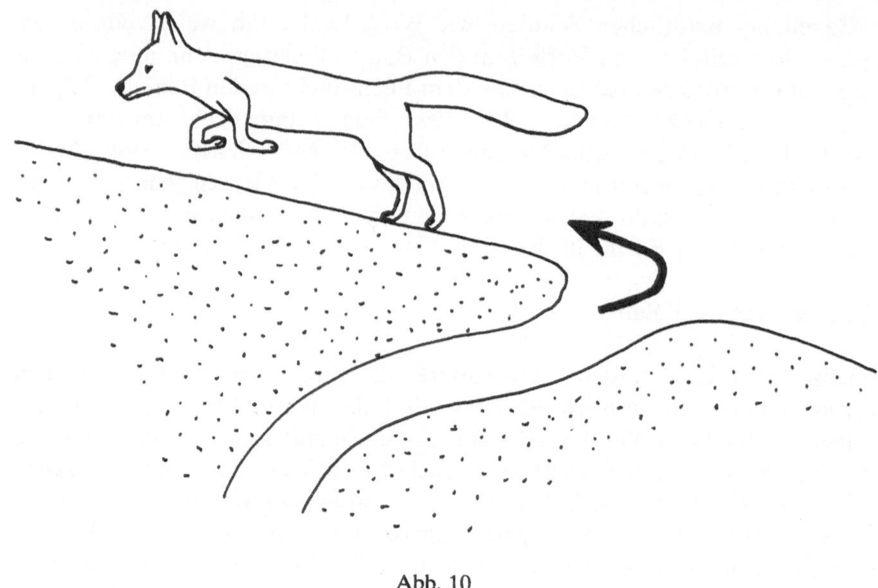

Abb. 10

zu benützen und damit die Distanz zum Beobachter zu vergrössern, mit anderen Worten, sie zeigten eine grosse Ortstreue auch innerhalb des Baufeldes.

FLUCHTBESTIMMENDE FAKTOREN

Die Schreck- und Fluchtreaktionen hängen von verschiedenen Faktoren ab. Diese, nachfolgend kurz skizziert, sind verantwortlich für die Komplexität und häufig zu beobachtende Divergenz im Reaktionsverhalten auf Umweltreize, bei verschiedenen und ein und denselben Individuen.

Individualität

Gerade bezüglich die Scheu und Vertrautheit kann bei Jungfüchsen desselben Wurfes beträchtliche Unterschiede festgestellt werden.

Art und Stärke der Umweltreize

Auf das "Klick" der Kamera reagieren die weitaus meisten Jungfüchse überhaupt nicht. Das Knacken eines Zweiges oder raschelndes Laub erregt jedoch sogleich ihre Aufmerksamkeit. Auch andere spontane Geräusche, wie das plötzliche Flattergeräusch eines vorbeifliegenden Vogels, lassen Jungfüchse heftig erschrecken und in den Bau flüchten. Sie haben aber meist keine nachhaltige Fluchtreaktion zur Folge, die Tiere erscheinen bald wieder vor dem Bau. Kombinierte Störreize akustischer und optischer Art, also ein sichtbares Wesen, das Geräusche verursacht, lassen Jungfüchse stärker erschrecken, als der einzelne Reiz für sich allein.

Erfahrungswert

Wenn wir die Scheuheit vor dem Menschen von Jung- und Altfuchs vergleichen, so stellen wir zwischen diesen beiden Altersstufen einen enormen Unterschied fest. Jungfüchse sind noch viel vertrauter, werden aber mit zunehmendem Alter immer scheuer. Offenbar lernen die Tiere in diesem Zeitraum durch Eigenerfahrung, aber auch über die Eltern, den Feind kennen. Je grösser die Feinderfahrung ist, umso erfolgreicher ist die Feindvermeidung und damit die Ueberlebenschance.

WARNVERHALTEN

Altfüchse, und zwar beide Geschlechter, warnen ihre Jungen, wenn sie einen Feind in der Nähe bemerken. Dabei sind zwei verschiedene Warnlaute voneinander zu unterscheiden:

Warnruf aus Distanz. Der Fuchs, der zum Bau zurückkehren will, bemerkt den Feind und lässt nun einen jaulend-bellenden Laut vernehmen, der ca alle 4 Sekunden wiederholt wird, und dies solange, bis der Feind verschwunden ist. Das Tier verharrt dabei in Sitz-oder Standstellung. Zwischendurch wechselt es schweigend den Standort, um den Feind neu zu lokalisieren. Fähe und Rüde scheinen nicht gleichzeitig zu warnen, sondern lösen sich ab. Bei zwei Paaren war das Männchen rufaktiver:

Bspl. Fähre warnt 63 mal von 0311–0325
Rüde warnt 286 mal von 0326–0420

Das Warnen vernahm ich bisher nur während der Dämmerungs- oder Nachtzeit. V. Braunschweig (pers. Mittl.) hörte es auch am hellichten Tage. Reaktion der Jungen: Sofortiger Abbruch des Spiels und anderer Tätigkeiten. Eilen zum Bau, aber häufig keine Flucht in diesen hinein, sondern Sitzenbleiben vor den Röhren und aufmerksames Lauschen. Nach mehreren, hier nicht näher erläuterten Beobachtungen, scheint das Warnverhalten der Eltertieren eine feindlehrende Wirkung auf die Jungen zu haben.

Einmaliger Schrecklaut. Der Altfuchs befindet sich mit den Jungen auf dem Bau. Der Feind nähert sich dem Bau, der Altfuchs stösst einen kurzen, spontanen Schrecklaut aus, was die sofortige Flucht der Jungen in den Bau zur Folge hat. Der Altfuchs flüchtet voraus in den Bau oder rennt in den Wald davon. Ein hoher Aktivitätsdrang lässt die Jungen jedoch bald wieder an der Erdoberfläche erscheinen.

DAS PHÄNOMEN DER NACHTVERTRAUTHEIT

Bei vielen Säugetierarten fällt auf, dass sie sich in der Nachtvertrauter benehmen, als tagsüber. Kuehme (1966) beobachtete die Nachtvertrautheit bei Löwen in der Serengeti. Ihre Fluchtdistanz gegenüber dem Auto war tagsüber rund vier mal grösser, als in der Nacht.

Eibl-Eibesfeldt (1950) bemerkte bei seinem zahmen Dachs, dass er als Jungtier auch tagsüber vertraut war, mit zunehmendem Alter jedoch während der Tageszeit immer scheuer wurde und erst in der Nacht seine Vertrautheit wieder zurückgewann. Zimen (1978) beobachtete bei freilebenden Wölfen in den Abbruzzen, dass die sonst so scheuen Tiere in der Nacht eine viel geringere Fluchtdistanz aufwiesen und mitten in menschlichen Siedlungen anzutreffen waren.

Beim Fuchs lassen sich entsprechende Beobachtungen machen. Jungfüchse werden bei zunehmender Dunkelheit vertrauter, ja gewisse Individuen können regelrecht zahm werden und beginnen mit dem Beobachter zu spielen. Zweifellos hat die Einschränkung der optischen Wahrnehmung ihren Einfluss. Doch braucht es im Durchschnitt auch wesentlich stärkere akustische Störreize, um Jungfüchse zur Flucht zu veranlassen, als tagsüber.

1. *Beobachtungsbeispiel.* 7. August. Auf einem Stoppelfeld nähert sich mir nachts ein Jungfuchs. Ich folge ihm im Abstand von ca 6 m, er zeigt vor meiner Gestalt und den Geräuschen keinerlei Furcht ebenso nicht vor Lampenlicht. Er verschwindet zwischen Häusersiedlungen in einem Garten. In der Morgendämmerung beträgt die Fluchtdistanz rund 60 m.
2. *Beobachtungsbeispiel.* 8. Oktober. Ich sitze nachts unter einem Apfelbaum auf freiem Feld. Ein nun ca 7 Monate alter Jungfuchs nähert sich dem Baum und frisst direkt vor meinen Füssen einen Apfel. Lampenlicht aus nächster Nähe und mein Sprechen in normaler Lautstärke beeindrucken ihn in keiner Weise.

Altfüchse mit grosser Erfahrung reagieren auf Lampenlicht mit sofortiger Flucht. Offenbar können sie das Licht mit dem Feind Mensch assoziieren. Füchse zögern aber keineswegs, im Schutze der Dunkelheit mitten durch menschliche Siedlungen zu gehen.

Springen an dem Bau

Die nächtliche Aktivität ist in der Tierwelt weit verbreitet. Gründe dafür sind: (i) Feindvermeidung (ii) Anpassung an Beuteaktivität (iii) Ausweichen vor Beutekonkurrenten. Damit ist die Nachtvertrautheit noch nicht erklärt. Es fällt aber auf, dass diejenige Tierarten, bei welchen die Nachtvertrautheit beobachtet werden konnte, in Höhlen, unterirdischen Bauen oder ähnlichen Lokalitäten geboren werden, kurz dort, wo es dunkel ist und die Tiere die Geborgenheit und Sicherheit erleben. Eine Hypothese könnte demnach lauten: Die Dunkelheit der Nacht vermittelt dem Tier deshalb das Gefühl der Sicherheit, weil sie das Aequivalent zur Dunkelheit der Geburtsstätte darstellt, denn nur ein sich sicher fühlendes Tier benimmt sich vertraut. Darus ableitend wäre die Frage näher zu untersuchen, ob Säugetiere, die in Höhlen geboren werden, vorwiegend nachtaktiv sind, vorausgesetzt, sie wären zur Nachtzeit nicht einem hohen Feinddruck ausgesetzt.

ZUSAMMENFASSUNG

Der Flucht vor Feinden geht das Sichern zur Orientierung voraus. Der in dieser Situation verunsicherte Fuchs zeigt dabei verschiedene Verhaltensweisen der Unsicherheit, wie Heben einer Vorderpfote, Kopfpendeln, wechselnde Stellung der Lauscher. Auch der Angstausdruck zeigt im extremen Fall charakteristische Merkmale.

Während Jungfüchse bei Gefahr immer in den Bau flüchten, ist dies bei adulten Tieren häufig nicht mehr der Fall, was als Anpassung an die Jagdmethoden des Menschen gedeutet werden kann.

Altfüchse warnen ihre Jungen akustisch vor Feinden. Es sind dabei zwei verschiedene Warnrufe zu unterscheiden: Warnen aus Distanz mit ständig wiederholtem bellartigem Schreilaut, und Warnen aus der Nähe, mit einmaligem, spontanem Schrecklaut.

Die auch beim Fuchs zu beobachtende Nachtvertrautheit könnte ein spezifisches Phänomen für in Höhlen geborene Säuger sein, indem die Dunkelheit der Nacht diesen Tieren dieselbe Sicherheit vermittelt, wie die Dunkelheit ihrer Geburtsstätte (Bau, Höhle u.a.).

LITERATUR

Blaser H. 1975. Beobachtungen zur Mutter-Kind-Beziehung beim Rotfuchs. *Feld, Wald, Wasser/Schweiz. Jagdzeitung*, Nr. 2, 39–47.

Eibl-Eibesfeldt I. 1950. Ueber die Jugendentwicklung des Verhaltens eines männl. Dachses (Meles meles L.) unter bes. Berücksichtigung des Spieles. *Z. f. Tierpsychol.*, 7: 327–355.

Kuehme W. 1966. Beobachtungen zur Soziologie des Löwen in der Serengeti-Steppe. *Z. Säugetierkde*, 31: 205–213.

Labhardt F. 1977. Paralleles Verhalten der Unsicherheit bei Fuchs (*Vulpes v.*) und Reh (*Capreolus c.*). *Feld, Wald, Wasser/Schweiz. Jagdzeitung*, Nr. 6: 2–4.

Tembrock G. 1957. Zur Ethologie des Rotfuches (*Vulpes vulpes*) unter bes. Berücksichtigung der Fortpflanzung. *Zool. Garten Leipzig*, 23: 289–532.

Zimen E. 1978. Der Wolf. Meyster Verlag, Wien, München.

15 FOXES, WOLVES AND CONSERVATION IN THE ABRUZZO MOUNTAINS

D. W. Macdonald,* L. Boitani† and P. Barrasso‡

INTRODUCTION

In 1973 the World Wildlife Fund established an enquiry into the status of the wolf, *Canis lupus*, in Italy. The wolf population numbered about 100 individuals confined to a few pockets in the central and southern Appenines (Zimen and Boitani, 1975). Approximately 25 wolves survive in the mountains between the Gran Sasso range in the north and the Parco National d'Abruzzo in the south.

Various aspects of the wolves' biology and conservation were studied over 4 years (Zimen and Boitani 1979; Zimen 1978, Macdonald and Boitani 1979; and Zimen and Boitani in prep.). A small part of this study concerned foxes, *Vulpes vulpes*, which were found throughout the wolves' range in Italy. It was thought that foxes and wolves might interact in two ways, both detracting from attempts to conserve the wolf. First, foxes were hunted and persecuted even more than were wolves in this region and there was a risk that significant numbers of wolves were succumbing to poison supposedly destined for foxes. Any attempt to influence the traditional policy of fox control by poisoning had to be based on better information on fox numbers, their use of the range and in particular the effect of garbage availability on their biology. Second, both foxes and wolves were seen to feed from rubbish tips and it was possible that wolf conservation was being thwarted by competition between these species. To investigate these issues we made a preliminary study of the fox's natural history in the context of the wolves' behaviour. In this paper we will summarize these preliminary findings and, more importantly, illustrate the general conservation problem of how the legal control of one species leads to the abuse of legislation to conserve another.

DIRECT INTERACTION BETWEEN FOXES AND WOLVES

Wolves occasionally kill foxes (e.g. Mech, 1970), although the victim may not be eaten. Italian wolves readily kill and eat dogs and may also prey upon foxes. Also, it is possible that wolves and foxes in the Abruzzo region compete for food, perhaps more so since overhunting by man exterminated

* Department of Zoology, South Parks Road, OX1 3PS, U.K.
† Instituto di Zoologia, Viale dell'Universita, 32, 00100 Roma, Italy.
‡ Parco Nazionale Gran Paradiso, Valnontey, 11012 Cogne (Aosta), Italy.

Table 1. Diet of Abruzzo Wolves (1973–76)

Percentage of frequency of occurrence in sample of wolf faeces						
Prey	Total $n = 220$		Summer $n = 84$		Winter $n = 136$	
	%	*n*	%	*n*	%	*n*
Garbage	16.8	(37)	10.7	(9)	20.6	(28)
Sheep/Goat	46.3	(102)	35.7	(30)	52.9	(72)
Cow/Horse	2.7	(6)	7.1	(4)	1.5	(2)
Dog	4.5	(10)	3.6	(3)	5.1	(7)
Small Mammal	3.6	(8)	8.3	(7)	0.7	(1)
Veg. matter	78.6	(173)	82.1	(69)	76.5	(104)
100% "matrix"	25.9	(57)	30.9	(26)	22.7	(31)
Other	10.0	(22)	17.9	(15)	5.1	(7)

the wolves' original prey (red deer, *Cervus elaphus* and roe deer, *Capreolus capreolus*) in the late nineteenth century. We posed two questions: 1) To what extent do the diets of foxes and wolves overlap?. 2) How do foxes and wolves react to each other during direct encounters and do they influence each other's movements?

1. *Diet*

A sample of 220 wolf and 425 fox faeces was collected from the Maiella range in the north-central part of the Abruzzo, where individuals of both species were radio-tracked. Table 1 presents the percentage frequency of occurrence of different food remains in the wolf faeces (Summer and winter samples differ at $p < 0.01$).

No cases of fox remains in wolf faeces were found, nor were carcasses of foxes killed by wolves discovered. Ten faeces contained remains of domestic dogs. The "100% matrix" class describes faeces that consisted solely of amorphous white powder whose origin could not be traced. The ungulate prey remains stemmed largely from feeding sites which we provisioned from the local slaughter-house. The wolves also preyed on domestic stock.

Figure 1 presents the results of analysis of 425 fox faeces collected between February and November 1976. The offal represented in the fox diet (*S* on Fig. 1) was derived from either garbage tips or bait sites established for observation and trapping of wolves. In addition to offal, small rodents were an important prey. Of these, 78% were wood mice (*Apodemus sylvaticus*) and 22% voles (*Microtus* or *Clethrionomys* spp.). Elsewhere, Microtine rodents have been found to be relatively more important in fox diet and, indeed, foxes seem to exhibit a preference for that genus (Macdonald 1977). We know nothing of the availability of different prey species.

The diet of some Abruzzo foxes also contained a lot of offal (although ungulate remains were uncommon). In contrast , only 8 of the 220 wolf faeces contained rodent remains. Similarly, foxes ate a lot of wild fruit (especially *Rosa canina*) while most of the vegetable matter in the wolf diet was garbage (the circumstantial evidence being that 149 of the 173 wolf

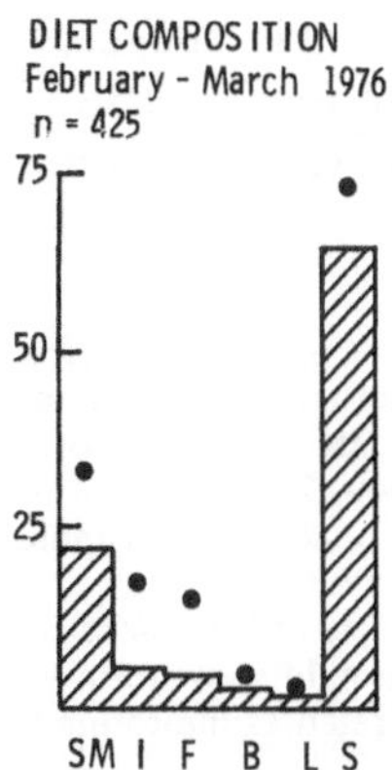

Fig. 1. Composition of the diet of foxes in the Abruzzo region, as represented by the contents of 425 faeces. The data are expressed on bars as percentage volume (corrected for dry weight) and by dots is percentage frequency of occurrence (see Lockie, 1959).

faeces containing fruit also contained garbage). The open garbage pits of the Abruzzo region represent ample food supplies adjacent to every mountain village. These data on wolf and fox diet suggested to us that the garbage tips represented the only likely area of competition for food.

In a similar situation, Grace (1976) suggests that foxes may compete with wolves for garbage. However, if competition occurs in the Abruzzo, it is unlikely to be because the quantity of garbage is limited, but rather because it is highly clumped in availability, so that several individuals of different species may be forced to depend on one dump where they compete directly. How do wolf and fox react to each other in direct encounters, particularly around garbage tips?

2. *Direct observations*

On each of 3 occasions when foxes and wolves were seen together the foxes seemed to be uneasy. For instance on 13th November 1975 E. Zimen observed a wolf, a fox and a cat feeding simultaneously from the refuse dump outside the village of Caramanico. The cat consistently moved out of the way of the fox and the wolf; the fox, in turn, avoided the wolf. Throughout the observation period the fox was "jumpy", feeding in quick bursts, frequently looking up, running around and watching the wolf which in contrast moved calmly about the dump. Occasionally the wolf and the fox were within a few metres of each other and the wolf then appeared to ignore the fox, which showed considerable interest in whatever the wolf was eating. Whenever the wolf moved, the fox ran to the place where the wolf had been feeding. Subsequently we found that all 3 carnivores had been feeding from a large heap of spaghetti!

3. *Snow tracking*

Fox and wolf tracks in the vicinity of baited trap sites and feeding places confirmed that members of both species ate at these sources of offal. Our interpretations of tracks and signs suggested that foxes avoided wolves, waiting for them to vacate the feeding site before venturing forth. Typical examples of our evidence from winter snow (i–iii) and summer field signs (iv) are as follows:

i) 24th March 1976: At least 2 wolves had been feeding at a bait site when a fox approached from the south at a trot. The fox had travelled some distance in a straight line towards the feeding site, which was apparently its destination. The drift of the snow suggested the wind blew from the fox towards the wolves. About 25 m short of the trap site, the fox stopped abruptly, spun round and ran off back the way it had come. We interpreted this to mean that the fox had reacted to the sounds of the feeding wolves.

ii) 6th March 1976: 5 or 6 different sets of fox tracks criss-crossed a bait at which wolves had also been feeding. Fresh snow from a mild snowstorm had accumulated in the wolf prints but not in the fox prints, which we interpreted as indicating that the foxes arrived after the wolves had left.

iii) 16th March 1976: A fox had walked slowly up the path through Valle Cupa. Two wolves had trotted up the same path. Both fox and wolf tracks broke into a run with the fox starting to run 50 m further up the path than the wolves. The fox cut off the path into the woodland and the wolves started to trot again. A possible interpretation is that the fox broke into a run as the wolves approached, whereupon the wolves gave brief chase.

iv) We collected both wolf and fox scats around the rubbish dump of Campo di Giove. During one day's tracking we found 10 wolf faeces together with abundant signs of feeding wolves. All these faeces and the majority of the field signs were found on the west side of the dump. On the same day we collected 66 fox faeces of which only 9% were found on the west side of the dump where wolf signs were concentrated. Possibly this resulted from the foxes avoiding the traditional feeding and resting places of wolves.

Tracks in the snow revealed two types of interaction between wolves and foxes: first, in scent marking, and second, in food caching. Foxes (Macdonald, 1979) and wolves (Peters and Mech, 1975) both urine mark on conspicuous objects and we found that they commonly marked the same site. For instance, on 21st February 1976 2 wolves and 3 foxes had visited and apparently urinated upon the same bush protruding above the snow cover. Furthermore, to reach the bush all 5 animals had travelled over 50 m in straight lines and from different directions. We also found fox droppings on top of wolf droppings and fox urine and droppings on excavated wolf cache sites (see Henry, 1977).

Murie (1936) reported foxes trailing wolves and looting their caches. This seems to be a common habit as many wolf caches in the vicinity of our trap sites had been investigated by foxes, and probably looted. Where fox tracks crossed fresh wolf tracks the fox might turn and follow the wolf trail, and sometimes they led to the wolf's cache. We have no evidence that foxes specialize in raiding each other's caches so it is interesting that they do loot wolf caches (see Macdonald, 1976). Perhaps wolf caches are easier to locate for a fox than are other fox caches but more probably the explanation is that the rewards are bigger with the large food items in wolf caches which merit the expenditure of time in finding them. Three wolf caches which we excavated contained large marrow bones of a size the fox could not have carried far. However, we have also seen occasional signs of wolves following fox tracks and excavating their caches. On April 13th 1976 E. Zimen followed the fresh tracks of a lone juvenile female wolf, who at that time lived in a small part of the range of her former pack but with only occasional contact with them. Close to one of the winter feeding sites, which was at that time not in use, the wolf crossed a set of fox tracks which were about 2–3 hours old. These fox tracks were leading from the feeding site and the wolf followed them closely. Within the next 2.5 km the wolf excavated and chewed two different bones which the fox had cached. The first bone came from an established cache which the fox had merely visited, and the second was from a freshly dug cache. Once she had uncovered the first bone the female wolf urinated into the hole.

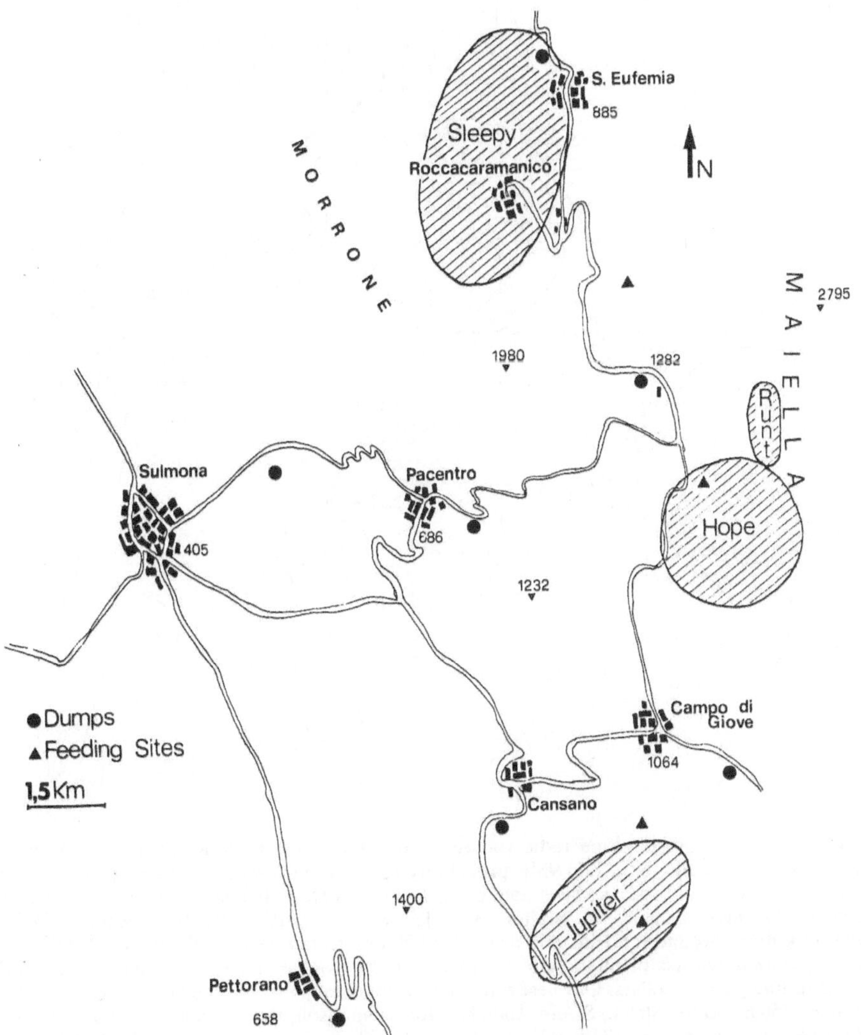

Fig. 2. Map showing the locations of fox home-ranges mentioned in the text. Wolf ranges were enormous compared to those of the foxes: members of one group of wolves travelled the area between Sulmona and Campo di Giove and between Pacentro and Pettorano, and as far again to the South. So, excluding the aggregating effect of the dumps, the chances of the wolf group being in a given fox's range are small.

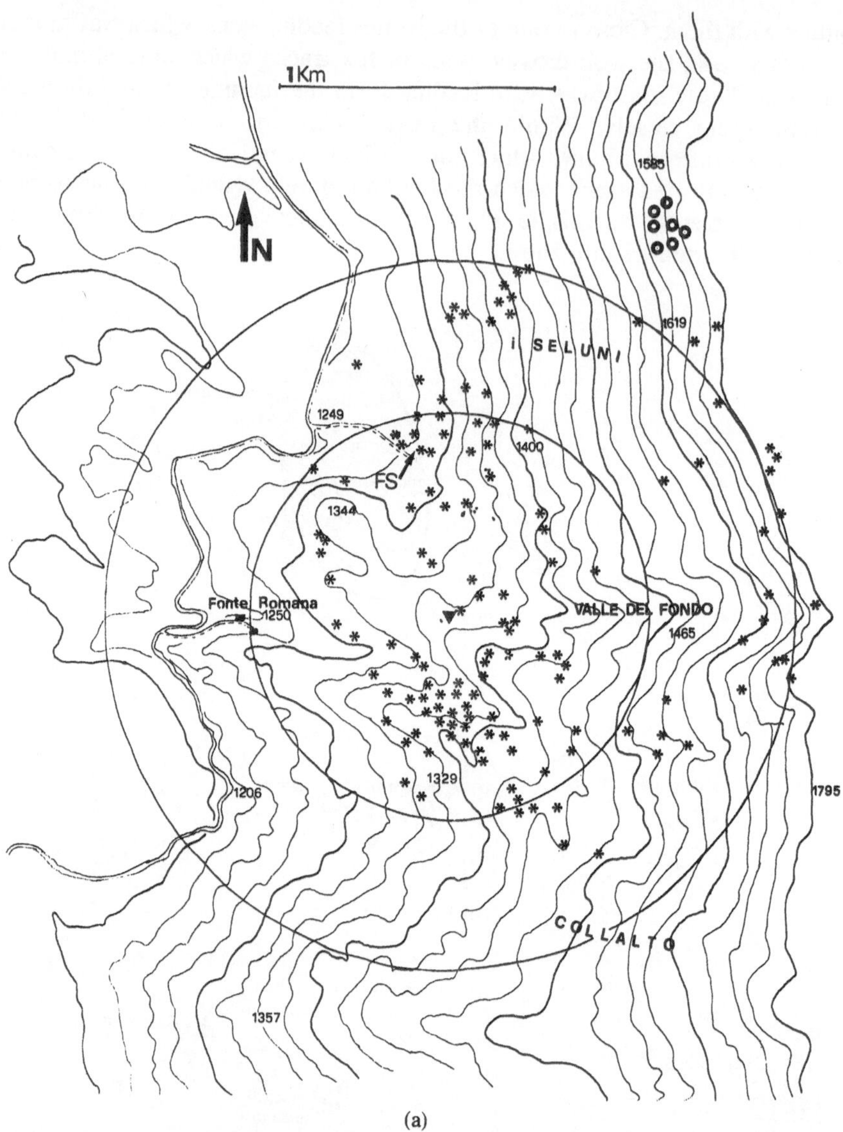

(a)

Fig. 3. Home ranges of three radio-collared foxes. Individual radio-locations are shown by stars. For comparison, 62% and 95% probability elipses are drawn around the radio-locations for each fox (as discussed in Macdonald et al, 1980). 3a) Hope (circles show area where Runt centred his movements). 3b) Jupiter. 3c) Sleepy (note particularly the concentration of radio-locations around the cliffs and around the dump, to the north at S. Eufemia). FS = sites where large quantities of offal were provided to attract wolves for observation and trapping, prior to fitting radio-collars. Our best estimates of home range size in the field for each fox was Jupiter 250 ha, Hope 80 ha., Sleepy 1000 ha. (for comparison, the minimum polygon estimates for these three foxes are: 431.9, 117.0, 1380.1; the 62.3% elipses are: 139.5, 58.1, 423.5; the 95% elipses are: 418.3, 174.4, 1270.7).

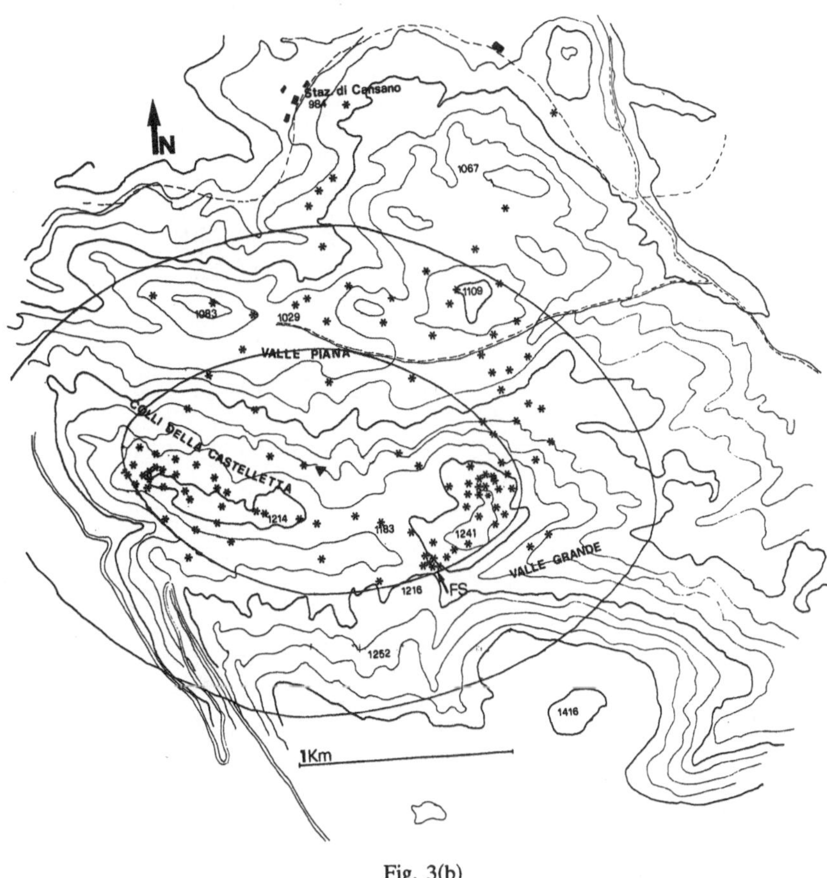

Fig. 3(b)

4. *Radio Tracking*

Out of 8 radio-tagged foxes, only 4 survived long enough to give results by radio-tracking. The mean home range size of 2 wolf packs was about 250 km^2 which is between 100–400 times the range size of foxes in the same area (see Fig. 2). Hence, in this respect the probability of a wolf being within a given fox's range is quite low. On the other hand wolves also depend largely on the food available at the dumps which they visit almost every night (Zimen and Boitani, 1979). So, for those foxes which have home ranges including one or more dumps or for those which make excursions to dumps (which is probably the majority), the probability of direct encounters is rather high (remembering that both species are largely nocturnal and follow a similar pattern of activity throughout the night). Where the two species did come into close proximity the following typify our observations:

i) On 6th May 1976 a fox, (Jupiter ♂), was followed on foot from 2130 hrs. He moved eastwards until 2215 hrs when he was within 60 m of a female wolf who was stationary. Jupiter ♂ circled downwind of the wolf and then moved rapidly south-westwards. By 2240 hrs he had resumed his previous pace but was moving away from the wolf.

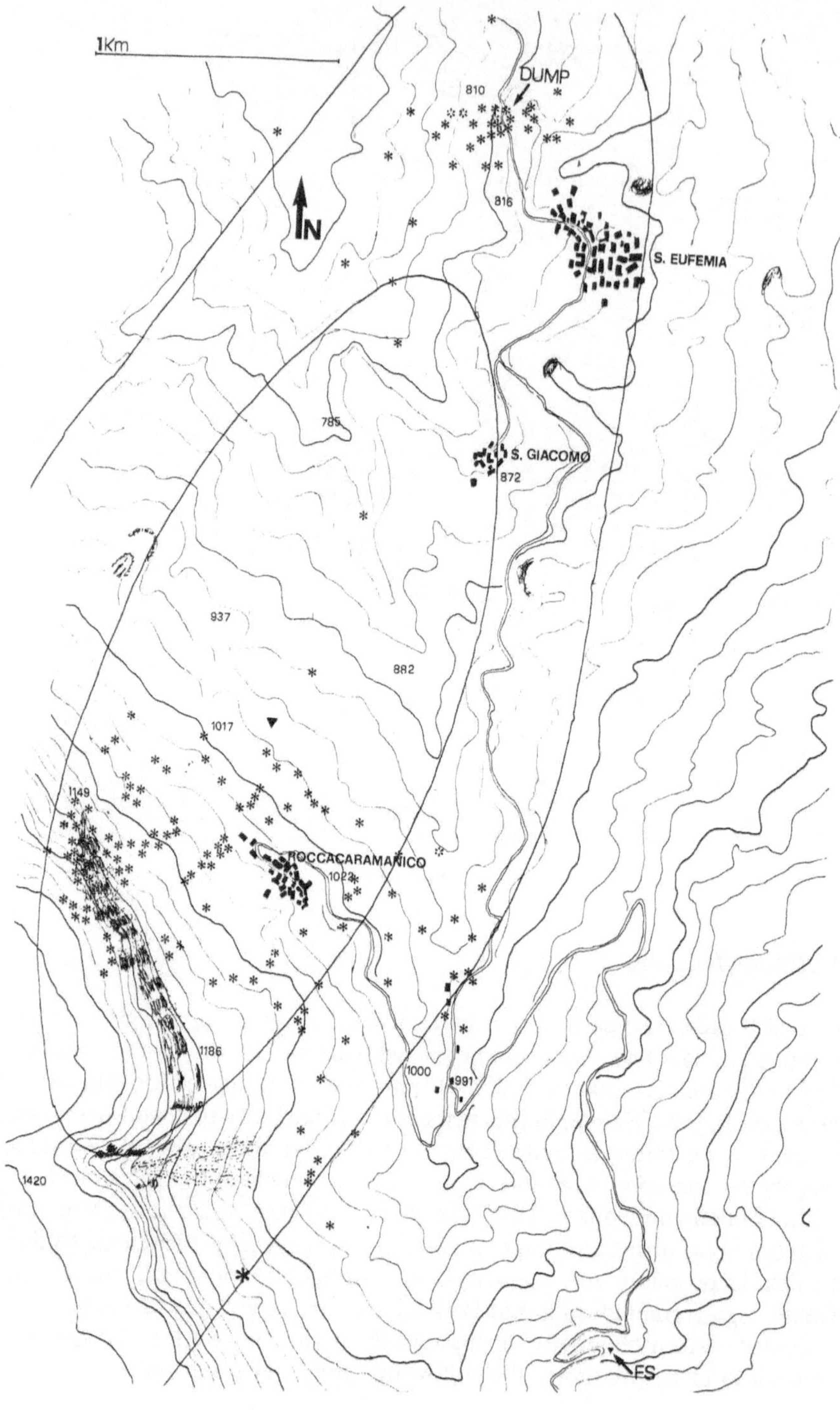
1Km
DUMP
810
816
N
S. EUFEMIA
785
S. GIACOMO
872
937
882
1017
1149
ROCCACARAMANICO
1022
1186
1000
991
1420
FS

Fig. 3(c)

ii) On 16th March from 2110 hrs Jupiter ♂ followed a generally westwards loop through his range until 2310 hrs when he was crossing below Colli della Castelletta. At 2320 hrs E. Zimen howled from further to the west (about 400 m) in imitation of a wolf (an imitation good enough to fool the wolves!). At once Jupiter ♂ changed his pace and disappeared rapidly to the north.

In summary, these and similar observations suggested that while there was some overlap in fox and wolf diet, this was unlikely to be a key factor in limiting either population. Food at the dumps was certainly at times superabundant and to avoid competition with wolves, foxes simply had to "wait their turn". This situation might have been quite different, and more to the foxes' disadvantage, had the wolf numbers not been so low for other reasons. By analogy S. Allen and A. Sargent (pers. comm.) have found that foxes decline where coyotes increase. Although foxes avoid direct contact with wolves, they do loot their caches. Where wolves are numerous, their kills may represent an important source of scavenged food for foxes. This was the case around carcasses at which Mech (1970) witnessed wolves killing foxes which ventured too close.

Fox ranges in relation to garbage tips

There are open garbage dumps in the vicinity of every village in Abruzzo, and throughout central and southern Italy; only recently are some being replaced by incinerators. Although the quantity of food may vary, for example during the tourist season, it is always substantial and often includes fresh offal from the local slaughter houses, in the form of fresh meat and bones. The importance of these dumps to carnivores may be increased by the mountainous landscape (up to 2900 m) which is covered by snow for several months a year, during which natural food availability is drastically reduced.

Apart from the high incidence of garbage and offal in fox diet, the data on the radio tagged foxes confirmed that dumps were important to them. In general terms two of the foxes were residents, while two were probably not:

a) Jupiter ♂ and Hope ♀ (Figs. 3a and b) were both resident animals of more than 1 year old, occupying ranges of approximately 250 and 80 ha respectively. In addition to one feeding place, set for the wolves and irregularly provided with food, Jupiter's range was in the vicinity of the dump at Cansano, while the dump at Campo di Giove, where Jupiter had been caught, apparently on an excursion (see Niewold, 1974), was about 5 km from the fox's main earth. In contrast, Hope's range was far from any dump (but included one wolf feeding place).

b) The other two foxes were quite different:

Runt ♂ was a yearling male in very poor condition when captured in March 1976. His small range neighboured that of Hope ♂ close to the wolf feeding place on the slopes of Monte Amaro: Runt ♂ seemed to depend for food on hurried sorties to the feeding place where he faced competition from resident foxes and from which he retreated to a range on the steep snow-covered slope. Indeed, after several heavy snowfalls in the absence of food at the feeding place Runt ♂ died. An autopsy revealed that his stomach contained only a few old bone chips from the feeding place.

Sleepy ♂ was a mature dog-fox (Fig. 3c) whose frequent trips (3–4 times a week) to the dump of Sant'Eufemia took him about 4 km from his daytime resting place on the cliffs above Roccacaramanico. At the dump and around the cliffs at Roccacaramanico he moved slowly, staying for up to an hour in the one spot, but in contrast he crossed the valley between the two areas quickly. We believed the valley was inhabited by other foxes. The social status of males like Sleepy is unknown, but our point is simply that the presence of the dump had an important effect on his movements. Faecal analysis confirmed not surprisingly, that foxes with more access to dumps (or feeding sites) ate more garbage or offal.

Figure 4(i) shows that diet varied for the foxes in different areas from which faeces were collected. For comparison, samples of faeces collected immediately adjacent to two dumps are also shown. However, these comparisons are biased by variations in the monthly sample size for each area. Fig. 4(ii) overcomes this bias (at the sacrifice of sample size) by representing only data from one month (January). The importance of offal in the diet of foxes who have access to dumps or feeding sites is clear, and in some cases

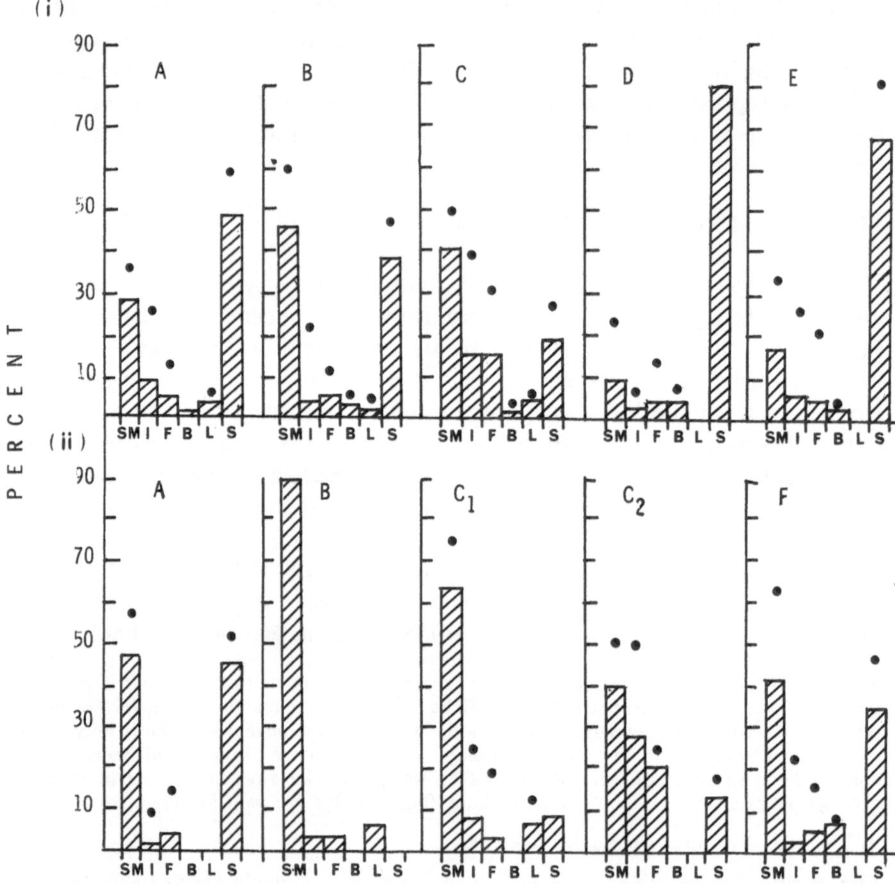

Fig. 4. Differences between the composition of fox faeces collected from different localities within the Abruzzo study area. Bars represent percentage volume of the faeces consisting of each prey (corrected for dry weight of each scat), dots represent percentage frequencies of occurrence. Figure 4(i) lumps together samples collected throughout 1976 whereas Fig. 4(ii) concerns only January (1977). The areas are: A) Jupiter ♂ home-range ($N = 133$, $n_{jan} = 36$), including woodland, alpine meadows, valleys and a garbage dump. B) Hope ♀ home-range ($N = 108$, $n_{jan} = 16$), entirely comprised of mountainside beech forest with access to a trap site. C) Cresta Maggiore ($N = 83$), a mountainous area far away from any dumps or feeding sites (C_1 = January 1977 ($n = 16$), C_2 = January 1978 ($n = 28$)). D and E) Garbage dumps ($N = 52$ and 64). F) Sleepy ♂ home-range ($n_{jan} = 27$), around isolated cliffs, but including almost nightly excursions to a dump.

was dramatic; for example, of 101 fox faeces collected around the dump of Campo di Giove during February–March 1976 75% (by volume) were remains of scavenged food.

Faeces in Jupiter's range comprised 49.2 and 27.0% offal and rodents respectively. In Hope's range (with the irregularly provisioned feeding site) offal and rodents comprised 39% and 45.7% of the diet. Faeces collected at Sleepy's daytime area, at the cliffs of Roccacaramanico comprised 42.4 and 34.8% rodents and offal respectively. By comparison, faeces collected at Cresta Maggiore, an area far from the nearest dump comprised 41.4% rodents and 19.4% offal and the diet was supplemented by a relatively high proportion of insects (14.7%) and fruit (15.4%). These two items were barely represented in the diet in the other areas, although present in the habitat.

The attraction of the garbage dumps for the foxes involved more than the offal to be found at these places. In one sample of 29 faeces collected in August–September at an earth inhabited by a family of foxes some 300 m above the dump at Campo di Giove a large proportion of the diet was comprised of rats, *Rattus rattus.* In fact, the composition of the faeces by volume were only 33.1% derived from offal attributed to the dump and 43.6% rodents. Of the rodents 35.8% were rats.

While our original interest in fox diet in this area was to both assess overlap with wolf diet and the frequency of depredation on game species (which were insignificant in fox diet) the dependence on dumps does raise another problem: it is paradoxical that considerable time and money is expended on killing foxes (see below) on the one hand, when on the other, conditions are made ideal for increasing the fox carrying capacity of the habitat by continued use of open dumps. This irony could become all the more notable as fox control is introduced in an attempt to stop the spread of rabies southwards through Italy (see Macdonald, 1980). However, if dumps were fenced to debar foxes it would be disastrous for the wolves. The solution proposed to the W.W.F. is the reintroduction of ungulate prey for the wolves (Zimen and Boitani, 1979) and plans are afoot to do this. In the absence of rabies or evidence of serious depredations on game there is nothing obviously undesirable about boosting fox numbers, nor of having them in attendance around each dump.

Fox hunting and poisoning

Italian hunting law permitted the use of poison to kill pest animals until a ministerial decree, protecting the wolf as an endangered species, also banned poison from November 22nd 1976. Until that date any game warden or hunter who wanted to use poison baits had only to request a permit from the local hunting committee: the permits were granted freely, on signature, each hunter only being obliged to sign for the number of poison baits received from the hunting committee and to return, at the end of the season, those left over. The law also required that the bait be placed at sunset, checked at sunrise and removed during daytime – what might be termed optimistic legislation! Of course in reality the baits were checked intermittently for a few days, they were rarely removed and any missing baits were simply discarded. The result was an enormous amount of poison spread in the environment with a potential killing power extended through several weeks or months. The most widely used type of poison was "Cyonan", a kind of cyanide, prepared in glass capsules and then effective until the glass is broken. Furthermore, the permit entitled each Italian hunter to have access to all the commune's land, so many hunters operated in the same area, each acting as if he was the only one there. Bounties paid by the hunting organizations for foxes further stimulated the use of poison.

During the hunting season of 1975–76 in the province of L'Aquila alone, (5,034 km^2) 5000 phials of Cyonan were issued to the hunters: only 129 were returned unused, together with 545 Hooded Crow (*Corvus cornix*) and 2,882 foxes, killed for a total expense of about $ 18,000 in bounties. The other 1,444 phials were declared missing. It is hardly surprising that poisoned wolves were found each year in L'Aquila province. These figures ignore illegal poisoning, mainly done with strychnine, which is used by shepherds and farmers in the attempt to protect their property from foxes, wolves and dogs. In the same year, 1975–76, in the neighbouring province of Pescara (1,225 km^2), ranging from sea level to over 2,700 m, 2,000 Cyonan phials were distributed to hunters. Eight hundred and thirty seven foxes were killed and not a single phial was returned. An additional 388 foxes were killed during the same season during 215 "battute di caccia", (fox drives with a line of men and dogs). This method of hunting is almost as dangerous for wolves.

The total of foxes killed in Pescara province was thus 1,225, which is one per square km. The impact of this kill on the fox population or on the survivors depredations on game was unknown. The release of game animals (hare, partridge and pheasant) in L'Aquila province alone costs about $ 150,000 per annum. The only tangible result of this massive campaign against foxes was an annual increase in the number of foxes and wolves killed, although studies have suggested that reducing fox numbers does increase the game bag (e.g. Spittler, 1974). In the end, data on fox and wolf behaviour had little relevance to the drawing up of the 1976 legislation which banned poison, which was proposed on political rather than scientific grounds. Although strychnine is readily available from pharmacists and illegal poisoning continues, it is greatly reduced and since the law came into force we have evidence of only 2–3 cases of Abruzzo wolves being poisoned each year. In 4 years of work in Abruzzo we know of only two cases when poison was used deliberately to kill a wolf, all the other cases were accidents associated with poison set for foxes.

ACKNOWLEDGEMENTS

We are grateful to the Italian branch of the World Wildlife Fund for sponsorship, and to E. Zimen for discussion of an earlier draft of this chapter.

REFERENCES

Grace, E. S. 1976. Interactions between men and wolves at an arctic outpost on Ellesmere Island. *Canadian Field Nature* 90 (2): 149–156.

Henry, J. D. 1977. The use of urine marking in the scavenging behaviour of the red fox (*Vulpes vulpes*). *Behaviour* 61: 82–105.

Lockie, J. D. 1959. The estimation of the food of foxes. *J. Wildl. Manag.* 23: 224–227.

Macdonald, D. W. 1976. Food caching by red foxes and some other carnivores. *Z. Tierpsychol.* 42: 170–185.

Macdonald, D. W. 1977. On food preference in the red fox. *Mammal Rev.* 7: 7–23.

Macdonald, D. W. 1979. Some observations and field experiments on the urine marking behaviour of the red fox, *Vulpes vulpes. Z. Tierpsychol.* 51: 1–22.

Macdonald, D. W. 1980. Rabies and Wildlife: a Biologist's Perspective. Oxford University Press, Oxford.

Macdonald, D. W., Ball, F. and Hough, N. G. 1980. The evaluation of home range size and configuration from radio tracking data. In: Amlaner, C. J. and Macdonald, D. W. (eds.) A Handbook on Biotelemetry and Radio Tracking. Pergamon Press, Oxford.

Macdonald, D. W. and Boitani, L. 1979. The management of carnivores: a plea for an ecological ethic. In: Patterson, D. and Ryder, R. (eds.) Animal Rights, Centaur Press, London.

Mech. L. D. 1970. The Wolf. Natural History Press, New York. 385 pp.

Murie, A. 1936. Following fox trails. *Univ. Mich. Misc. Publ.* 32: 7–45.

Niewold, F. J. J. 1974. Irregular movements of the red fox (*Vulpes vulpes*) determined by radio-tracking. *Proc. XI Int. Congr. Game Biol.*. Stockholm (1973), pp. 331–337.

Peters, P. P. and Mech, L. D. 1975. Scent marking in wolves. *Am. Sci.* 63: 628–637.

Spittler, H. 1974. Zur Populations dynamic des Fuchses (*Vulpes vulpes L.*) in Nordrhein-Westfalen. *Proc. XI Int. Congr. Game Biol. Stockholm* (1973), pp. 167–174.

Zimen, E. 1978. Der Wolf, Mythos und Verhalten. Meyster, Munchen. 333 pp.

Zimen, E. and Boitani L. 1975. Number and distribution of the wolf in Italy. *Z. Saugetierkunde* 40: 102–112.

Zimen, E. and Boitani, L. 1979. Status of the wolf in Europe and possibilities of conservation and reintroduction. In: Klinghammer, E. (ed): The behaviour and Ecology of Wolves. Garland Press, New York.

16 EPIDEMIOLOGY OF FOX RABIES

A. I. Wandeler*

Rabies has been with mankind since time immemorial. Its history is lost in antiquity (Steele, 1975). Rabies is an important disease of foxes, but also of other animal species. Due to the character of the zoonosis in densely populated areas of North America and Europe, and thanks to WHO's effort to coordinate research, more is known about rabies in foxes than in any other animal species (except laboratory rodents). But rabies in other species causes greater economic loss and more tragedy.

Epidemiology describes disease patterns in populations and animal communities. The observed epidemiological phenomena are explained by properties of the disease agent, the host, and host populations. Wildlife diseases are especially difficult to investigate. Adequate sampling in populations of unknown size and structure becomes a problem. This is also true for rabies.†

THE WORLD SITUATION

Today rabies is widespread on all continents except Australia. The red fox (*Vulpes vulpes*) is responsible for maintenance and spread of the disease in northern and eastern North America, northern Asia, and in Europe. The gray fox (*Urocyon cinereoargenteus*), where he exists, is also involved (Winkler, 1975). In arctic areas *Alopex lagopus* replaces Vulpes in ecological as well as in epidemiological respects (Crandell, 1975). In the American Midwest and in California rabies is predominant in skunks (*Mephitis mephitis*) (Parker, 1975). The raccoon (*Procyon lotor*) is the main host and vector for rabies in Florida and southern Georgia (McLean, 1975). In Africa domestic and wild **canids** (Röttcher and Sawchuk, 1978) and mongooses (Snyman, 1953) are responsible for the disease. Mongooses (*Herpestes sp.*) introduced to some Caribbean islands for snake control are now the main reservoir there (Everard et al., 1974). In South and Central America vampire bat (*Desmodus rotundus*) rabies is of economic importance because of its transmission to cattle (Baer, 1975c). In North and Central America insectivorous bats are supporting a rabies epizootic which is independent from terrestrial carnivore rabies (Baer, 1975b). In urban as well as in rural areas of Africa, Asia, South- and Central America rabies in stray dogs is of major concern. Several hundred human rabies cases reported every year to WHO are mostly caused by dog bites (Steck, 1978).

Rabies incidence and regional prevalence are changing at different rates in different areas. In Europe rabies used to be a serious problem in past

* Virology Division, Veterinary Bacteriology, University of Berne, Berne, Switzerland.

† Many statements made in this brief review are results of not yet published investigations made in collaboration with Prof. Dr. F. Steck, Dr. Ch. Stocker, U. Häfliger, and P. Bichsel.

centuries. It is peculiar that there are only very few reports from before 1900 mentioning rabies in wildlife (Köchlin, 1835). Possibly this is because rabid foxes were of minor concern when compared to the more important dog rabies. The widespread occurrence of urban rabies stimulated Louis Pasteur to devote himself to rabies research. His well-known success was the first approach to therapeutic immunization of man after exposure. Rabies disappeared from Central Europe at the turn of the century, possibly due, among other suspected reasons, to the fact that the stray dog problem was brought under control at that time. During World War II a new epizootic originated in northeastern Europe and spread west- and south-westward. The frontwave penetrated into the Federal Republic of Germany in 1950, reached Denmark in 1964, Austria, Belgium and Luxembourg in 1966, Switzerland in 1967, France in 1968, the Netherlands in 1974 and Italy in 1977 (Wandeler et al., 1974; Kauker, 1975; Toma, 1978). Similar epizootic waves of fox rabies have been observed in North America (Friend, 1968; Johnston and Beauregard 1969; Tabel et al., 1974).

THE FOX RABIES EPIZOOTIC

The spread of fox rabies has been described and analyzed in detail by numerous authors (Moegle et al., 1974; Tabel et al., 1974; Kauker, 1975; Bögel et al., 1976; Toma and Andral, 1977; for older publications see Winkler, 1975). The most important features of fox rabies are:

1. The first rabies cases registered in newly invaded areas are almost always foxes.
2. The frontwave moves into new areas with a speed of approximately 25–60 km/year. Much higher rates of movement have been observed in Canada, when rabies moved from the arctic southward (Johnston and Beauregard, 1969; Tabel et al., 1974).
3. Rivers, lakes, and higher mountains function as natural barriers. Rivers are usually crossed where bridges are available.
4. Intensive fox control may also stop the spread of rabies (Müller, 1971).
5. The case density in the frontwave center is very high. In areas with good surveillance up to 2 rabid foxes per km^2 per year are recovered.
6. Foxes constitute the majority (60–85%) of all diagnosed rabies cases.
7. In foxes grouped according to conditions under which the animals were collected (found dead, killed because of abnormal behaviour, killed with normal behaviour, killed by traffic, trapped) the proportion of rabid ones is always higher than in similarly grouped categories of other species (Wandeler et al., 1974).
8. The frontwave epizootic usually lasts not more than one or two years. Rabies may become enzootic and stay for more years under special topographic conditions.
9. Where rabies itself and fox control reduce fox population density below a certain level, rabies disappears not only in foxes, but also in all other species.

10. An area that becomes free of rabies may be reinvaded after a few years from adjacent endemic regions. This is explained by a rapid recovery of fox populations during the rabies-free years.

Some remarks must be made about quantitive aspects, which were somewhat overemphasized in several papers. To give classical epidemiological parameters, like prevalence rate or incidence rate on rabies is impossible at present. We do not know size and densities of fox populations. We also have no proof that our samples of this universe are representative and unbiased. Nevertheless there are good indications that there is a strong relationship between the rabies epizootic and fox population densities. In addition to the statements made above, it should be noted that only the fox and the badger, but no other species, are reduced in population density by the event of rabies.

There remains the question: How is it possible that a single species can be responsible for spread and maintenance of a fatal disease? Parasites deleterious enough to threaten elimination of hosts, also threaten to eliminate themselves. The adaptation of rabies virus to the biology of its carnivore host is unique in nature. Properties of the virus and the unusual pathogenesis determine a good part of the epidemiology.

The virus

Rabies virus is a rhabdovirus. Other animal rhabdoviruses are: Vesicular stomatitis virus, bovine ephemeral fever virus, Marburg virus, and a few less important isolates from vertebrates and insects (Kundson, 1973; Fenner, 1976). They all have in common a bullet-shape, a membrane envelope with surface projections, and a helical ribonucleoprotein capsid (Murphy, 1975). The genome is negative strand RNA (Sokol and Koprowski, 1975). The dimensions of rabies virus particles are: length 180 nm, diameter 75 nm. The antigen inducing rabies neutralizing antibodies in a host is a glycoprotein, one of the three major proteins associated with the envelope (Dietzschold et al., 1978).

A small group of different viruses isolated from African mammals, man, and insects is more closely related to rabies than the rest of the rhabdoviruses (Murphy, 1975; Fenner, 1976). Lagos bat, Mokola, Duvenhage, Obodhiang, and Kotonkan viruses, and rabies virus, share common or closely related nucleocapsid antigens (Shope, 1975). These neurotropic viruses are assigned to different serotypes of the rabies serogroup. Classic rabies virus is the most widespread among the rabies-related viruses.

In contrast to many other common viruses, rabies shows only very little antigenic variation (Shope, 1975). Only monoclonal antibodies produced in hybridoma cell cultures do sometimes discriminate between different isolates and strains with different passage histories (Wiktor and Koprowski, 1978). Different variants of rabies virus are characterized according to biological properties such as virulence, pathogenicity, invasiveness, incubation period, and duration and kind of clinical symptoms (Baer et al., 1977). All these properties are difficult to characterize and to reproduce experimentally. Not only origin, but also passage history, virus titer, defective interfering virus particles and other interfering components in the inoculum and their formation in the initial phases of infection, may well influence the outcome of an otherwise standardized inoculation. There is indication that many so-called strains or substrains are genetically not uniform. Rabies virus genetics is still poorly understood. All explanations of epidemiological particularities on the basis of properties of assumed virus variants are therefore highly hypothetical.

Rabies virus, like all other viruses, needs living host cells in order to replicate and survive. Extracellular virus is vulnerable to the detrimental influence of physical and chemical agents. It is sensitive to heat, ultraviolet light and lipid solvents. It is relatively resistant to drying, freezing and thawing, and relatively stable at pH values between 5 and 10. Rabies virus keeps its infectivity when frozen at low temperature. Stabilizing factors (e.g. certain polypeptides) may prevent the rapid inactivation of a virus suspension.

The tissues of an animal that has died from rabies loose their infectivity according to their initial virus content and to the environmental influence. All infectious virus may be gone within a few days in a body lying above ground in summer; while infectivity can be maintained for weeks in a carcass protected by cool dirt or in snow and ice.

Pathogenesis

After experimental intramuscular infection of different street rabies isolates, first evidence of viral replication is found in nearby striated muscle cells (Murphy, 1977; see this paper for older references). This is an important source of virus for the invasion of the peripheral nervous system. The time the virus spends in these muscle or in other cells before entering the nervous system varies greatly; it may be a few days or several weeks. Murphy et al. (1973) have shown that the nerve endings of neuromuscular and neurotendinal spindles are a probable site for virus entry into the nervous system. The transport of the virus genome from the periphery to the central nervous system is within the axoplasm of nerve fibers, most likely by retrograde axonal transport. In experimental infection the very rapid transport from the periphery to the central nervous system can be stopped by cutting the nerve innervating the site of inoculation, or by amputation of the inoculated leg (Baer, 1975a, Baer and Cleary, 1972, Dean, Evans, and McClure 1963). Transport of virus by bloodstream from the site of initial replication to the final target organs (central nervous system, salivary glands) does not occur in rabies pathogenesis. Even late in the course of the disease when many organs yield infectious virus, viremia is observed very seldom. By its peculiar route within nerve fibers the virus does not stimulate any host immune response, and if it is already stimulated, intracellular i.e. intraaxonal virus is protected from antibody and immunocyte attacks.

Sensory nerves of the limbs and of the body enter the spinal cord via dorsal root ganglia. This is also the route rabies virus genom takes in its centripetal movement. Certainly, entry of virus into the central nervous system is also possible in cranial nerves, if the virus has entered the body at accordingly innervated sites. After reaching the spinal cord, its ascent to the brain is only a matter of hours. Again it takes only a short period till all parts of the brain are involved; the neocortex becoming infected relatively weakly and late. The virus replicates primarily in neurons. Replication leads to the formation of cytoplasmic inclusion bodies, consisting of large amounts of viral nucleocapsid material. The cellular inclusions are seen as Negri bodies by light microscopy. In infected neurons the virus buds primarily from endoplasmic reticulum membranes. Apart from signs of virus replication not much cell damage can be seen by electron microscopy. But the infected cells are functionally altered. The clinical course of the disease is correlated with the spread and localisation of virus within the central nervous system.

The infection of spinal cord and brain soon leads to a centrifugal movement of virus to peripheral organs. Again, the transport of virus genome is inside axons i.e. in the axoplasm of neurons. Despite the fact that rabies virus is primarily neuronotropic, many peripheral organs do support virus replication. In late stages of the disease virus can be demonstrated in skin, bucal and nasal mucosa, adrenals, pancreas, striated muscles, myocardium and occasionally also in other places. Most important for disease transmission is virus replication in salivary glands. Here, virus budding comes primarily from the plasma membrane of the cell surface, and not from internal membranes as in neurons. This way infectious virus is shed directly into the ductuli of the gland and excreted with the saliva.

Transmission and susceptibility

Rabies is usually transmitted by bite, i.e. by intratissue inoculation. Rabies virus penetration through intact skin does not occur, but infection through mucous membranes is occasionally possible. Airborne infection has been observed only under exceptional circumstances; for example, in caves harbouring large numbers of rabid bats (Constantine, 1967) or in laboratories when virus aerosol was created purposely or accidently (Winkler et al., 1972; Winkler et al., 1973). The susceptibility of all species so far exposed to rabies aerosol seems to be quite high (see Winkler 1975b for references). Infection by ingestion is also possible (Soave, 1966; Fischman and Ward, 1968; Correa-Giron et al., 1970; Bell and Moore, 1971; Ramsden and Johnston, 1975). In this case, the virus has to penetrate oral mucosa (e.g. by infecting taste buds); in the stomach it is inactivated by the gastric acid. All species tested so far are much less susceptible to the oral than to the intramuscular route of inoculation with street rabies virus. In our own experiments foxes were approximately 100,000 times more resistant to oral challenge than to intramuscular inoculation with Swiss fox isolates (to be published).

Street rabies virus infectivity is expressed as 50 per cent endpoint of a titration of mouse intracerebral lethal doses ($MICLD_{50}$). One $MICLD_{50}$ is the calculated dilution of a virus suspension which kills half of the intracerebrally inoculated weanling mice in a titration experiment. Suckling or weanling mouse brain is the most susceptible and reliable test system for street rabies virus.

The virus dose needed to successfully infect 50 per cent of the inoculated foxes was less than 5 $MICLD_{50}$ in our own experiments (to be published) with Swiss fox isolates inoculated into neck muscles. It was less than 5 $MICLD_{50}$ in experiments with American fox and skunk isolates (Sikes 1962, Sikes 1966, Parker and Wilsnack, 1966) inoculated into cervical muscles also. Black and Lawson (1970) in Canada observed a 50 per cent endpoint for foxes at about 16 $MICLD_{50}$ when the virus originating from foxes is injected into gluteal muscles. In contrast to these results is an experience described by Winkler et al. (1975). They were unable to induce disease in foxes with 250 $MICLD_{50}$ of a fox virus given into the gluteal muscle. These differences may reflect the importance of the inoculation site for susceptibility.

The susceptibility of other species may partly explain epidemiological observations. Sikes (1962, 1966) demonstrated marked species differences in susceptibility to intramuscular injection of American street rabies isolates. The calculated virus doses necessary to successfully infect 50% of the inoculated animals was as follows: foxes 5 $MICLD_{50}$; skunks 500 $MICLD_{50}$; raccoons 1,000 $MICLD_{50}$, and opossums more than 80,000 $MICLD_{50}$. In Parker and Wilsnack's (1966) experiments foxes and skunks were equally susceptible to a skunk isolate. Unfortunately similar data for the European **mustelids** are still missing.

Incubation period

The incubation period between exposure and first clinical symptoms is quite variable. It depends on the site of virus entry into the body (cells supporting virus replication, quantity and quality of nerve endings, distance to CNS) and on properties of the inoculum (origin of virus, virus titer). In a study published by Sikes (1962) the shortest incubation period of a fox inoculated with fox salivary gland virus was 12 days; the longest was 109 days. Richards (1962) reported incubation periods of 4 to 181 days in foxes inoculated with a variety of different isolates from foxes, skunks, and dogs. Foxes inoculated with skunk salivary gland virus showed incubation periods of 14 to 57 days (Parker and Wilsnack, 1966). Atanasiu et al. (1970) observed incubation periods in foxes inoculated with fox and bovine isolates ranging from 10 to 105 days. The time span between inoculation and death was 15 to 153 days in experiments published by Black and Lawson (1970). Three studies (Sikes, 1962; Parker and Wilsnack, 1966; Black and Lawson, 1970) report that the incubation period was inversely related to the amount of virus inoculated.

Morbidity period and clinical symptoms

The period from the first observed clinical sign until death occurs is relatively short in foxes. Sikes (1962) reported morbidity periods of 1 to 3 days, Parker and Wilsnack (1966) 1 to 17 days, Richards (1962) 5 to 13 days, and Atanasiu et al. (1970) 1 to 9 days. In all studies short morbidity periods are common, while long ones are the exception.

The clinical symptoms are quite variable (Richards, 1962; Winkler, 1975a; Sykes-Andral, 1976). The observable disease in experimentally infected foxes may begin with loss of appetite or sudden aggressiveness. Hyperactivity, loss of activity pattern, tremor, hypersensitiveness, and increased and abnormal sexual behaviour are other early symptoms. These symptoms rarely last longer than just one or a few days. Hydrophobia, so typical for the disease in humans, is rarely seen in animals. Convulsions and paralysis (often ascending) are common late symptoms. They lead to death almost without exception.

The disease in free-ranging foxes is poorly understood. Commemoratives sent with animals to diagnostic laboratories do not tell the whole story (Sykes-Andral, 1976). Of all the foxes found rabid by the Swiss rabies center, only about $\frac{1}{3}$ had been sent there because they showed aggressiveness, or had quarrelled with domestic animals. Completely unknown are alterations of other very important aspects of fox ethology, especially of social and territorial behaviour. As long as these are unknown, interpretations of the mechanisms leading to the spread of the disease within a fox society will remain hypothetical.

Organ distribution of virus and virus excretion

In rabid foxes, rabies virus antigen can be demonstrated with fluorescent antibody in almost all organs (Debbie and Trimarchi, 1970). In our own experimentally infected foxes (to be published) we regularly found substantial amounts of infectious virus only in the brain, salivary glands, and nasal mucosa. But it was found only irregularly at low titers in the adrenals and pancreas, while skeletal muscle, intestines, heart, lungs, kidneys, liver, blood and cerebrospinal fluid almost never yielded any demonstrable amount of infectious virus. The concentration of infectious virus in most peripheral tissues is usually too low to guarantee transmission by oral uptake. Naturally infected foxes, badgers and stonemartens show about the same virus concentration in their brains, the median titers being around 100 $MICLD_{50}$ in 0.03 ml tissue homogenate. In 93% of 816 rabid foxes and in 83% of 82 rabid badgers we found virus in the salivary glands, with median titers of $10^{3.4}$ and $10^{3.5}$ $MICLD_{50}$, respectively. In contrast, of 36 rabid stonemartens only half of them had infected salivary glands and a low median titer of $10^{1.5}$ $MICLD_{50}$ (Wandeler et al., 1974). In Sikes' (1962) study, experimentally infected foxes had demonstrable virus in their saliva only for 1 to 3 days before death. Foxes given a low virus inoculum had a longer incubation period and the greatest percentage of saliva virus excretion during morbidity. The same was found in another experiment by Parker and Wilsnack (1966). Virus excretion a few days before clinical symptoms became apparent has been found occasionally in dogs (Vaughn et al., 1965), cats (Vaughn et al., 1963), and skunks (Sikes 1962). This period of preclinical virus excretion may be longer in bats (Baer and Bales, 1967; Bell et al., 1969). Unfortunately, adequate data on the duration of virus excretion by infected foxes are missing. It is very unlikely that virus excretion begins long before clinical symptoms become obvious. There is no salivary gland infection without previous involvement of the brain. In 635 foxes shot in rabies endemic areas in Switzerland, whose brains were negative for rabies, we could not detect any virus in salivary glands (Wandeler et al., 1974).

Antibodies and nonfatal rabies

A measurable immune response of an infected host to rabies virus arises, if at all, only very late in the course of the disease. Murphy (1977) has specified the possible mechanisms keeping rabies antigen from stimulating the host defence mechanisms at initial stages of the infection. Whether local and systemic immune responses are involved in terminal phases of the illness is still a point of controversy.

Rabies is usually fatal. Survivors with sequelae do occur under certain experimental conditions (Bell 1964). Complete recovery is very rare in observed naturally and experimentally infected animals. Survivors of clinical disease do have high neutralizing antibody titers not only in serum, but also in cerebrospinal fluid. Bell et al. (1971) did not find any animals which met these criteria in a large sample of a dog population in rabies endemic Argentina. Quite a few investigators found in sera of wild mammals (Sodja et al., 1971; Everard et al., 1974; McLean, 1975; Price and Everard, 1977; Carey and McLean, 1978) and in birds (Gough and Jorgenson, 1976) a high frequency of what they interpreted as antibody against rabies. In our own studies we compared in neutralization tests blood samples from 61 rabies-negative foxes from rabies endemic areas with samples from 75 foxes from an area free from rabies for the past 100 years. The frequency and quantity of rabies neutralizing activity was not significantly different in the two groups (Wandeler et al., 1974). We had the same experience with blood samples of **rodents** and **insectivores** (to be published). But we never found any rabies inhibiting activity in sera of wild caught foxes after they had been held in captivity for several weeks. Inhibitors against rabies virus were also demonstrated in Japan in sera of rabbits which never had had any contact with rabies virus (Sekine and Yoshino, 1974).

Caution is therefore recommended in the interpretation of neutralizing and hemagglutinating activity of wild animal sera. We feel that only very few of the rabies negative foxes in our material had been actively immunized. Whether they were survivors of clinically apparent disease is an open question. Ramsden and Johnston (1975) found antibody in some foxes and skunks which remained healthy after being fed with infected mice. It has also been suspected that intratissue inoculation of street rabies may occasionally lead to limited peripheral infection without the involvement of the CNS (see Bell, 1975, for references). This is the mechanism inducing immunity after injection of attenuated rabies viruses (Flury LEP and HEP, ERA) into peripheral tissues.

Latency

Rabies in its incubation period meets the criteria for latency. Rabies virus stays in a latent stage in the periphery for variable lengths of time before it moves to the central nervous system. Nobody knows the mechanisms keeping the virus genome intact, but silent in peripheral tissue. There also arises the question whether the sudden invasion of the nervous system can be initiated by external factors. Soave et al. (1961) and Soave (1964) suggest such a possibility; but their experiments still need further confirmation.

The whole spectrum of varying incubation periods is of great influence on the course of the epizootic. But in Europe it is rather unlikely that very long incubation periods account for the reappearance of the disease in an area after some years of absence.

Reservoir

A rabies reservoir outside the main carnivore host has often been postulated, for theoretical reasons as well as for emotional motivations. Johnson (1966) thought that sporadic cases in wildlife would indicate such a reservoir.

A reservoir host should be a symptomatic or asymptomatic carrier, able to transmit the disease intra- and interspecifically. From observations made in the course of epizootics, symptomatic carriers other than those already known can be excluded. This statement is also true for rodents (Winkler, 1966).

Asymptomatic carriers were assumed among small **mustelids**, **rodents**, and **insectivores**. We did not detect any signs of carrier state in **mustelids** (Wandeler et al., 1974), **rodents** and **insectivores** (to be published). But other investigators (Sodja et al., 1971; Sodja and Matouch, 1972a+b; Schneider and Schoop, 1972) found that the brain (negative for rabies by immunofluorescence), salivary glands, and/or brown fat of wild caught **rodents** inoculated intracerebrally into weanling mice may occasionally result in the isolation of rabies after several blind passages. Unfortunately there is no way to prove that the final isolates really originate from the wild **rodents**. The virus in its occult form is not transmitted, except maybe vertically to offspring (by an unknown mechanism). Once it is apparent after several blind passages, it is indistinguishable from classical rabies and remains that way. In laboratories where rabies virus is not propagated, investigators with the same method of blind-passaging for any virus in **rodents**, have never isolated rabies. We therefore suspect that at least our own rabies isolate from a blind-passaged *Apodemus* salivary gland results from a contamination. Even if the conclusion is valid, that rabies virus can occur in an occult form in **rodents**, it is hard to imagine that this strange phenomenon is of any importance for the fox rabies epizootic.

The spread of rabies within a fox population

By making reasonable assumptions on fox population parameters and using the data on rabies in foxes it is possible to show in computer simulation that the single species fox is well able to support and maintain a rabies epizootic (Smart, 1970; Preston 1973). In contrast to the simplifying assumptions necessary for models stands the complicated, dynamic, not fully understood society of foxes. An adult pair (Sargeant, 1972) or one male plus several females (MacDonald, 1977) seasonally share a family territory with their young of the year. The half year-old subadults may emigrate and disperse, being transient, non-territorial foxes for some time. Dispersal movements of subadults, breeding beginning in mid-winter, and territory-defence throughout the year may all be important factors of rabies epidemiology. There are numerous possibilities for infectious contacts. An aggressive sick fox may attack a healthy family member, a healthy territory owner may attack a sick intruder, or a healthy fox may penetrate into the territory of his sick neighbour. A healthy animal has to expose himself to the bites of a sick one in order to become infected. These hypothetical examples should show that an at random transmission to all social categories is not very likely. As a matter of fact, age and sex composition of rabid foxes does not represent the population structure and changes throughout the year. In a collective of rabid foxes young individuals are underrepresented in summer, then their occurrence increases inversely to the expected declining proportion in the living population. (Wandeler et al., 1974).

CONCLUSIONS

The evaluation of virological and serological findings has raised at least as many questions as it has answered. With caution some partially hypothetical conclusions can be drawn.

Rabies virus shows adaptation to different hosts and may be divided into variants with different biological properties. But rabies epidemiology seems to be independent from antigenic variation, antigenic drift or shift, so important for other infectious diseases. The relative antigenic uniformity may be due to the lack of selection through immune individuals in the host populations.

The fox is one of the most susceptible species. He is easily infected by bite. Infection by oral route is possible, but certainly not the rule in nature. After an incubation period which usually lasts a few weeks, a fox becomes sick and possibly aggressive. A very high proportion of the sick foxes excretes virus in the saliva and is capable of infecting other animals. The interpretation of our own serology data shows that most infected foxes die from rabies. Immune individuals are therefore rare and do not influence the further course of the epizootic. Other authors (Carey and MacLean, 1978) take the opposite view. Immunity or any other acquired state of resistance would favour the survival of older immune foxes in rabies epidemic areas. An according shift in age structure has not been found in Canada (Johnston, personal communication) or in Switzerland.

In most areas of Europe surveillance is good enough to reveal an almost continuous movement of the epizootic, forwards and backwards, with varying speed, and staying in different areas for different lengths of time. Two systems determine the character of the epizootic. One system is the virus within the host, the other the host population. After a long incubation period the virus causes a short period of morbidity and virus excretion, and a disease characterized by behavioural and nervous disorders. For transmission by bite there must be contacts between individuals. These depend on the social structure of the population, and on the social use of space, which may be altered by density. In spring and summer rabies has to be transmitted from territory to territory or within territories; in fall and winter the disease may be brought over larger distances by dispersing subadults and young adults.

REFERENCES

Atanasiu, P., Guillon, J. C., and Vallée, A. 1970. Contribution à l'étude de la rage expérimentale du renard. *Ann. Inst. Pasteur* 119: 260–269.

Baer, G. M. 1975a. Pathogenesis to the central nervous system. In Baer, G. M. (ed.): The natural history of rabies, vol. I: 181–198. New York.

Baer, G. M. 1975b. Rabies in nonhematophagous bats. In Baer, G. M. (ed.): The natural history of rabies, vol. II: 79–97. New York.

Baer, G. M. 1975c. Bovine paralytic rabies and rabies in the vampire bat. In Baer, G. M. (ed.): The natural history of rabies, vol. II: 155–175. New York.

Baer, G. M. and Bales, G. L. 1967. Experimental rabies infection in the Mexican freetail bat. *J. Infect. Dis.* 117: 82–90.

Baer, G. M. and Cleary, W. F. 1972. A model in mice for the pathogenesis and treatment of rabies. *J. Infect. Dis.* 125: 520–527.

Baer, G. M., Cleary, W. F., Diaz, A. M., and Perl, D. F. 1977. Characteristics of 11 rabies virus isolates in mice: Titers and relative invasiveness of virus, incubation period of infection, and survival of mice with sequelae. *J. Infect. Dis.* 136: 336–345.

Bell, J. F. 1964. Abortive rabies infection. 1. Experimental production in white mice and general discussion. *J. Infect. Dis.* 114: 249–257.

Bell, J. F. 1975. Latency and abortive rabies. In Baer, G. M. (ed.): The natural history of rabies, vol. I: 331–354. New York.

Bell, J. F., Gonzalez, M. A., Diaz, A. M., and Moore, G. J. 1971. Nonfatal rabies in dogs: Experimental studies and results of a survey. *Am. J. Vet. Res.* 32: 2049–2058.

Bell, J. F. and Moore, G. J. 1971. Susceptibility of *carnivora* to rabies virus administered orally. *Am. J. Epidemiol.* 93: 176–181.

Bell, J. F., Moore, G. J., and Raymond, G. H. 1969. Protracted survival of a rabies-infected insectivorous bat after infective bite. *Am. J. Trop. Med. Hyg.* 18: 61–66.

Black, J. G. and Lawson, K. F. 1970. Sylvatic rabies studies in the silver fox (*Vulpes vulpes*). Susceptibility and immune response. *Can. J. Comp. Med.* 34: 309–311.

Bögel, K., Moegle, H., Knorpp, F., Arata, A., Dietz, K., and Diethelm, P. 1976. Characteristics of the spread of a wildlife rabies epidemic in Europe. *Bull. World Health Organ.* 54: 433–447.

Carey, A. B. and McLean, R. G. 1978. Rabies antibody prevalence and virus tissue tropism in wild carnivores in Virginia. *J. Wildl. Dis.* 14: 487–491.

Constantine, D. G. 1967. Rabies transmission by air in bat caves. U.S. Dep. of Health, Education and Welfare, Public Health Service Publication No. 1617. 51 pp.

Correa-Giron, E. P., Allen, R., and Sulkin, S. E. 1970. The infectivity and pathogenesis of rabiesvirus administered orally. *Am. J. Epidemiol.* 91: 203–215.
Crandell, R. A. 1975. Arctic fox rabies. In Baer, G. M. (ed.): The natural history of rabies, vol. II: 23–40. New York.
Dean, D. J., Evans, W., and McClure, R. 1963. Pathogenesis of rabies. *Bull. World Health Organ.* 29: 803–811.
Debbie, J. G. and Trimarchi, Ch. V. 1970. Pantropism of rabies virus in free-ranging rabid red fox *Vulpes fulva*. *J. Wildl. Dis.* 6: 500–506.
Dietzschold, B., Cox, J. H., and Schneider, G. 1978. Structure and function of rabies virus glycoprotein. *Developments in Biological Standardization* 40: 45–55.
Everard, C. O. R., Baer, G. M., and James, A. 1974. Epidemiology of mongoose rabies in Grenada. *J. Wildl. Dis.* 10: 190–196.
Fenner, F. 1976. Classification and nomenclature of viruses. *Intervirology* 7: 1–115.
Fischman, H. R. and Ward, F. E. 1968. Oral transmission of rabies virus in experimental animals. *Am. J. Epidemiol.* 88: 132–138.
Friend, M. 1968. History and epidemiology of rabies in wildlife in New York. *New York Fish and Game J.* 15: 71–97.
Gough, P. M. and Jorgenson, R. D. 1976. Rabies antibodies in sera of wild birds. *J. Wildl. Dis.* 12: 392–395.
Johnson, H. N. 1866. Sporadic cases of rabies in wildlife: Relation to rabies in domestic animals and character of virus. Proc. National Rabies Symposium, NCDC, Atlanta, 25–30.
Johnston, D. H. and Beauregard, M. 1969. Rabies epidemiology in Ontario. *Bull. Wildl. Dis. Ass.* 5: 357–370.
Kauker, E. 1975. Vorkommen und Verbreitung der Tollwut in Europa von 1966–1974. Sitzungsber. Heidelberger Akad. Wiss.; Math.-naturwiss. Klasse, Jg. 1975/2: 49–84.
Knudson, D. L. 1973. Rhabdoviruses. *J. gen. Virol.* 20: 105–130.
Köchlin, J. R. 1835. Ueber die in unseren Zeiten unter den Füchsen herrschende Krankheit und die Natur und Ursachen der Wutkrankheit überhaupt. Zürich.
MacDonald, D. 1977. The behavioural ecology of the red fox. In Kaplan, C. (ed.): Rabies – The facts. 70–90. Oxford.
McLean, R. G. 1975. Raccoon rabies. In Baer, G. M. (ed.): The natural history of rabies, vol. II: 53–77. New York.
Müller, J. 1971. The effect of fox reduction on the occurrence of rabies. Observations from two outbreaks of rabies in Denmark. *Bull. Off. Int. Epiz.* 75: 763–776.
Murphy, F. A. 1975. Morphology and morphogenesis. In Baer, G. M. (ed.): The natural history of rabies, vol. I: 33–61. New York.
Murphy, F. A. 1977. Rabies pathogenesis. Brief review. *Arch. Virol.* 54: 279–297.
Murphy, F. A., Bauer, S. P., Harrison, A. K., and Winn, W. C. 1973. Comparative pathogenesis of rabies and rabies-like viruses. Viral infection and transit from inoculation site to the central nervous system. *Lab. Invest.* 28: 361–376.
Moegle, H., Knorpp, F., Bögel, K., Arata, A., Dietz, K. und Diethelm, P. 1974. Zur Epidemiologie der Wildtiertollwut: Untersuchungen im südlichen Teil der Bundesrepublik Deutschland. *Zbl. Vet. med.* B 21: 647–659.
Parker, R. L. 1975. Rabies in skunks. In Baer, G. M. (ed.): The natural history of rabies, vol. II: 41–51. New York.
Parker, R. L. and Wilsnack, R. E. 1966. Pathogenesis of skunk rabies virus: Quantitation in skunks and foxes. *Am. J. Vet. Res.* 27: 33–38.
Preston, E. M. 1973. Computer simulated dynamics of a rabies-controlled fox population. *J. Wildl. Manage.* 37: 501–512.
Price, J. L. and Everard, C. O. R. 1977. Rabies virus and antibody in bats in Grenada and Trinidad. *J. Wildl. Dis.* 13: 131–134.
Ramsden, R. O. and Johnston, D. H. 1975. Studies on the oral infectivity of rabies virus in carnivora. *J. Wildl. Dis.* 11: 318–324.
Richards, St. H. 1962. Rabies study data summary, 1953–60. North Dakota Game and Fish Department, Bismark. 29 pp.
Röttcher, D. and Sawchuk, A. M. 1978. Wildlife rabies in Zambia. *J. Wildl. Dis.* 14: 513–517.
Sargeant, A. B. 1972. Red fox spatial characteristics in relation to waterfowl predation. *J. Wildl. Manage.* 36: 225–236.
Schneider, L. G. and Schoop, U. 1972. Pathogenesis of rabies and rabies-like viruses. *Ann. Inst. Pasteur*, 123: 469–476.
Sekine, N. and Yoshino, K. 1974. Inhibitors against rabies virus present in normal rabbit sera. *Arch. Virusforschung* 45: 89–98.

Shope, R. 1975. Rabies virus antigenic relationships. In Baer, G. M. (ed.): The natural history of rabies, vol. I: 141–152. New York.

Sikes, R. K. 1962. Pathogenesis of rabies in wildlife. I. Comparative effect of varying doses of rabies virus inoculated into foxes and skunks. *Am. J. Vet. Res.* 23: 1041–1047.

Sikes, R. K. 1966. Wolf, fox, and coyote rabies. Proc. National Rabies Symposium, NCDC, Atlanta. 31–33.

Smart, Ch. W. 1970. A computer model of wildlife rabies epizootics and an analysis of incidence patterns. MSc Thesis, Virginia Polytechnic Institute, Blacksburg, Virginia. 125 pp.

Snyman, P. S. 1953. Rabies in the Union of South Africa. *Bull. Epiz. Dis. Afr.* 1: 94–97.

Soave, O. A. 1964. Reactivation of rabies virus in a guinea pig due to stress of crowding. *Am. J. Vet. Res.* 25: 268–269.

Soave, O. A. 1966. Transmission of rabies to mice by ingestion of infected tissue. *Am. J. Vet. Res.* 27: 44–46.

Soave, O. A., Johnson, H. N., and Nakamura, K. 1961. Reactivation of rabies virus infection in a guinea pig with adrenocorticotropic hormone. *Science* 133: 1360.

Sodja, I., Lim, D., and Matouch, O. 1971. Isolation of rabies virus from small wild rodents. *J. Hyg. Epidemiol. Microbiol. Immunol.* 15: 271–277.

Sodja, I. and Matouch, O. 1972. Adaptation to the laboratory mouse of rabies-like viruses isolated from wild rodents. *Acta virol.* 16: 147–152.

Sodja, I. and Matouch, O. 1972. Pathogenicity of "wild-mouse" rabies-like strains for laboratory animals. *Acta virol.* 16: 153–158.

Sokol, F. and Koprowski, H. 1975. Structure-function relationships and mode of replication of animal rhabdoviruses. *Proc. Natl. Acad. Sci. USA* 72: 933–936.

Steck, F. 1978. Epidemiologie der Tollwut. *Münch. med. Wschr.* 120: 271–274.

Steele, J. H. 1975. History of rabies. In Baer, G. M. (ed.): The natural history of rabies, vol. I: 1–29. New York.

Sykes-Andral, M. 1976. La rage du renard. *Revue méd. vét.* 127: 1641–1674.

Tabel, H., Corner, A. H., Webster, W. A., and Casey, C. A. 1974. History and epizootiology of rabies in Canada. *Can. Vet. J.* 15: 271–281.

Toma, B. 1978. Evolution de la rage – en Europe, – en France. In: La rage. Informations techniques des services veterinaires, Nos. 64–67: 13–20, 21–26.

Toma, B. and Andral, L. 1977. Epidemiology of fox rabies. *Adv. Virus. Res.* 21: 1–36.

Vaughn, J. B., Gerhardt, P., and Newell, K. W. 1965. Excretion of street rabies virus in the saliva of dogs. *J. Am. Med. Ass.* 193: 363–368.

Vaughn, J. B., Gerhardt, P., and Peterson, J. C. S. 1963. Excretion of street rabies virus in saliva of cats. *J. Am. Med. Ass.* 184: 705–708.

Wandeler, A., Wachendörfer, G., Förster, U., Krekel, H., Schale, W., Müller, J., and Steck, F. 1974. Rabies in wild carnivores in Central Europe. I. Epidemiological studies. *Zbl. Vet. Med.* B 21: 735–756.

Wandeler, A., Wachendörfer, G., Förster, U., Krekel, H., Müller, J., and Steck, F. 1974. Rabies in wild carnivores in Central Europe. II. Virological and serological examinations. *Zbl. Vet. Med.* B 21: 757–764.

Wiktor, T. J. and Koprowski, H. 1978. Monoclonal antibodies against rabies virus produced by somatic cell hybridization: Detection of antigenic variants. *Proc. Natl. Acad. Sci. USA* 75: 3938–3942

Winkler, W. G. 1966. Rodent rabies. Proc. National Rabies Symposium, NCDC, Atlanta, 34–36.

Winkler, W. G. 1975. Fox rabies. In Baer, G. M. (ed.): The natural history of rabies, vol. II: 3–22. New York.

Winkler, W. G., Baker, E. F., and Hopkins, C. C. 1972. An outbreak of non-bite transmitted rabies in a laboratory animal colony. *Am. J. Epidemiol.* 95: 267–277.

Winkler, W. G., Fashinell, T. R., Leffingwell, L., Howard, P., and Conomy, J. P. 1973. Airborne rabies transmission in a laboratory worker. *J. Am. Med. Ass.* 226: 1219–1221.

Winkler, W. G., McLean, R. G., and Cowart, J. C. 1975. Vaccination of foxes against rabies using ingested baits. *J. Wildl. Dis.* 11: 382–388.

17 CHARACTERISTICS OF THE SPREAD OF A WILDLIFE RABIES EPIDEMIC IN EUROPE

K. Bögel* and H. Moegle†

During the past 10 years much data on fox ecology has been obtained within the WHO/FAO Coordinated Research Programme on Wildlife Rabies in Europe. Scientists in a number of European countries have studied the interaction between fox populations, rabies, and control measures under different topographical conditions. Moreover, the role of other wild carnivores and the effect of rabies and fox population control on these species, as well as on small game and rodents, have been investigated (Moegle et al., 1971, Müller, 1971, Wandeler et al., 1974(a) and (b)). In some European countries these research projects were supplemented by ecological studies on the composition and annual turnover of stable fox populations after reduction by rabies and control measures (Bögel et al., 1974, Lloyd et al., 1976, Wandeler et al. (c), 1974). Special studies using electronic data processing dealt with the characteristics of the spread of wildlife rabies in the former administrative area of Südwürttemberg-Hohenzollern and two adjacent districts (Stockach and Uberlingen) of the Land Baden-Württemberg, Federal Republic of Germany (Bögel et al., 1976).

Whereas all results of the WHO coordinated research were assessed in a comprehensive way at the 2nd European Conference on the Surveillance and Control of Rabies in 1977, this presentation summarizes data of particular interest to those concerned with fox ecology.

The improvement of rabies surveillance for adequate postexposure treatment in man and development of control measures called for intensive ecological studies of fox population dynamics and behaviour, which, however, had often to be confined to limited and relatively small study areas. Territorial behaviour could be investigated by observation of individual animals in a fox family. Details on contacts between foxes of adjacent territories and on contact rates during migration remained largely unknown to us. More knowledge, in this respect, could be helpful in our efforts to improve rabies control operations. Another area which we are presently investigating concerns natural barriers for the spread of rabies. Again fox ecology and behaviour is the key to the answer to many open questions.

However, as in many areas of disease control, much data on the ecological basis of patterns, difficult to study in nature, have been obtained through epidemiological investigation, including successes and failures of control measures. Theories and plans for countrywide strategies for the control of

* Veterinary Public Health, World Health Organization, Geneva, Switzerland.
† Referat Veterinärwesen, Regierungspräsidium, Nauklerstrasse 47, 74 Tübingen, Federal Republic of Germany.

epidemics have often been assessed by a test-and correct procedure under field conditions. Large scale studies on the reservoir species of diseases often include investigation of epidemiological patterns. Conclusions drawn from such observations have subsequently to be supported by ecological studies in selected and relatively small areas.

Without doubt there is, particularly in wildlife, an important mutual contribution to ecology by direct studies of the animal species involved and of epidemics spread by these species under different conditions. Distances between cases recorded and the recovery of populations following reduction by rabies and control measure are typical examples.

This leads us to the specific question of the spread of rabies and its relation to fox ecology.

Wildlife rabies invaded Baden-Württemberg in 1957, coming from the north and spreading in a large front to the south (Fig. 1). In our first study we tried to explore the relation between the density of the fox population and the frequency of rabies. For an estimate of the fox population density we use the number of annually shot foxes per km^2 given by the hunting statistics. Experience in various parts of Europe shows that this "indicator value" can be applied despite all reservations, if it refers to areas of about 2,000 km. For the State of Baden-Württemberg, we plotted the indicator values of fox population density and the frequency of rabies over the course of several years. The area was subdivided into zones of homogeneous topographic character with an average size of 5,000 km^2.

Figure 2 shows the frequency of rabies and the fox density in the north-eastern part of Baden-Württemberg, first invaded by rabies.

There is evidence that rabies alone caused in this area a sharp decrease of the fox population, a decrease of more than 50%. Gassing of fox dens caused a further reduction to a level of 0.2 shot foxes per km^2. Rabies

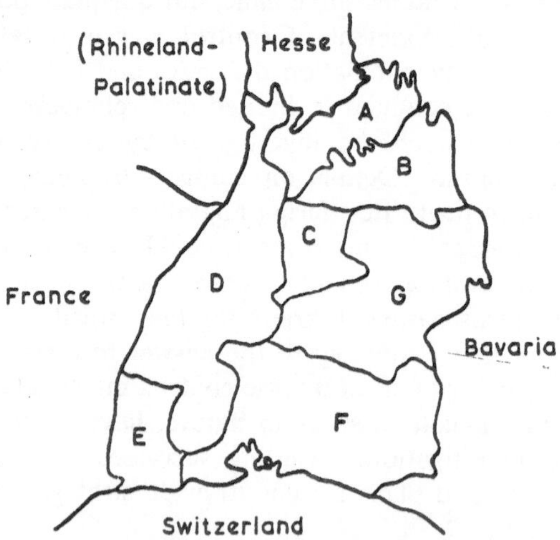

Fig. 1. Surveillance of the Fox Population in Rabies Control-Study Areas in Baden-Württemberg.

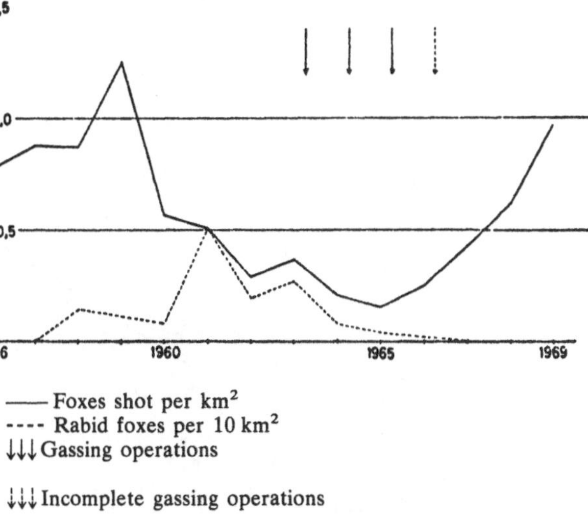

Fig. 2. Area A – Eastern part of North Baden (3 168 km^2).

disappeared from the area. The fox population recovered in the following years, and in 1971 new outbreaks of rabies occurred in the area. Nevertheless, owing to gassing operations, rabies-free intervals of several years could be recorded.

As the main result of this study we defined for our region a level of fox population density of 0.2 to 0.3 foxes per km^2 – deduced from hunting records and actual population data – at which rabies disappears.

Another method of comparison between fox density and frequency of rabies is shown in Fig. 3. We compared the fox population density and the frequency of rabies in all 62 counties of Baden-Württemberg. For this purpose we determined for each county the indicator values of the population density at the time of the first outbreak of rabies, whereas the frequency of rabies is calculated as average frequency over the first 3 years of the infection (Fig. 3).

There is a positive correlation between rabies frequency and the indicator value of the fox population density, the correlation-coefficient being 0.52. The line of regression cuts the x-axis at a population density value of about 0.3. This indicates that rabies does probably not enter an area with a lower density of the fox population.

In further studies we investigated the pattern of the spread of rabies in our area. We therefore used the possibilities of electronic data processing at WHO, Geneva. We recorded all 2,822 cases of animal rabies in our district – a district with a size of 11,000 km^2 – from 1963, if when rabies entered the area in the north, until rabies reached the southern border in 1971.

We collected the following data: date of diagnosis in the laboratory, animal species, and localization of the cases defined by the coordinates of the centre of the community where the animal was found.

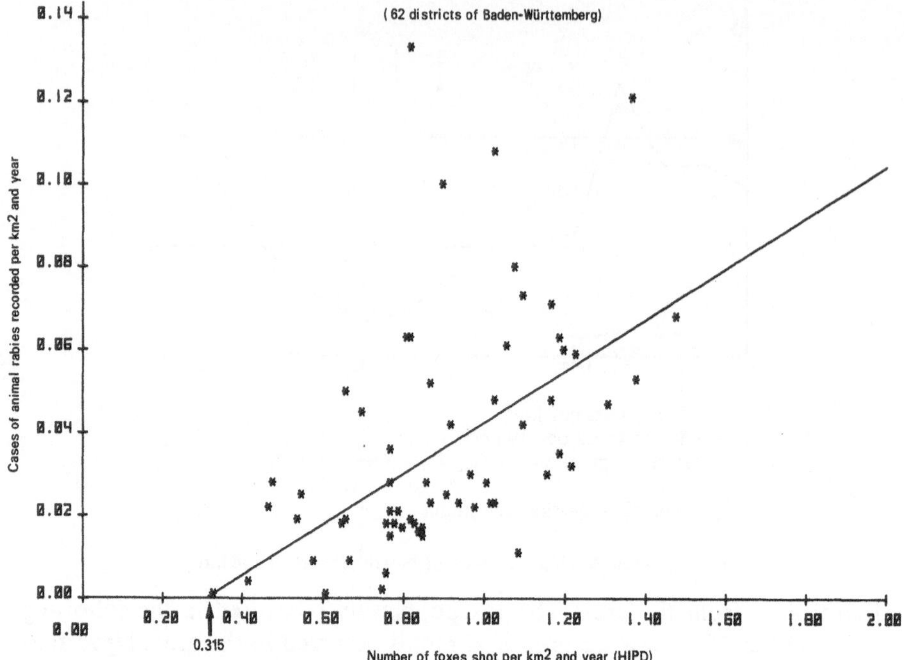

Fig. 3. Regression of the frequency of animal rabies with respect to the hunting indicator of fox population density (HIPD). (62 counties of Baden-Württemberg).

The first question we tackled concerned the distance of new cases of rabies from the preceding cases, and the time interval between the cases. A simple question, but it took us some time to develop a computer-adapted method which would define a frontline of an epidemic and provide a satisfactory answer to our questions.

Considering the clearly preferred direction of spread from north to south, we could define frontlines of the disease by linking the most southern cases of rabies on a map. Figure 4 shows a computer print of the study area. Each character means a community. The lines indicate the frontlines of the disease at different times. The analysis of our calculation procedure revealed that it was not necessary to define a new frontline every time a new case occurred. We fixed the frontlines for each month. Thus we could check the distance of each new case from the frontline of the preceding month, and define the mean distances of these cases from the frontline at a distinct time.

Table 1 shows the distances of new cases of rabies from the frontline. 93.4% of all new cases occurred at a distance of less than 10 km beyond the frontline. The greatest distance of a new case from the frontline was 20.5 km.

The slopes of mean and maximum distances of new cases from the frontline showed a characteristic cyclic pattern over the year (Fig. 5). The annual cycle of the frequency of wildlife rabies is well known, so it was reasonable to compare both cycles. Both curves show a similar course. However, the distance of new cases from the frontline begins to increase

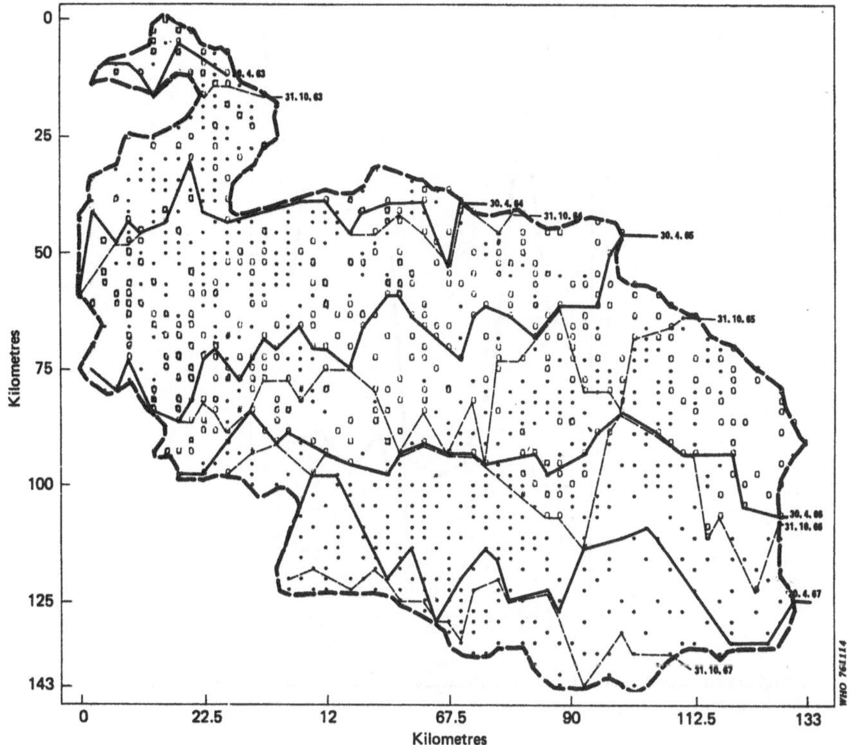

Fig. 4. Southward spread of rabies in the study area in the southern part of the Federal Republic of Germany. (Frontline drawn on a computer print-out for the situation on 31 October 1966.)

clearly before the increase of rabies frequency in the autumn (Fig. 6). The increase of frequency in the autumn, therefore, appears to be the result of the increasing velocity of spread, not its cause. Field observations support this statement.

We had three zones of different density of fox population and of different frequency of rabies in our study area (Table 2), namely the sub-areas:

A. *Black Forest:* low density of fox population
B. *Suebian Jura:* medium density
C. *Pre-alpine zone:* high density

Table 1. Distance of new rabies cases from the frontline of the previous month

New cases	Distance from the frontline in km				
	0–5[a]	5–10	10–15	15–20	20
Number	238	115	20	5	3
Percentage	66.4	27.0	4.7	1.2	0.7
	93.4				

[a] Upper limit included.

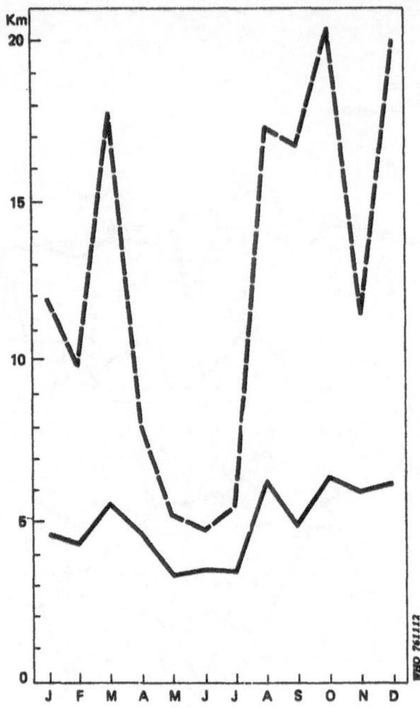

Fig. 5. Observed mean (solid line) and maximum (broken line) distances of new rabies cases from the frontline of the preceding month.

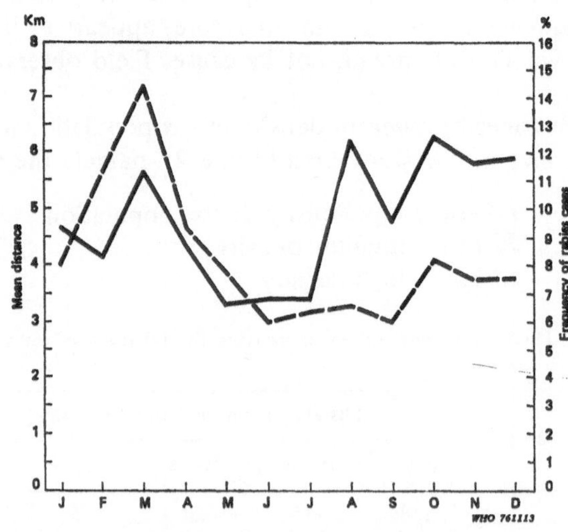

Fig. 6. Mean distance of rabies cases recorded ahead of the monthly determined frontline (solid line) and rabies cases as a percentage of all 2,822 cases recorded from 1963 to 1971 (broken line).

Table 2. Frequency and distance of rabies cases in zones of different indicator values for the density of fox populations

	A "Black Forest"	B "Swabian Jura"	C "Pre-alpine area"
HIPD[a]	0.7	1.1	1.5
Frequency of rabies[b]	0.044	0.051	0.065
Distance of new cases[c]	4.98	4.71	4.70

[a] HIPD = Hunting indicator of Population Density: foxes shot per km^2 per year.
[b] Average number of recorded cases of rabies in animals per km^2 and per year.
[c] Mean distance in km of new cases ahead of the monthly determined frontline.

The differences between the rabies frequencies of these areas are statistically significant. There is no difference in the mean distances of new cases beyond the frontline. This result is also verified by several field observations and was postulated in 1970 by French authors, based on theoretical considerations.

Another question of interest is the role of the different animal species in the spread of rabies. From the distribution of the different species before and behind the frontline, we can deduce that no other species than the fox apparently plays a significant role in the spread of the disease before the frontline. Ahead of the frontline foxes accounted for 97.2% of all animals found rabid as compared to 79.8% behind the frontline.

Let us try to recapitulate:

1. The frequency of animal rabies and the density of the fox population are correlated, the co-efficient being 0.52. Rabies disappears from an area or will not enter when the fox density (estimated by hunting records) falls short of 0.2–0.3 foxes shot annually per km^2.
2. The spread of wildlife rabies shows a distinct pattern. The mean distances of cases per month does not seem to depend on fox population density, nor on the frequency of rabies. The increase of the distance of new cases from the frontline precedes the increase in rabies frequency; the increasing distance between cases appears to be the cause of the increase of case frequency and not its result.
3. Animals other than the fox do not play a significant role in the spread of the disease beyond the frontline.

Our findings suggest that a more detailed recording and computer analysis will, in the future, become an important contribution to two areas of public health significance, namely:

The strategy of rabies control and fox protection, taking into consideration the role of foxes in the frontwave of an epidemic; and

The improvement of post-exposure treatment in man, taking into consideration the seasonal pattern of distances between rabies cases in nature.

SUMMARY AND MAIN CONCLUSIONS

WHO/FAO coordinated research on wildlife rabies in Europe has contributed much to our understanding of fox ecology. Studies on the turnover and recovery of fox populations and the migration and territorial behaviour of individual foxes were supplemented by data on distances between observed cases by time and space, and on the seasonal pattern of case frequency and the spread of the epidemic into new territories.

Further research on fox behaviour appears to be useful in order to improve and support new strategies which are being developed for the control of the disease through the establishment of healthy fox populations. Measures will be well aimed and limited by space and time, according to the epidemiological pattern. The further reduction of the risk of exposure to man and protection of the animal populations (which are the principal victims of rabies) shall remain the major objectives of the internationally coordinated research.

REFERENCES

Bögel, K., Arata, A. A., Moegle, H. and Knorpp, F. 1974. Recovery of reduced fox populations in rabies control. *Zbl. Vet.-Med.* B21: 401–412.

Bögel, K., Moegle, H., Knorpp, F., Arata, A., Dietz, K. and Diethelm, P. 1976. Characteristics of the spread of a wildlife rabies epidemic in *Europe. Bull. Wld Hlth Org.* 54: 433–447.

Lloyd, H. G., Jensen, B., van Haaften, J. L., Niewold, J. J., Wandeler, A., Bögel, K. and Arata, A. A. 1976. Annual turnover of fox populations in Europe, *Zbl. Vet.-Med.* B 23: 580–589.

Moegle, H., Knorpp, F. and Bögel, K. 1971: Einfluss der Begasung der Fuchsbaue auf die Fuchsdichte und die Wildtollwut in Baden-Württemberg. *Berl. Münch. tierärztl. Wschr.* 84: 437–441 (English summary: WHO Wkly Epidem. Rec. No. 5 (1972) 49.

Müller, J. 1971: The effect of fox reduction on the occurrence of rabies. Observation from two outbreaks of rabies in Denmark. *Bull. Off. int. Epiz.* 75: 763–776.

Wandeler, A., Wachendörfer, G., Förster, U., Krekel, H., Schale, W., Müller, J. and Steck, F. (1974a): Rabies in wild carnivores in Central Europe. I. *Epidemiological studies. Zbl. Vet.-Med.* B 21: 735–756.

Wandeler, A., Wachendörfer, G., Förster, U., Krekel, H., Müller, J. and Steck, F. (1974b): Rabies in wild carnivores in Central Europe. II. Virological and serological examinations. *Zbl. Vet.-Med.* B 21: 757–764.

Wandeler, A., Wachendörfer, G., Förster, U., Krekel, H., Schale, W., Müller, J. and Steck, F. (1974c): Rabies in wild carnivores in Central Europe. III. Ecology and biology of the fox in relation to control operations. *Zbl. Vet.-Med.* B 21: 765–773.

18 SHORT REPORT ON MATERIALS AND METHODS USED IN A STUDY OF THE EFFECT OF RABIES ON THE DYNAMICS OF FOX POPULATIONS IN FRANCE, WITH SOME PRELIMINARY RESULTS

M. Artois* and L. Andral*

INTRODUCTION AND AIMS

The aims, methods and format of analyses of data obtained in a study of the population dynamics of the fox of relevance to the occurrence of rabies in foxes in France are described, together with some preliminary results.

The purpose of this report is to describe the materials and methods used in the pursuit of this study from 1975–78. Analyses of data are as yet incomplete but some preliminary results are given.

It is not known whether or not rabies is transmitted as readily and persistently by foxes of different age groups and sex. In order to promote a more effective fox control campaign in rabies-free areas ahead of the disease front–for prevention of spread of rabies–investigations on the role of foxes of different ages and sex in rabies transmission were undertaken. In addition, selectivity of the currently available control methods on the different components of a fox population were studied.

In order to provide an understanding of the population dynamics of the fox, the relative abundance, the age structure, sex ratios and reproduction of fox populations were examined in three areas, viz

a. The rabies-free area in advance of the rabies infected zone.
b. The area where rabies in foxes was prevalent in the newly infected frontal zone.
c. The area in the wake of the frontal zone where the disease was less intense but intermittently persistent.

MATERIAL

The heads of wild and domestic animals suspected of rabies are sent from all parts of France, except the region around Paris, to CNER* for rabies diagnosis. Animals implicated in a biting or scratching incident are diagnosed elsewhere. The National Defence Office has seconded "vétérinaires

* Centre National d'Etudes sur la Rage, Ministère de l'Agriculture, Direction de la Qualité, Services Vétérinaires, B.P. n° 9-54220 Malzeville, France.

biologistes" to the Veterinary Service Offices in each department throughout France. Many have assisted CNER by conducting autopsies on entire fox carcases sent to their laboratories, before severing the head for despatch to CNER. In addition they have examined some foxes which were not suspected of being rabid, and they have also collected information on the annual number of foxes killed by Federal Gamekeepers.

During the period of study from 1975–78 inclusive, about 2,000 foxes have been examined for rabies diagnosis and age determination (among the whole of foxes submitted for rabies diagnosis).

METHODS

Abundance and fluctuations in number of foxes. The number of foxes known to have been killed in 12 surveillance areas have provided a measure of relative abundance, in some cases, over the entire 4 year period of study.

Unfortunately, none of the surveillance areas has been traversed by the advancing rabies front, and the work will need to be extended to observe the effects of rabies on abundance of foxes in newly infected areas. This method based upon an index of number of adults killed in quadrats covering different habitats allows, however, interesting comparisons between different regions.

Rabies diagnosis. Diagnosis is made in brain tissue by the fluorescent antibody test using Ammon's horn impression slides, and by the inoculation test using laboratory albino mice.

Age determination. Heads received for rabies diagnosis have been aged by the cementum annulation technique in decalcified canine teeth according to the method described by Jensen and Nielsen (1968).

Sex determination. Except where a fox has been examined at autopsy by our collaborators at the Departmental Veterinary Service Offices, the sex of heads received is dubious or not known. To resolve this examination of cells of the brain, tissue for sex chromatin (found only in females) has been carried out, using Harris haematoxylin on Ammon's horn impression slides.

METHODS OF ANALYSES AND CORRELATION OF RESULTS

Rabies, age structure and sex ratios. For each Region foxes diagnosed as positive or negative for rabies are classed into 4 age groups, 0–1; 1–2; 2–3; and over 3 years old. Where possible these data are further subdivided according to sex. Seasonal fluctuations in the sex ratios of rabid and nonrabid foxes are noted and possible bias of methods of capture upon sex. The sex ratio of embryos is also noted.

Reproduction. From 1976–78 inclusive, 249 vixens have been examined. Placental scars were visible in 180 vixens; 57 vixens were pregnant and of these 22 were in the last third of the 52–53 day gestation period. Data are analysed to determine the monthly proportion of females pregnant. Fecundity is noted by counts of corpora lutea, fertility by the number of embryos counted from implantation to full term, and birth litter size is estimated from the number of embryos observed in the later stages of gestation.

Variation in mean weights of ovaries are measured to reflect sexual activity in non-pregnant vixens, and placental scars are used as information complementary to that obtained by direct observations of pregnant vixens. Onset of pregnancy and dates of parturition are obtained by extrapolation from the stages of gestation estimated by the weights of the embryos.

Unlike Englund (1970) who classed placental scars into 6 groups according to their intensity, only 3 classes of intensity were noted in this study. The mean numbers of corpora lutea and embryos and placental scars have been compared. As far as possible all the available data on reproduction are examined according to year, region and age of the vixen. The proportions of nonproductive (sterile or barren) vixens was estimated from the proportions of adult vixens without placental scars in the post-breeding season.

BRIEF OUTLINE OF SOME RESULTS

Relative abundance. Preliminary analyses reveal that population levels estimated from bounties and hunting records are surprisingly low varying from one fox per 150 to 1,000 ha. These data are incompatible with estimates made in other ways, e.g. by spot light counts at night, and a re-examination of the data is required

Age of foxes and rabies. Rabies is less prevalent among foxes less than 1 year old than among older animals. The dispersal movements of these young animals at the onset of winter, are, however, of considerable importance in the long distance progression of the disease, and it seems that they are of greater significance to the seasonal spread of the disease rather than to its prevalence in an area.

Reproduction. Preliminary results indicate that most vixens give birth in the last fortnight in March, the mean number of (dark) placental scars representing live births is 4.5 and the proportion of non-productive vixens is low, at 5% in 1976. These and other data are very similar to results obtained by other workers in western and central Europe. All data are or will be subjected to statistical analyses.

CONCLUSIONS

Preliminary analyses indicate that the results of this study will complement and add significantly to investigations conducted in Europe (Lloyd et al., 1976) and Ontario (Johnston and Beauregard, 1969). In addition, the effects of some methods of control on fox populations, and rabies on population structures will be elucidated. Such information will be of value in rabies modelling and computer simulation programmes. (Copies of the final report will be available on request).

ACKNOWLEDGEMENTS

I am indebted to H. G. Lloyd (Ministry of Agriculture, Fisheries and Food, Powys, U.K.) for help in the translation of this paper and also to E. Bertolotti et M. J. Duchéne.

REFERENCES

Englund J. 1970. Some aspects of reproduction and mortality rates in Swedish foxes (*Vulpes vulpes*), 1961–63 and 1966–69. *Viltrevy*,1970, 8, n° 1; 2–82.

Fairley J. S. 1970. The food, reproduction, form, growth and development of the fox *Vulpes vulpes* L. in north-east Ireland. *Proc. R. Irish Acad.*, 1970, 69 B; 103–137.

Jensen B. and Brunberg Nielsen L. 1968. Age determination in the red fox (*Vulpes vulpes* L.) from canine tooth section. *Danish Rev. Game Biol.*, 1968, 5 n° 6; 3–15.

Johnston D. H. and Beauregard M. 1969. Rabies epidemiology in Ontario. *Bull. Wildlife Diss. Assoc.*, 1969, 5; 357–370.

Lloyd H. G., Jensen B., Van Haaften J. L., Niewold F. J. J., Wandeler A. Bogel K., and Arata A. A. 1976. Annual turnover of fox populations in Europe. *Zbl. Veter. Med.*, 1976, 23 580–589.

19 EPIZOOTIOLOGY AND CONTROL OF RABIES IN CENTRAL EUROPE

G. Wachendörfer and J. W. Frost*

THE SITUATION OF RABIES IN CENTRAL EUROPE

Since the beginning of World War II the epizootic of sylvatic rabies has spread from a focus south of Kaliningrad: (a) southwards into Czechoslovakia, Hungary, Yugoslavia and Romania; (b) westwards through the Democratic and Federal Republics of Germany into Denmark (invaded for the third time in Sept. 1977), Belgium, Luxembourg, Austria, Switzerland, France, the Netherlands, and Northern Italy; (c) eastwards into USSR (Fig. 1).

In the period 1972 to 1976 in 11 central European countries 63,672 cases in animals were reported: 82 per cent in wildlife and 18 percent in domestic animals (Fig. 2). Foxes accounted for 73.6 per cent of all cases. In this area the fox maintains the chain of infection. Other wild carnivores, if at all, play only a secondary role. Dogs (3.1 per cent) and cats (4.4 per cent) were important transmitters between wildlife and man. Since 1972 there has been a sharp increase of the epizootic in most of the affected countries, especially in Austria, Belgium, the Democratic and Federal Republics of Germany, and France. Nineteen cases were reported in man, of which 10 were infected outside of central Europe (Wachendörfer, 1977).

Some figures exist, which throw light on the so called "dark number" of cases in wildlife. In the Federal Republic of Germany the highest incidence of rabies was observed in 1976. In the state of Hesse an average of 11.7 cases per 100 km^2 was recorded (Fig. 3). It is estimated that the infected fox populations are reduced by the disease to 80–40 per cent (WHO-Report, 1973; Wandeler et al., 1974a; Stellmann et al., 1974). Presuming an average population density of 3 to 4 foxes per km^2, a factor of approximately 10 to 20 seems to be justified in calculating the number of actual cases in wildlife. At least during an acute outbreak, this gives more precise information concerning the true incidence of rabies under field conditions (Farrenkopf, 1977; v. Braunschweig, 1976; Wachendörfer, 1977).

THE PRESENT STATUS OF EPIDEMIOLOGICAL RESEARCH

The ecological equilibrium in central Europe has been upset for decades in favour of the fox. From hunting records analysed in Denmark it can be concluded that since the beginning of the century the fox population has increased two- to fourfold. For this exist various reasons, such as for

* State Veterinary Investigation Laboratory, Frankfurt/Main, FRG.

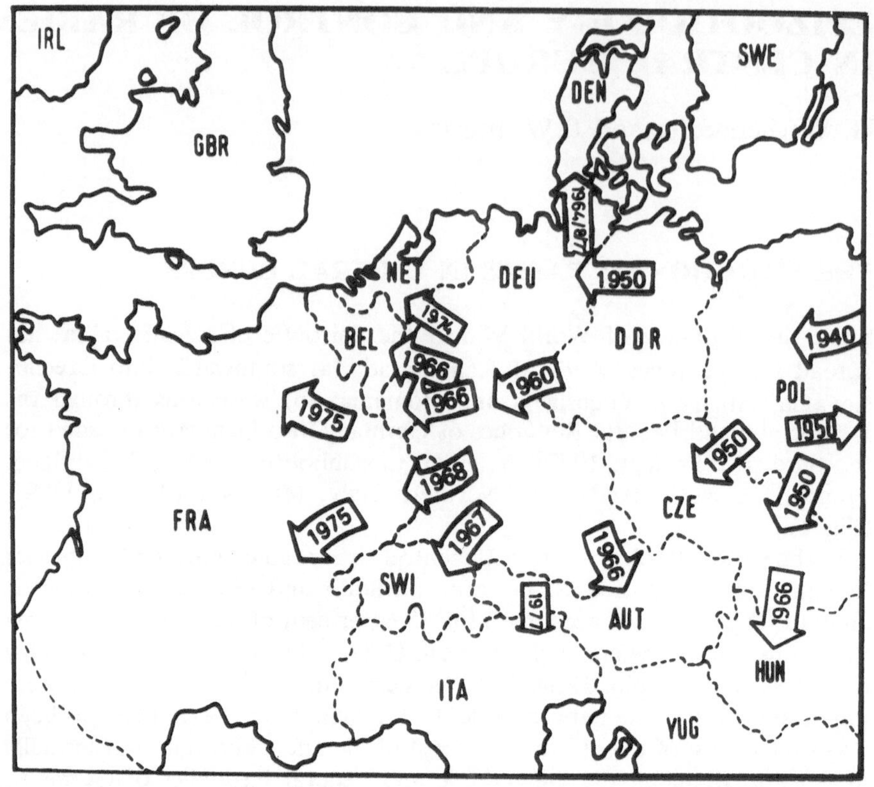

Fig. 1. Spread of sylvatic rabies in Europe 1940–1977.

instance the elimination of the natural enemies of the fox, the decrease in fox hunting – especially after World War II – and changes in agriculture (Steck, 1975). Thus a population density was attained providing ideal conditions for the perpetuation of the infection.

In the past years as a result of the 1st European WHO-Conference on the Surveillance and Control of Rabies in 1968 the epizootiological situation of sylvatic rabies in central Europe intensively was studied within the framework of a multinational FAO/WHO coordinated research program. In this report only the most eminent results can be presented which were obtained by working groups in Denmark, Czechoslovakia, Switzerland, and the Federal Republic of Germany.

It was confirmed that only the red fox maintains the chain of infection. Other wild-living animals including all species of carnivores living in central Europe, play only a secondary role or none at all. In study areas the red fox accounted for 62.2 to 78.7 per cent of recorded rabies cases found in animals (Fig. 4). Dogs and cats represented only 0.4 to 8.4 per cent of the cases and are considered as an overspill from sylvatic rabies. Sequences of

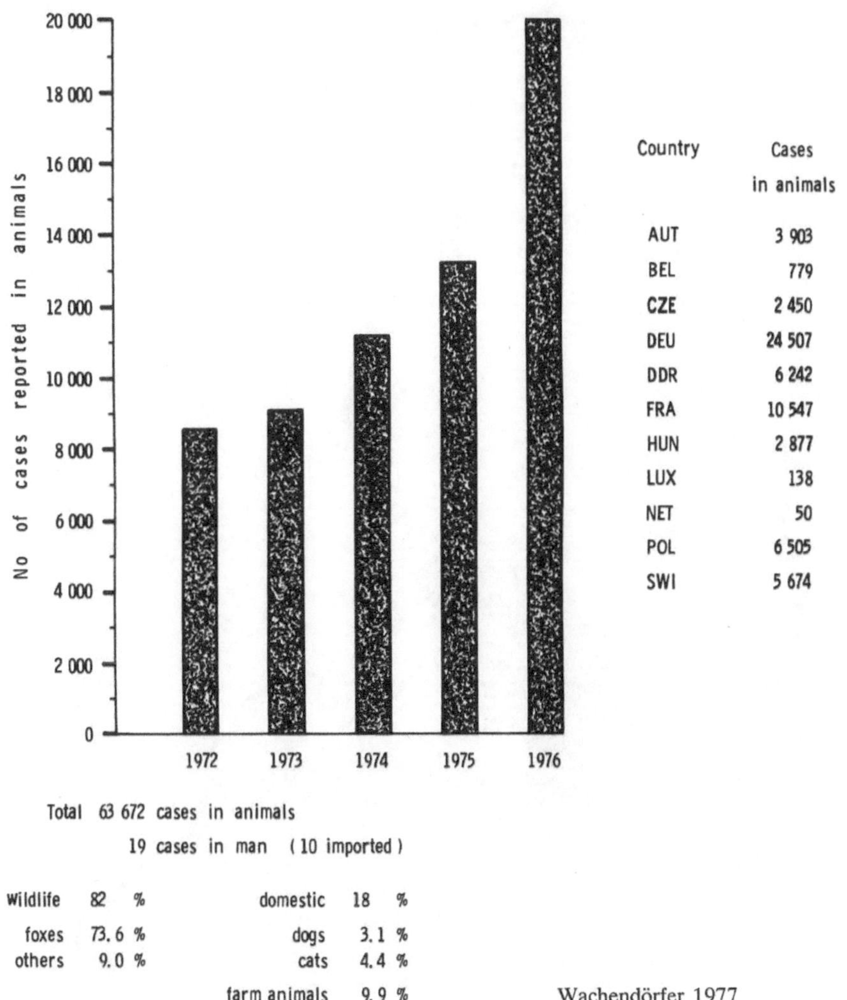

Fig. 2. Sylvatic rabies in Central Europe 1972–1976. Pattern I.

infection in dogs or cats are very rare in our region and if they occur consist only of very few cases (Steck, 1975). The compulsory vaccination of dogs in infected countries like Switzerland and Denmark prevents rabies in this species almost completely and thus reduces human exposures significantly (Fig. 5). Cases of rabies in other species, except in ungulates (especially cattle and deer), are seldom. These and other non-predatory animals are not involved in distributing the disease. To our present knowledge bats – in contrast to the Americas – are not included in the cycle of infection.

Now let us briefly mention some additional facts emphasizing the role of the fox as sole expansor of the epidemic: If the population density is below 0.2 to 0.3 foxes hunted per km^2 and year, rabies will extinguish or will not

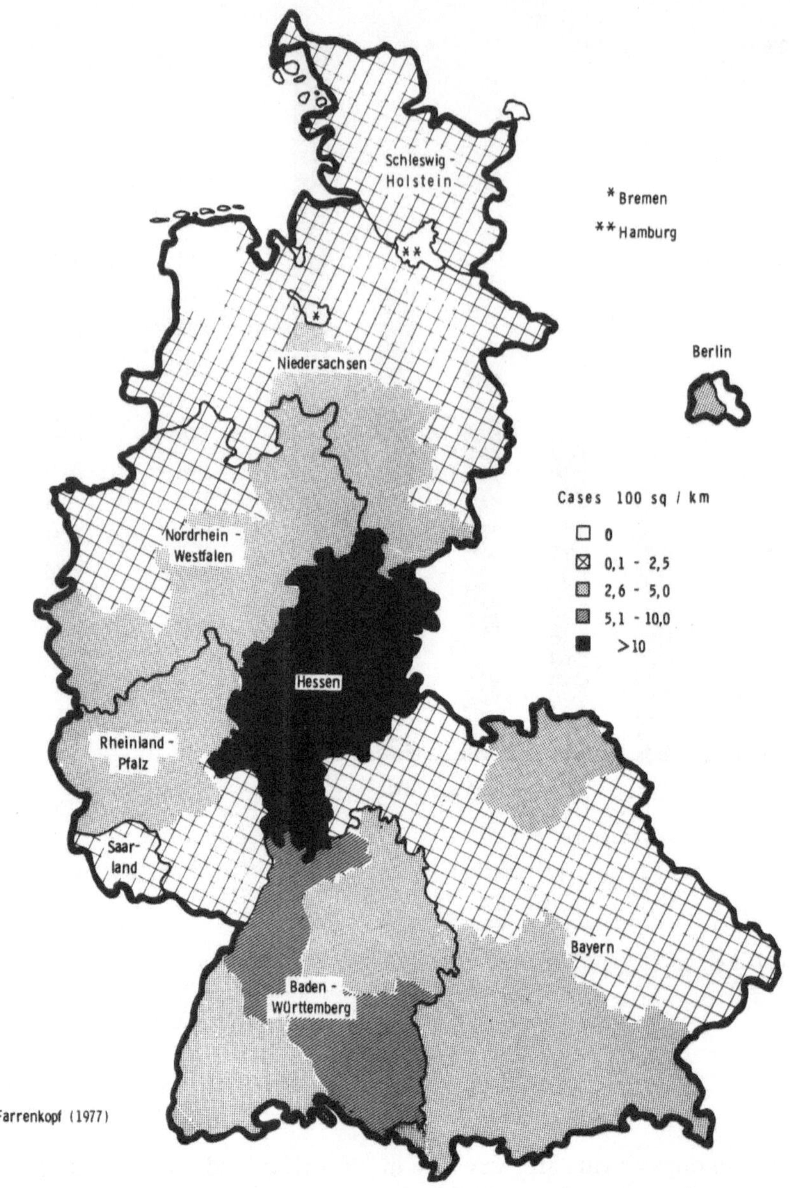

Fig. 3. Distribution of rabies in The Federal Republic of Germany 1976. Density of cases recorded.

invade such an area (Moegle et al., 1971, 1974, 1977; Wandeler et al., 1974c; Bögel et al., 1976).

A computer analysis of case data from Baden–Württemberg resulted in information on speed, tendency of spread and seasonal variation confirming the role of the fox in epidemiology (Bögel et al., 1976). In this study more than 97 per cent of rabid animals ahead of the monthly drawn front line

	Total	Period
Denmark	237	1964/65, 1969/70 [1]
Hesse	14 143	1954 - 1977 [2]
Switzerland	8 494	1967 - 1977 [3]

Wildlife in %	Denmark	Hesse	Switzerland	Domestic in %	Denmark	Hesse	Switzerland
Foxes	78.4	62.2	78.7	Cattle	6.0	6.4	3.1
Roe-deer	2.5	12.8	4.6	Cats	5.1	8.4	3.4
Mustelids	3.0	2.7	6.7	Dogs	0.4	4.4	0.4
Other Wildlife	----	0.8	0.2	Other domestic Animals	4.6	2.3	2.9
Total	83.9	78.5	90.2		16.1	21.5	9.8

1) Müller, Copenhagen

2) Communicable Disease Statistics of GFR

3) Steck, Bern

Fig. 4. Distribution of rabies among wild and domestic animals in Denmark, Hesse and Switzerland.

were foxes compared with about 80 per cent behind (Fig. 6). It could also be shown by this analysis, that the percentage of infected mustelids is definitely higher at the end of the epizootic (Bögel et al., 1976; Moegle et al., 1977). However, there exist hints for rare and limited cycles of infection within the mustelids. These mustelids cycles come to an end after fox rabies is eliminated (Bögel et al., 1977). An explanation for this self-limitation is the fact, that mustelids excrete significantly less infectious virus with their saliva than the fox.

The rabies epizootic follows a clear seasonal pattern, which results from the social behaviour and population dynamic of the fox. Two peaks can be discerned (Fig. 7), the spring peak, being caused by the mating period with its high contact rate. The shallower peak in autumn seems to be due to the high population density and to the young foxes striving for their own home range. Cases in other species follow this pattern with a small lag period with the exception of cattle, where the spring peak is missing because of stabling during winter time (Kauker et al., 1963; Rojahn, 1977).

The observed increase of rabies already in August suggests, that cubs of less than 5 month of age may play a role in epidemiology (Bögel et al., 1976). The mean distance of migration for the frontline of the epizootic is approximately 5 km per month (Moegle et al., 1977).

For the perpetuation of the infection multiplication of rabies virus in the salivary glands and the subsequent excretion with saliva are essential (Dierks et al., 1969). Optimal conditions for the survival of the virus exist apart from ecological factors, if the diseased animal excretes infectious virus for a certain time in amounts sufficient for the provocation of new infections (Sikes, 1962). These prerequisites are fulfilled in particular by the fox.

Fig. 5. Species of animals responsible for post-exposure treatment in man.

Area/Period	Dogs	Cats	Cattle	Other domestic animals %	Foxes	Mustelids	Roe-deer	Other wildlife	Total Absolute figures = 100%
Southern Hesse 1953–1975	22.4	17.2	15.1	1.9	23.7	—	16.4	2.6	232
Northern-Westphalia 1972	16.0	14.8	13.0	27.8	17.8	1.8	8.3	0.6	169
Switzerland 1967–1977	2.0	75.0	—	—	12.0	11.0	—	—	254

Wachendörfer and Frost 1978

Species	Observed cases ahead frontline* n = 426 %	Observed cases behind frontline* n = 2396 %
Foxes	97.2	79.8
Roe deer	1.4	7.3
Badgers	0.5	2.3
Martens	--	2.1
Other wildlife	--	0.2
Cats	0.7	2.3
Dogs	0.2	2.1
Other domestic animals	--	4.0

* Frontline determined for each preceding month

Bögel et al. (1976) modified

Fig. 6. The role of different animal species in the spread of rabies. Study area (size 11,273 sq/km) in the southern part of the Federal Republic of Germany, 1963–1971.

Virological studies in Switzerland demonstrated, that virus production in salivary glands was only accompanied by infection of the brain. In 93 per cent of the rabid foxes and in 83 per cent of rabid badgers the virus could also be isolated from the salivary gland (Wandeler et al., 1974b). On the other hand, only in 50 per cent of rabid stonemartens infectious virus could be isolated from the salivary glands, the median titre being 2 $\log_{10} MLD_{50}$ lower than in foxes and badgers. Thus the high susceptibility of the fox for rabies virus (Sikes, 1966) together with the regular formation of large amounts of infectious virus in the salivary glands may be the most important factors in maintaining sylvatic rabies in our fox population.

Now we must question ourselves, why only the fox acts as vector and virus reservoir during the present epizootic and other wild-living predators do not. This problem was studied by our working group in form of adaptation trials. A rabies virus isolate from the fox was passaged eight times by the intramuscular route in weasels (*Mustela nivalis*) and European hamsters

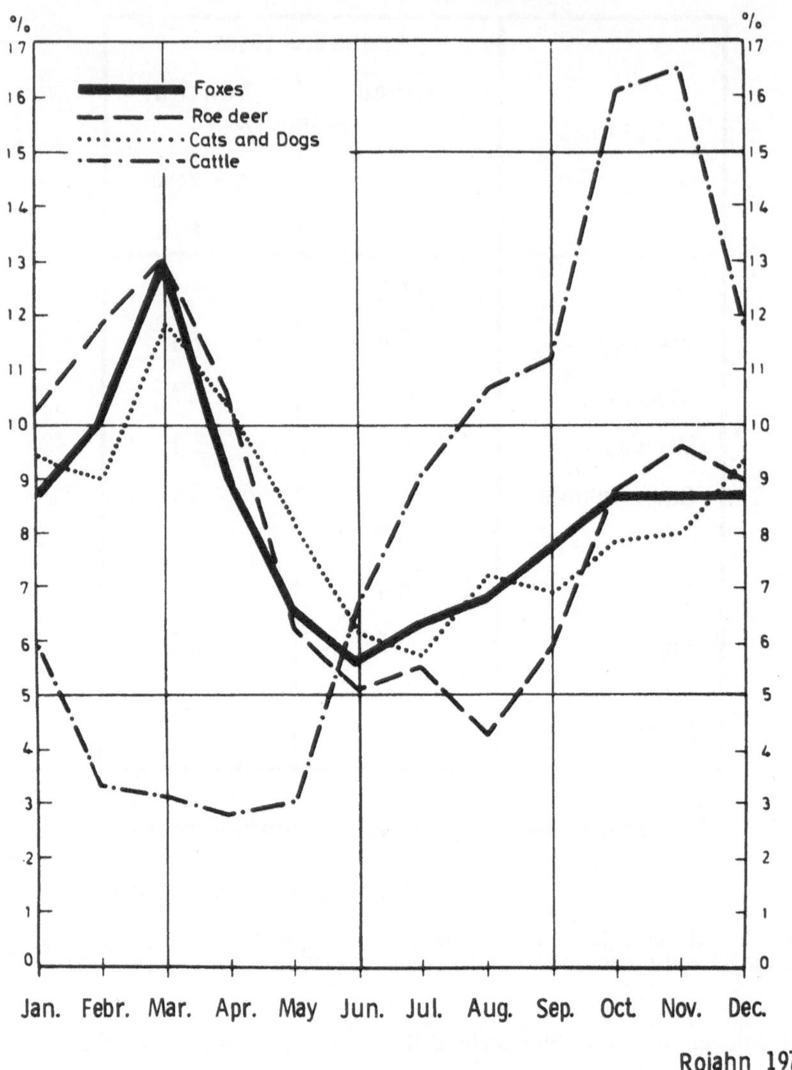

Fig. 7. Frequency of rabies in different animal species per month. Federal Republic of Germany, 1955–1975.

(*Cricetus cricetus*) (Fig. 8). The fast increase of infectious virus in the brains of the weasels and the centrifugal spread into salivary glands indicated a rapid adaptation of the fox isolate to this species. From the fifth passage on, however, duration and quantity of virus formation decreased suggesting insufficient conformation with the new host. Possibly this is one of the reasons why under field conditions independant cycles have not developed in this species.

The European hamsters proved to be rather resistant to infection with rabies virus. Infectious titres from salivary glands were usually low and in 56

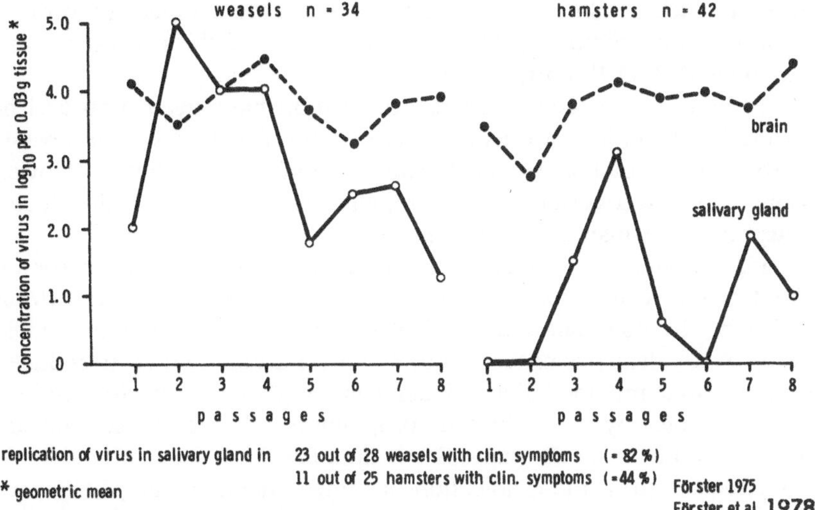

Fig. 8. Concentration of a fox isolate of rabies virus in brain and salivary gland of weasels and European Hamsters during 8 passages.

per cent no virus at all could be detected. This is the reason why hamsters have no epidemiological relevance.

Intensive search for rabies in small rodents has already shown, that they are not directly involved in the rabies epizootic. Rabid mice with high titres of infectious virus in brain, salivary glands or brown fat, as it is in the case in other animals and in laboratory infected mice, have not been detected among the many rodents examined in France, Hesse, and Switzerland (WHO-Report, 1972; Steck, 1975; Förster et al., 1977).

On the other hand, in Czechoslovakia (Sodja et al., 1971; 1972) and in the Federal Republic of Germany (Schneider et al., 1972; Schoop, 1977) virus isolates were obtained from rodents after serial blind passages in suckling mice. These isolates belonged to the serotype I of the rabies virus group and seem to reflect latent infections. Many aspects of these so called "rodent strains" remain obscure (Steck, 1975), in particular pathogenesis and spread. In large scale contact trials with laboratory mice virus transmission could not be detected (Weinhold, 1977).

Our group studied adaptation and peripheral spread of rodent strain W56/71 (Schneider et al., 1972) in ferrets (*Mustela putorius domesticus*) and European hamsters (Förster, 1975; Förster et al., 1978). Both species developed high titres of infectious virus in the central nervous system, but only in one animal minute amounts of virus could be isolated from the salivary glands. This rodent strain is characterized by an extreme neurotropism and thus differs biologically from the fox isolates.

For latent rabies virus infections in other wild-living species in Europe no evidence has been found (Wandeler et al., 1974b; Wachendörfer et al., 1978).

Since at present only the fox serves as reservoir for rabies virus and as expansor of the disease, the density of susceptible individuals must be reduced to such a value, that infectious contacts are avoided. In most parts of central Europe the only practical means to achieve this consists in gassing of fox dens (WHO-Report, 1973).

Other methods to reduce the fox population have been used and have their merits. Trapping and distribution of poison baits have the advantage that the number of killed animals can be checked immediately. However, poisoning and to a certain degree trapping, are only used with reluctance because of the danger for other animals.

Normal hunting or even rewarding hunting can be carried out for years without reducing the fox population by more than 25 per cent (Müller, 1977). Only in mountain areas in Switzerland shooting was found efficient. Gassing of fox dens has proved to be the most useful method to reduce the fox population, provided that 75 per cent or more of the fox litters are known and destroyed. Reduction programmes should be conducted by trained personnel from the particular region. Even if these measures have so far not succeeded in the eradication of rabies, they have helped to reduce the overall incidence, especially in domestic animals and man (WHO-Report 1972, 1973, 1978).

A very intriguing alternative to gassing of fox dens was postulated and demonstrated in laboratory trials by orally vaccinating foxes against rabies (Black and Lawson, 1970; Baer et al., 1971). As has been shown in field studies, about 70 per cent of a given fox population can be reached by baits (Johnston, 1975; Manz, 1975; Manz and Berger, 1978).

In the past few years several research groups in Canada, the USA, in France, Switzerland and in the Federal Republic of Germany investigated into this subject. Various live virus vaccines were orally applicated to foxes and the immune response checked. Since 1972 these studies are coordinated by WHO.

Debbie (1974) used embryonated chicken eggs infected with vaccinal rabies virus and was able to provoke an immune response in 50 per cent of the foxes. The working group of Andral in Nancy reported good results with the live vaccine from the Institut Pasteur, Paris, by just instilling it into the buccal cavity of foxes. However by the same vaccine in baits – the vaccine was injected into chicken heads – not a single fox was protected against test infection. Obviously the virus was inactivated very fast by autolytic processes within the bait. Winkler and coworkers (1976a) filled two variants of ERA vaccine in plastic straws, which were then hidden in sausages and fed to foxes. With the commercial ERA vaccine one third of the animals developed antibodies against rabies virus. When using a higher titring, laboratory manufactured vaccine, about 80 per cent of the foxes proved to be protected.

In Switzerland Steck and his coworkers as well as our group had promising results with the ERA-derived vaccine virus grown in BHK-21 cells using chicken heads as bait and small plastic containers or thin walled glass vials, respectively, to protect the vaccine. Thus this aspect of oral vaccination seems to be solved, apart from problems deriving from mass production for large scale application in the field.

Another point, and this still poses a serious obstacle, is the problem of the residual pathogenicity of the vaccines tested so far. The use of vaccine strain Flury LEP is impossible because of its pathogenicity for animals younger than three months. The highly adapted strain Flury HEP, which was believed to be completely innocuous, proved to cause vaccinal rabies in mice (*Apodemus sylvaticus*), in rats (*Rattus norvegicus*) as well as in mustelids (*Mustela erminea* and *Mustela nivalis*) (Förster et al., 1976). The very immunogenic ERA vaccine virus revealed residual pathogenicity for most of the wildliving rodents tested to far (Förster et al., 1976; Winkler et al., 1976; Farrenkopf, 1977; Lohrbach, 1977) and gave indications for an adaptation to muskrats (*Ondatra zibethica*) (Wachendörfer et al., 1978) which became extremely aggressive.

Up to now none of the available rabies live virus vaccines is safe enough to free application in the field. At present a cloned variant of the Flury HEP strain is under investigation in our laboratory and to date no adverse effects were observed in small rodents (*Microtus arvalis, Mus musculus, Apodemus sylvaticus*), in hamsters (*Cricetus cricetus*), rats (*Rattus norvegicus*), muskrats (*Ondatra zibethica*), in mustelids (*Mustela putorius, Mustela erminea, Mustela nivalis*), and in racoons (*Procyon lotor*) and foxes (*Vulpes vulpes*). However, some safety and efficacy testing still remains to be performed before a final evaluation of this vaccinal strain can be given.

SUMMARY

The fox (*Vulpes vulpes*) as vector and reservoir of rabies virus in central Europe is ascertained by a large volume of scientific data. Other species, especially mustelids, are only secondarily involved in the epidemic and so far no independent cycles of transmission have developed. Rodents, despite the fact that rabies-like virus strains have been isolated in several areas from these species, play no actual role in epidemiology.

Chains of infections in dogs and cats are rare events and only consist of a short series of cases. However, these animals act as a link between wildlife and man.

To the present knowledge the necessary interruption of infectious contacts can only be achieved by reduction of the fox population. As an alternative to gassing of fox dens the possibility of orally vaccinating foxes against rabies is discussed. Various problems, such as choice of bait, production of a highly effective vaccine, its protection within the bait, are solved or at least pose no serious problems.

Studies on residual pathogenicity of rabies virus strains suitable for the oral vaccination demonstrated fatal vaccinal rabies for non-target species which eventually could consume the baits. The present state of innocuity testing of live rabies virus vaccines is reported.

REFERENCES

Andral, L. 1977. WHO report of consultations on oral vaccination of foxes against rabies. Nancy 12–13 Nov. 1977, VPH/78.7.

Baer, G. M., Abelseth, M. K., and Debbie, F. G. 1971. Oral vaccination of foxes against rabies. *Am. J. Epidemiol.* 93: 487–490.

Black, J. G. and Lawson, K. F. 1970. Sylvatic rabies studies in the silver fox (*Vulpes vulpes*). Susceptibility and immune response. *Can. J. comp. Med.*, 34: 309–311.

Bögel, K., Moegle, H., Knorpp, F., Arata, A., Dietz, K. and Diethelm, P. 1976. Characteristics of the spread of a wildlife rabies epidemic in Europe. *Bull. Wld. Hlth. Org.* 54: 433–447.

Bögel, K., Schaal, F. and Moegle, H. 1977. The significance of martens as transmitters of wildlife rabies in Europe. *Zbl. Bakt. Hyg. I. Abt. Orig. A* 238: 184–190.

Braunschweig, A. von 1976. Tollwut ein jagdliches Problem. Paper presented at the meeting of the German Veterinary Assoc., Hannover 28–29 Oct. 1976.

Debbie, J. G. 1974. Use of inoculated eggs as a vehicle for the oral vaccination of red foxes (*Vulpes vulpes*). *Infect. Immun.* 9: 681–683.

Dierks, R. E., Murphey, F. A., and Harrison, A. K. 1969. Extraneural rabies virus infection. Virus development in fox salivary gland. *Amer. J. Pathol.* 54: 251–273.

Farrenkopf, R. 1977. Passageversuche zur Unschädlichkeitsprüfung des Impfstammes ERA-BHK-21 an Bisamratten (*Ondatra zibethica*) und Feldhamstern (*Cricetus cricetus*) - Ein Beitrag zur oralen Immunisierung von Füchsen gegen Tollwut. *Vet. Med. Diss., Gießen.*

Förster, U. 1975. Zur Frage der Adaptationsfähigkeit von zwei in Mitteleuropa isolierten Tollwutvirusstämmen an eine domestizierte und zwei wildlebende Spezies - Ein Beitrag zur Epidemiologie der Tollwut. Habilitationsschrift, Medizinische Fakultät, Frankfurt am Main.

Förster, U., Wachendörfer, G., Schnettler, R. und Weber, J. 1976. Unschädlichkeitsprüfungen von Tollwut-Lebendvakzinen an wildlebenden Säugern. 11. Kongr. Bericht: 257–262.

Förster, U., Wachendörfer, G. und Krekel, H. 1977. Untersuchungen zum Nachweis von Tollwutvirus bei Nagern und Insektivoren - Ein Beitrag zur Epidemiologie der Tollwut. *Berl. Münch. Tierärztl. Wschr.* 90: 335–337.

Förster, U., Schale, W. und Wachendörfer, G. 1978. Zur Frage der Adaptationsfähigkeit von zwei in Mitteleuropa isolierten Tollwutvirusstämmen an eine domestizierte und zwei wildlebende Spezies - Ein Beitrag zur Epidemiologie der Tollwut. *Zbl. Vet. Med.* B 25: 826–834, 841–848.

Johnston, D. H. 1975. The principles of wild carnivore baiting with oral vaccines. WHO consultation on oral vaccination of foxes, Frankfurt am Main, 1–3 Sept. 1975.

Kauker, E. und Zettl, K. 1963. Zur Epidemiologie der sylvatischen Tollwut in Mitteleuropa und zu den Möglichkeiten ihrer Bekämpfung. Vet. Med. Nachr. 2/3: 181–204.

Lohrbach, W. 1977. Unschädlichkeitsprüfungen von zwei Modifikationen des Tollwutimpfvirus "Stamm ERA" bei Wanderratten (*Rattus norvegicus*) - Ein Beitrag zur oralen Immunisierung von Füchsen gegen Tollwut. *Vet. Med. Diss.*, Gießen.

Manz, D. 1975. Markierungsversuche an Füchsen im Revier als Vorbereitung für eine mögliche spätere perorale Vakzination gegen Tollwut. XI. Meeting of the German Vet. Med. Assoc. 10–12 Apr. 1975 - published in: *Fortschr. Vet. Med.* 11: 263–269 (1976).

Manz, D. und Berger, J. 1978. Feldversuche zur Köderaufnahme durch Füchse in Abhängigkeit von der Jahreszeit. *Zbl. Vet. Med.* B 25: 157–160.

Moegle, H., Knorpp, F. und Bögel, K. 1971. Einfluß der Begasung der Fuchsbaue auf die Fuchsdichte und die Wildtollwut in Baden-Württemberg. *Berl. Münch. Tierärztl. Wschr.* 84: 437–441.

Moegle, H., Knorpp, F., Bögel, K., Arata, A., Dietz, K. und Diethelm, P. 1974. Zur Epidemiologie der Wildtiertollwut. Untersuchungen im südlichen Teil der Bundesrepublik Deutschland. *Zbl. Vet. Med.* B 21: 647–659.

Moegle, H. und Knorpp, F. 1977. Zur Bekämpfung der Tollwut bei Wild. *Prakt. Tierarzt* 58: 105–112.

Rojahn, A. 1977. Vorkommen der Tollwut in der Bundesrepublik Deutschland. Berl. Münch. Tierärztl. Wschr. 90: 269–273.

Schneider, L. G. and Schoop, U. 1972. Pathogenesis of rabies and rabies-like viruses. *Ann. Inst. Pasteur* 123: 469–476.

Schoop, U. 1977. Praomys (Mastomys) natalensis: An African mouse capable of sustaining asymptotic rabies infection. *Ann. Microbiol.* (Inst. Pasteur) 128: 289–296.

Sikes, R. K. 1962. Pathogenesis of rabies in wildlife. I. Comparative effect of varying doses of rabies virus inoculated into foxes and skunks. *J. Amer. Vet. Res.* 23: 1041–1047.

Sikes, R. K. 1966. Wolf, fox and coyote rabies. Proc. Nat. Rabies Symp. CDC Atlanta pp. 31–33 U.S. Department of Health, Education and Welfare.

Sodja, J., Lim, D., and Matouch, O. 1971. Isolation of rabies-like virus from murine rodents. *J. Hyg. Epidemiol. Microbiol. Immunol.* 15: 229–230.

Sodja, J. and Matouch, O. 1972. Adaptation to the laboratory mouse of rabies-like viruses isolated from wild rodents. *Acta virol.* 16: 147–152.

Steck, F. 1975. Rabies in central Europe. WHO report of consultations on the WHO/FAO coordinated research programme on wildlife rabies in Europe, Frankfurt am Main 11–12 Dec. 1975, VPH/76.2.

Stellmann, C. et Beranger, G. 1974. Epizootologie de la rage en France de 1968 à 1972 selon un modele biomathematique. Incidence sur les modalités de prophylaxie (vaccination des bovins). *Rév. Méd. veter.* 125: 45–62.

Wachendörfer, G., Förster, U., Menzel, W., Krekel, H., Scharfen, E., Frost, J. W. und Osthoff, F. 1978. Serologische Untersuchungen bei wildlebenden Spezies Mitteleuropas zum Nachweis neutralisierender Antikörper gegenüber Tollwutvirus. *Berl. Münch. Tierärztl. Wschr.* 91: 357–360.

Wachendörfer, G., Farrenkopf, R., Lohrbach, W., Förster, U., Frost, J. W. und Valder, W. A. 1978. Passageversuche mit einer Varianten des Tollwut-Impfstammes ERA bei wildlebenden Spezies (*Ondatra zibethica* und *Rattus norvegicus*) - Ein Beitrag zur oralen Immunisierung von Füchsen gegen Tollwut. *Dtsch. Tierärztl. Wschr.* 85: 273–285.

Wachendörfer, G. 1978. Rabies in the European region. In: Surveillance and control of rabies - report on a conference, Frankfurt am Main 15–19 Nov. 1977: 36–50 WHO Copenhagen.

Wandeler, A., Wachendörfer, G., Förster, U., Krekel, H., Müller, J., and Steck, F. 1974. Rabies in wild carnivores in central Europe. I. Epidemiological studies. *Zbl. Vet. Med.* B 21: 735–756.

Wandeler, A., Wachendörfer, G., Förster, U., Krekel, H., Müller, J., and Steck, F. 1974. Rabies in wild carnivores in central Europe. II. Virological and serological examinations. *Zbl. Vet. Med.* B 21: 757–764.

Wandeler, A., Müller, J., Wachendörfer, G., Schale, W., Förster, U., and Steck, F. 1974. Rabies in wild carnivores in central Europe. III. Ecology and biology of the fox in relation to control operations. *Zbl. Vet. Med.* B 21: 765–773.

Weinhold, E. 1977. Versuche zur natürlichen Übertragung der Tollwutinfektion bei Mäusen. Tätigkeitsbericht des Bundesgesundheitsamtes, Berlin.

WHO-Report of the informal discussions on the WHO/FAO coordinated research programme on Wildlife Rabies in Europe, Nancy 3–5 July 1972.

WHO 1973. Expert committee on rabies, 6th report, WHO Techn. Rep. Ser. No 563, Geneva.

WHO-Report 1978.

Winkler, W. G., Shaddock, J. H., and Williams, L. W. 1976. Oral rabies vaccine: Evaluation of its infectivity in three species of rodents. *Am. J. Epidemiol.* 104: 294–298.

Winkler, W. G. and Baer, G. M. 1976. Rabies immunization of red foxes (*Vulpes fulva*) with vaccine in sausage baits. *Am. J. Epidemiol.* 103: 408–415.

20 FOX SOCIAL ECOLOGY AND RABIES CONTROL

EDITOR'S CLOSING REMARKS

Erik Zimen*

As Editor of these proceedings I cannot resist the temptation to comment on some of the more controversial topics discussed. Predominant were – how could it be otherwise – questions related to rabies control, i.e. on methods of estimating fox population density, of reducing fox densities by increased hunting pressure, and above all, fox den gassing.

We were fortunate in having at the Workshop representatives from many fields of fox-related research, with their various interests – the fox itself, rabies control, and animal welfare related thereto. Fluctuating subgroups reflected these different interests: one was composed mainly of veterinarians connected with the WHO Rabies Research and Control program, one mainly of zoologists immersed in behavioural and population ecology, one was of hunters concerned with fox predation on game species, and finally a most heterogeneous group united in opposing gassing as a means of rabies control. Obviously the preconditions for controversial discussion were ideal.

Opinions were in fact divided on many topics; but apart from some short rounds towards the end of the meeting, when seemingly "old sparring partners" squared up over fox den gassing, discussion was surprisingly rational. I believe this mutual understanding was partly because the vets, and others engaged in rabies control, have themselves done a lot of fox ecology research recently. These were also the participants looking for alternative methods, including oral vaccination, to lessen the spread of rabies, and who, when forced to use conventional methods, do not propagate over-simplified solutions to complicated questions. On the other hand, their obligation is primarily to human welfare, which rabies still threatens despite all vaccine improvement since Pasteur's time. Here is the stuff of conflict with those emotionally averse to the eradication of wildlife by any means, not excluding poison and gas – a most understandable attitude if we think in terms of recent European history.

But confining ourselves to rational factors, it now seems that not all rabies control schemes were based on a thorough understanding of the complicated interrelations of those ecosystems into which actions have been carried. In many cases it seems most doubtful whether the desired effects were finally achieved. There are even signs that the remedial measures taken favoured

---

* Lehrstuhl für Biogeographie, Universität des Saarlandes, D-6600 Saarbrücken, G.F.R.

the spread of the disease in some ways, and the frequency of its outbreak, perhaps delaying its end or favouring fresh outbreaks in infected areas.

The Workshop was not intended to evaluate different methods of rabies control, but we did hope to resolve some of the many still-open questions in fox behaviour and ecology, as a contribution to a rational habitat-related anti-rabies program. Biologists have stood aside too long, often criticizing the vets instead of working alongside them to solve those problems which are our common concern.

Fox den gassing

On this question I do not want to spend much time. Apart from its ethical implications, I think the effect of gassing has been exaggerated, both on fox populations and non-target species such as the badger (*Meles meles*), whose recent decline probably mainly results from rabies, not gassing. Observations during a recent rabies epidemic around Saarbrücken indicate that the reduction effect of gassing on the fox population is almost negligible compared to the effect of the disease itself (Zimen, in prep.). Gassing, in rabies-infected areas, can only kill off some foxes earlier than rabies would have done, and so somewhat reduce the danger to man. Large-scale annual gassing operations continued over many years, as practised some years ago in parts of Germany and Switzerland, were abandoned as ineffectual – putting all other, mainly humane, reasons aside. I think fox den gassing is out, except in some very specific situations, although it may be some years before the last veterinary officer in charge realizes this.

The hunting indicators of fox population density

One controversial question was whether the number of foxes reported killed in an area, in the official hunting statistics, in any way reflects actual fox population density. The concept of the so-called hunting indicator of fox population density (HIPD) was criticized on the following points:

- The unreliability of hunting statistics. In some countries these are most incomplete; in others only a small proportion of the foxes killed are supposed to be registered (Stubbe, chapter 3), while our research indicates that hunters in West Germany tend to exaggerate numbers of foxes killed, wherever or whenever the higher figures may aid their high social pressure for predator control – a pressure which is often increased when rabies hits the area.
- Hunting efficiency and methods are highly dependent upon the habitat, and its accessibility to the hunter, as well as on the age structure of the fox population, food availability, etc. (see Englund, chapter 9).
- The zeal of a hunter for fox control depends largely on the main species he hunts; zeal decreases with increasing prey size, that is, with decreasing competition with the fox.

- The time of year fox control is practised has a definite influence on fox population dynamics.

Thus the same low density could be indicated by the HIPD, for example, for two very different areas – one with low hunting pressure but high fox density, the other where high hunting pressure leads to a reduced population far below carrying capacity, with a corresponding low recruitment. It becomes clear that the HIPD is a very crude measure for population density, at best, and even a comparison of different countries' hunting statistics for such large areas as Europe (Stubbe, chapter 3) will result in very speculative estimates of fox populations. The same applies for single regions over a number of years (Braunschweig, chapter 8).

On the other hand there is a clear correlation between the number of foxes shot annually in an area and rabies frequency (Bögel and Moegle, fig. 3, chapter 17). It is most probable that the number of rabid animals depends on fox abundance; it seems therefore that the HIPD at least tends to reflect fox population density. For rapid, large-scale rough estimates of relative fox densities the use of hunting statistics is still the only available method; instead of rejecting it altogether because of its unquestionable drawbacks, we should try to improve it. In future, researchers should compare hunting statistics with statistics attained by other methods for small areas, to find reliable correction factors for different habitats and hunting methods – factors which could be used along with the HIPD.

Fox population dynamics

One of the most interesting phenomena in fox ecology is the great variation in reproductivity. Populations can be classed, roughly, in three categories:

1. Populations with large annual variation in the proportion of reproducing females, as well as in the number of pups per litter,
2. Populations with rather stable, low reproduction rate,
3. Populations with rather stable, high reproduction rate.

Wide variation seems to go with drastic changes in food availability (coniferous taiga of central and northern Sweden: Englund, chapter 9, Lindström, chapter 11), while stable conditions result when there is a diversity of food resources not fluctuating over the years (deciduous forests and areas of central Europe put to agricultural use, Englund 1970; Stubbe, chapter 3; Lloyd et al. 1976) and North America (Storm et al. 1976; Johnston pers. comm.). Lindström in his paper (chapter 11) discusses the possible causation for the observed annual fluctuation in reproduction within one unstable population and Englund (chapter 9), on the possible reasons for population stability or instability.

Stable populations may have either consistently high or low reproduction, apparently according in the main to human persecution. The very high rate of reproduction observed in North American populations can be correlated with the present high pelt price and consequent intensive fox shooting and trapping. In central Europe fox persecution is also intense, but probably not as extreme as in America. Thus in central Europe almost all females breed, but average litter size is lower. In traditionally pacified areas, on the other hand, as around Oxford, reproduction is continuously reduced as the number of breeding females is limited (Macdonald chapter 10).

Macdonald (chapter 10) discusses the possible proximate and ultimate causation of breding inhibition in vixen, and Lloyd (chapter 2) summarizes his experience on the spatial organization of fox populations, and the effect of human-induced mortality on the system. In unaffected populations the territorial system is probably rather stable, and vacant male or female territories rare whenever density reaches the capacity limit. With only slight chances of finding territories of their own in which to breed and raise their own pups, young females will then increase their inclusive fitness by staying in their home territories and helping the α female, often their mother, to raise hers.

Slowly our understanding of basic fox behavioural and population ecology takes shape. However, the important question of how young females assess population density, and through it their chances of breeding success, and adjust their behaviour accordingly, is still completely obscure.

Two explanations, not necessarily exclusive, seem theoretically possible. A young female experiences population density at first hand in late summer and autumn, in making short incursions into neighbouring territories. In case of frequent encounters with other foxes, aggressive or otherwise, she may return home; else she may continue to search for a vacant spot, making longer and longer excursions, or dispersing altogether.

The other possibility would be that populations in areas of traditionally high human hunting pressure tend to become r-selected, while unharmed populations tend to be K-selected. High dispersal tendencies are typical for r-selected populations (Geist 1979), and this includes general pup dispersal independent of density (Fig. 1).

Should female dispersal be a matter of individual adaptation, we could expect future research to show some variation in female reproductivity in neighbouring areas with different hunting pressure. From our present information it seems that in central Europe, both hunter motivation and fox susceptibility are low in mainly forested but high in mainly open areas (see below). Thus in forested regions, dispersal of young females should be less common, making for larger group size and lower reproduction and population turn-over rates. In experimentally pacified areas too, reproduction rates should decline quickly. If dispersal is a genotypical adaptation of whole populations, local variations in reproductive rates should be less distinct.

Whatever factors may actually alter female reproductive strategies, hunting pressure seems drastically to influence fox spatial organization, reproduction and annual population statistics. Thus hunting may actually increase, or at least does not necessarily decrease, fox numbers within an area for any year (Fig. 2). Furthermore hunting may unsettle spatial organization, and so cause social instability too. This may boost territorial trespassing, leading to

Expected environmental and social factors favouring fox	r-Strategy	K-Strategy
Climate	Seasonal and/or unpredictable	Constant and/or predictable
Habitat structure	Diverse	Monotonous
Succession	Less mature	More mature
Food resource	Diverse	Monotonous
Food distribution	Scattered	Evenly
Food availability	Abundant	Restricted
Spatial distribution of population	Scattered;vacuums	Evenly distributed;no vacuums
Mortality	High; non-directed; density independent; catastrophic	Low; age-and sex-directed; density dependent
Population size	Variable in time; non-equilibrium; below carrying capacity of environment; unsaturated communities or portions thereof	More constant in time; equilibrium; at or close to carrying capacity; saturated communities
Persecution by man	Strong	Weak

Expected adaptive characteristics of fox	r-Strategy	K-Strategy
Individual size	Large	Small
Life expectancy	Short	Long
Maturity	Early	Late or easy socialy controled
Litter sizes	Large	Small
Male reproductive strategy	Maximating copulations; promiscuous; limited parental investment	Tendency to monogamy; stong parental investment
Female reproductive strategy	Opportunistic	Restricted to dominant territory holders, others help raise kin pups
Pair bonds	Weak	Strong
Group sizes	Small, tendence to solitary living	Larger
Group stability	Weak, temporary	Stable
Intraspecific competition	Weak	Strong
Territoriality	Weak, seasonal	Strong, stable
Population turnover	Fast	Slow
Heterosis	Favoured	Selected against
Resource utilization through	Dispersal	Maintenance

Fig. 1. Some possible correlates of *r* and *k* selected fox populations (adopted from Pianka 1970 and Geist 1979).

more social contracts, perhaps aggressive, both between neighbours and within territories. In such populations, foxes would be obliged to re-establish their territory borders repeatedly, and their hierarchical rights within the group.

All these speculations may seem purely academic; but their implications are important for adequate rabies control methods. Up to now it has seemed all too logical to reduce fox numbers in order to cut rabies frequency, and thus the danger to human lives. Hunters in all rabies-infested countries are therefore told over and over again to do all they can to reduce fox

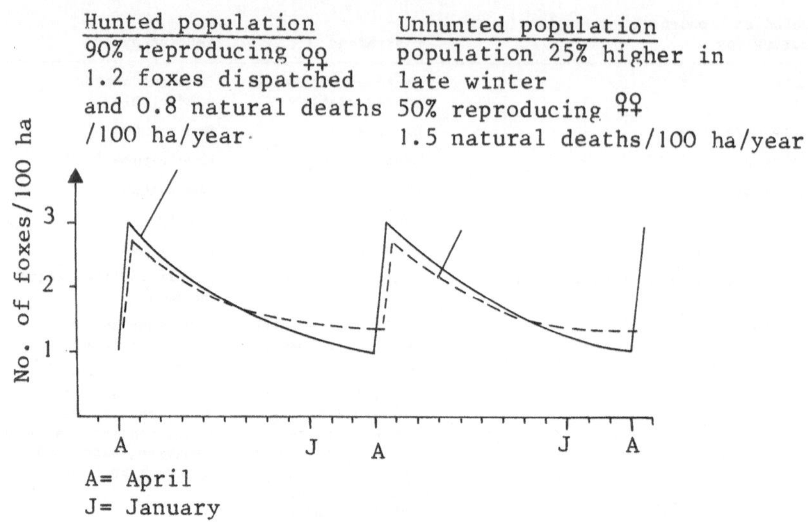

Fig. 2. Development of fox populations in Central Europe

Hunted populations: Wide-range estimates give density in late winter as 1 fox/100 ha or, assuming parity of the sexes, 0.5 vixen/100 ha, of which an average 90% whelp. The average number of cubs in a litter is about 4.5. The annual increase is accordingly (0.45 × 4.5) = 2.00/100 ha. Spring density is therefore 3 foxes/100 ha, of which an average of 1.2 is shot. To obtain the late winter density of 1 fox/100 ha, 0.8 foxes/100 ha must die natural deaths. (This model is based on data from Lloyd et al. 1976).
Unhunted populations: Wide-range calculations for Central Europe predict a reduction due to hunting of about 25% of fox numbers, so that in an unhunted population in late winter there should be about 1.3 foxes or 0.65 vixen/100 ha. Given a stable social organization, in the long run some 50% of all vixen would whelp annually. This represents an annual increase of (0.325 × 4.5) = 1.45/100 ha. Numbers in spring would be 2.75 foxes/100 ha. If population density stays constant 1.45 foxes/100 ha must die natural deaths annually. (Basis for this prediction was given by Macdonald (chap. 10), von Braunschweig (chap. 8), Englund (chap. 9) and Lindström (chap. 11)).

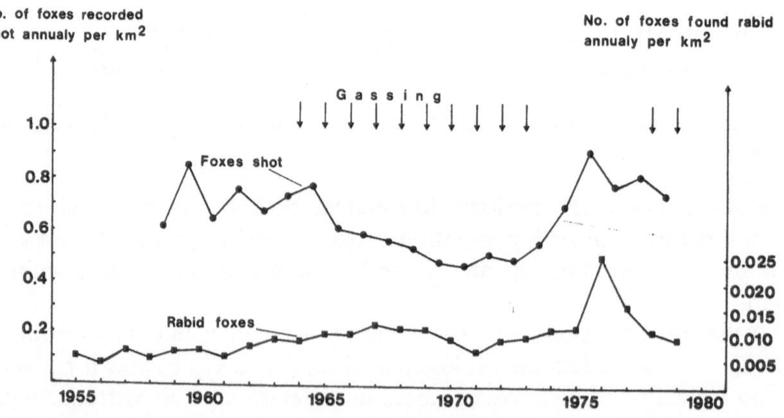

Fig. 3A. Annual registered no. of foxes shot and found rabid in the Fedral Republic of Germany (area hunted = 236.741 km^2). (Data from DJV Handbuch, 1980: Mainz).

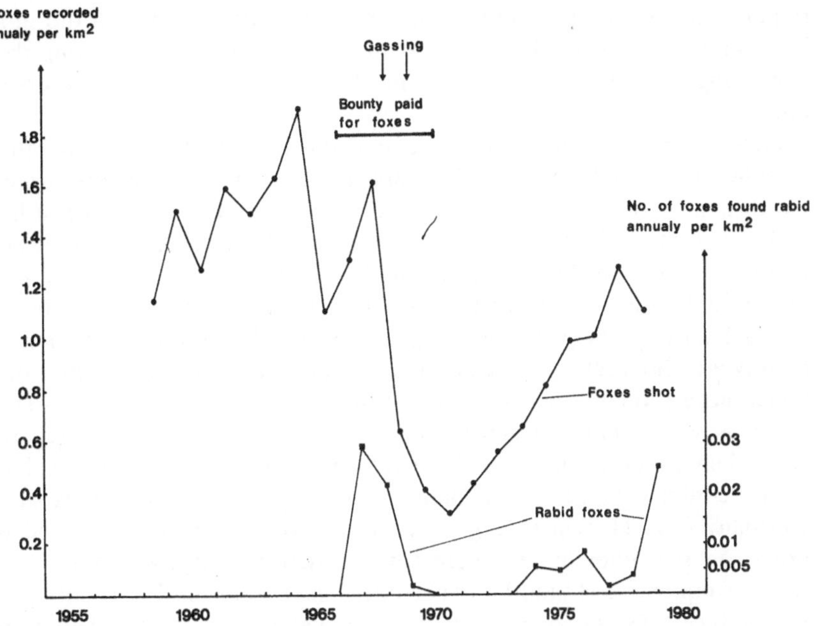

Fig. 3B. Annual registered no. of foxes shot and found rabid in the Saarland of southwestern Germany (area hunted = 2.468 km²).

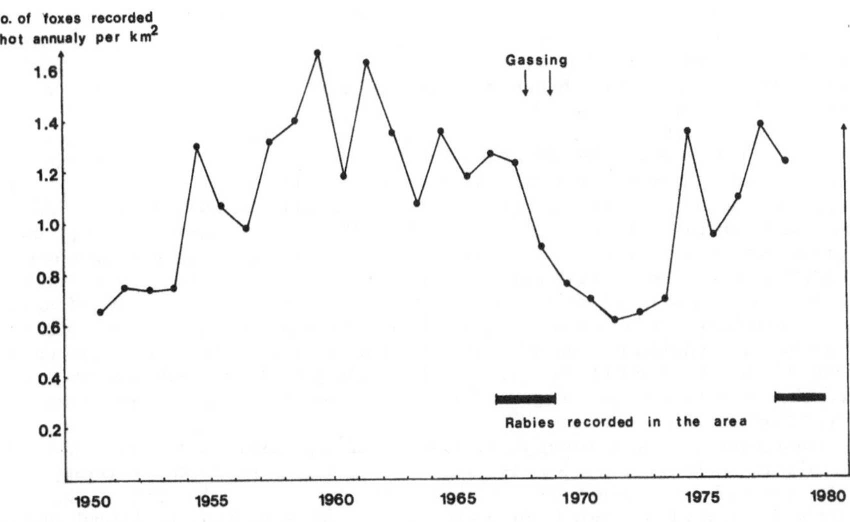

Fig. 3C. Annual registered no. of foxes shot in Kreis Homburg of southeastern Saarland (area hunted = 220 km²).

populations, not only in infected areas but everywhere and all the time. In fact, hunting statistics show a slight increase in foxes taken during the last years (Fig. 3), although such data must be regarded with due reserve (see above).

Hunters in some areas may have been successful in limiting the fox population to far below carrying capacity, as in the northwestern parts of Nordrhein-Westfalen and Niedersachen in Germany, a mainly open habitat with small-game shooting. Rabies frequency is lower here than elsewhere in Germany (Fig. 3 in Wachendörfer and Frost, chapter 19), perhaps a result of successful hunting efforts. In the more forested areas of Germany however it seems highly unlikely that hunting in any way limits fox populations; it is merely a compensatory mortality factor (von Braunschweig, chapter 8). It is much more probable that the present increase of hunting pressure – still too weak as yet to effect a substantial reduction – favours rabies, due to the socio-demographic influences discussed above. This would partly explain the high frequency of rabies in many of the more forested regions of West Germany (e.g. Hessen in Fig. 3, chapter 19). In very large, closed forest areas, on the other hand, where fox populations are less disturbed, lower rabies frequencies should be expected; in fact, data on the recent epidemic in southwest Germany indicate low frequencies in large coherent forest areas (Zimen, in prep.). Of course, densities may be lower here, owing to the probably lower carrying capacity, thus too accounting for the lower rabies frequencies.

Still, one must ask whether, in all honesty, vulpine rabies in its present European and North American form is not partly a man-made problem. Only intense causal studies in fox population biology can answer this question in detail. Some evidence exists however: forest clearance for

---

From approximately 1950 to 1970, when rabies was spreading over the FRG, the annual numbers of dispatched foxes decreased only slightly in spite of the steady increasing rabies toll (Fig. 3A). To bring this about hunting must have intensified. An increasing hunting pressure in the late 1950s and early 1960s becomes clearly apparent when the area surveyed is much smaller and population fluctuations do not overlap (Fig. 3C). The sharp rise in the numbers of foxes shot in the FRG in the mid 1970s on the other hand can hardly be attributed to still a greater hunting pressure. Here the underlying cause is likely to be the recovery of fox populations after the epidemic, also reflected in the increase of rabid foxes found (Fig. 3A). The spectacular increase of rabies cases in the mid 1970s (Fig 3A) was ascribed to the legal ban of gassing operations in 1974, just as the decline of rabies in the late 1960s was thought to result partly from large-scale gassing (Wachendörfer 1979, Bögel and Moegle, chap. 16). This must not necessarily be correct; to a biologist it seems more likely that fluctuations in rabies frequency stem from increased fox population instability due to heavy periodic mortality outbreaks. Hereby, quantitatively, rabies-caused mortality by far outroles all effects of human induced mortalityc(Zimen, in prep.). Thus, not gassing or hunting but rabies itself decreases infected populations. In the sense of Errington (1967) gassing in such populations is only of compensatory nature.

On a nation-wide scale, strong fluctuations in local population densities (Fig. 3B and 3C) overlap and so level out (Fig. 3A). But a certain large-scale correspondence in density and so also in rabies frequency is to be expected, i.e. a decline of rabies after the initial wave had swept the whole country by the early 1970s, followed by a second increase after most populations had recovered in the mid 1970s. Moreover, boosted hunting and gassing mortalities in presently non-infected populations has probably intensified fox population instability, and consequently rabies-frequency fluctuations (see text and Fig. 1).

agriculture ever since the Middle Ages, in Europe and eastern North America, must have favoured the fox, whose numbers are probably higher nowadays than ever before. In addition, human hunting may have favoured *r*-selected features and thus contributed to the fox's enormous adaptability. Man today is the most outstanding single environmental factor for most animal species, which must either adapt to him or disappear.

Perhaps the natural history of the fox, including our obvious inability to control either him or vulpine rabies, is yet another case of human incompetence in dealing with the extremely intricate and sensitive interrelationships of ecosystems. I own to feeling a great sympathy for the fox, who will survive in spite of man, but it is not only on his behalf that I urge that more ecological research be carried out to complement our understanding, before even apparently simple and logical schemes are sprung upon the natural community, of which man himself is part.

REFERENCES

Englund, J. 1970. Some aspects of reproduction and mortaility rate in Swedish foxes 1961–63 and 66–69. *Viltrevy* 8: 1–82.

Errington, L. P. 1967. On predation and life. Iowa State Univ. Press. 277pp.

Geist, V. 1979. Life strategies, human evolution, environmental design. Springer-Verlag. New York. 495 pp.

Lloyd, H. G., Van Haften, J. L., Niewold, F. J. J., Wandeler, A., Bögel, K. and Arata, A. A. 1976. Annual turnover of fox populations in Europe. *Zbl. Vet. Med. B.* 23: 580–589.

Pianka, E. R. 1970. On *r* and *K* selection. Amer. Nature. 104: 592–597.

Storm, G. L., Andrews, R. D., Phillips, R. L., Bishop, R. A., Siniff, D. B. and Tester, J. R. 1976. Morphology, reproduction, dispersal and mortality of midwest red fox populations. Wildlife Monographs No. 49, 82 pp.

Wachendörfer, G. 1979. Zur Epidemiologie und Bekämpfung der Tollwut in Mitteleuropa. Z. Säugetierkunde 44: 36–46.